Carbon-based Membranes for Separation Processes

Ahmad Fauzi Ismail • Dipak Rana
Takeshi Matsuura • Henry C. Foley

Carbon-based Membranes for Separation Processes

Springer

Ahmad Fauzi Ismail
Advanced Membrane Technology Research Centre (AMTEC)
Materials and Manufacturing Research Alliance
University Teknologi Malaysia, 81310, Skudai, Johor
Malaysia
afauzi@utm.my, fauzi.ismail@gmail.com

Takeshi Matsuura
Department of Chemical and Biological Engineering
University of Ottawa
161 Louis Pasteur, Ottawa, Ont. K1N 6N5
Canada
takeshi_mtsr@yahoo.ca,
matsuura@eng.uottawa.ca

Dipak Rana
Department of Chemical and Biological Engineering
University of Ottawa
161 Louis Pasteur, Ottawa, Ont. K1N 6N5
Canada
rana@eng.uottawa.ca

Henry C. Foley
Department of Chemical Engineering
The Pennsylvania State University
University Park, PA 16802
USA
hcf2@psu.edu

ISBN 978-0-387-78990-3 e-ISBN 978-0-387-78991-0
DOI 10.1007/978-0-387-78991-0
Springer New York Dordrecht Heidelberg London

Library of Congress Control Number: 2011930411

Printed on acid-free paper

Springer is part of Springer Science+Business Media (www.springer.com)

Foreword

Industrial culture has brought with it magnificent improvements in human health and well being over the last two centuries. At the same time, these advances in the human condition have come at a cost. All too often in the past, the overall impact of an industrial process or product was not fully accounted for; waste and by-products were considered to be merely zero cost disposables. Profit margins thus were apparently higher than we know in hindsight that they should have been. Whether it was the eighteenth century mill making wool and pouring waste sulfuric acid into a river or the twentieth century coal power plant emitting megatons of carbon dioxide as well as lesser amounts of sulfur and mercury, the real overall costs of production, which must include the environmental impact (or remediation) and public health effects, were not taken into account. Thus the waste product problems were not considered to be important, indeed, for a long time they were not considered problems.

At this juncture, early in the twenty-first century, much has changed. We now think in terms of process and product life cycles and take into account the full cradle-to-grave costs of production. But even more, we know that as human populations grow over the next 50–100 years, enormous pressure will be placed upon all our systems of production, delivery and consumption. Carbon dioxide in the atmosphere is leading to slow but sure global warming with unknown consequences. Fresh, clean water will become an increasingly difficult commodity to supply cheaply to large segments of the population. Energy needs will only rise with time as developing nations emerge as first world industrial powers. So what has this to do with carbon membranes the reader may ask? Simply put, many of the most pressing technological problems that we face will require solutions that involve energy efficient separations and carbon membranes will provide the solutions that we seek.

Carbon membranes are still in their infancy as a technology, yet the promise they hold is enormous. Already we know that nanoporous (0.5–1.0 nm average pore size) carbon membranes show an especially high affinity for carbon dioxide transport, a property that will undoubtedly be of utility in carbon capture and sequestration. They are robust enough to withstand use in aqueous media and at either high or low pH. When engineered with mesopores (1.0–3.0 nm), they can be used to provide ultrafiltration of water and other process fluids. In combination with catalysts, they are able to combine reaction and separation, thereby providing a viable means to selectively open the reaction zone and thereby overcoming equilibrium limitations of the closed system. Carbon membranes have been shown to be useful in a spectrum of separations as important and, yet, as delicate as, for example, the separation of oxygen and nitrogen. In other instances they have been shown to be compatible with blood, can be used to separate proteins and may offer opportunities for biomedical advances. When placed on ceramic or metallic supports, they are able to withstand high pressures.

If all this is known, then what is left to do? The fact is that most of these demonstrations have been at the laboratory level and not much beyond that level. Preparation of these materials is still an art and not a science. The science begins once a useful membrane has been prepared in the lab; prior to that point, despite much effort, the steps of preparation remain art. Can this be overcome? Certainly it can be. Similarly, it is clear that being able to manufacture materials readily, reproducibly and at low cost, remains a barrier to adoption and application.Then there is the science associated with the synthesis of the pore structures. Still too little is known about the details of the pore structure in many carbon membranes, and this has tended to limit the science that seeks to understand the mechanisms of separation—especially nanoscale kinetic separations. Eventually, synthesis of regular pore structures must be the goal, so that we can have carbon membranes tailored for each application with pore structures having optimal orientation for regular transport and low tortuosity, with pores that are sized for the separation to be done and on support media that lend themselves to ready incorporation into a module.

To get to this point with carbon membranes will take more dedicated research, the development of new ideas and with luck a few breakthroughs in synthesis and preparation. The present volume is a terrific starting point for the scientist or engineer who is embarking upon research in the area. Collected in one place for the first time is all that one needs to know to synthesize and test carbon membranes. The time saving offered by the collection of information into this volume is enormous and will prove to be useful not just for one newly entering the field but for the seasoned researcher as well.

State College, PA

Henry C. Foley

Preface

The concept of carbon membrane is not necessarily novel. Ash et al. compressed nonporous graphite carbon into a plug and called it carbon membrane in early seventies. The practical usefulness of carbon membrane was however realized in the beginning of eighties for the first time by the work of Koresh and Soffer who pyrolyzed many thermosetting polymers to produce carbon molecular sieve membranes. Since then attempts have been made to use carbon membranes for gas separation, nanofiltration and other membrane separation processes.

It was the realization that there were performance limits in polymeric membranes in gas separation which prompted research on carbon membrane. In 1991 Robeson set upper bounds in the selectivity-permeability plots of several gas-pairs by compiling experimental data for a large number of polymeric materials. Although the boundary lines have been shifted to the desirable direction after nearly 20 years' research efforts, the achievement has not yet been truly spectacular. Attention of membrane research community was then focussed on inorganic materials, such as silica, zeolite and carbon, which exhibited molecular sieving properties. Remarkable improvements have been made in terms of the selectivity-permeability plot but the exploitation of these materials for the practical application remains underachieved primarily due to their poor processibility.

In order to combine the superb molecular sieving effects of inorganic materials and the desirable mechanical and processing properties of polymers considerable efforts have been made recently to fabricate composite membranes, also called mixed matrix membranes (MMMs), in which inorganic particles are incorporated in host polymeric membranes. With respect to carbon material, attempts to fabricate MMMs were further encouraged by the recent progress in nano-technology in general and carbon nano-tubes in particular. Using nano-sized particles instead of micro-sized particles, the compatibility between the carbon particle and the host polymeric membrane seems to be enhanced. Moreover, the mass transport through the nano-tubes is remarkably different from the micro- and macro-sized materials, as evidenced by the recent discovery of carbon nano-tube membrane for seawater desalination by reverse osmosis.

The authors intend to make a historic overview of carbon-related membranes in this book. It will cover the development of carbon related membranes and membrane modules from its onset to the latest research on carbon mixed matrix membranes. After the review of the progress in the field, they also intend to show the future direction of R&D.

Chapter 1 is the introduction of the book. The general historic overview of the carbon and carbon-related membranes is made in this chapter.

Chapters 2–7 are dedicated to carbon membranes. A comprehensive literature review on carbon membrane transport, carbon membrane and carbon membrane module preparation is made in these chapters.

In Chap. 2, the unique feature of carbon membrane transport is briefly outlined.

Chapter 3 is for the carbon membrane configuration. Currently, all flat sheet, tubular, capillary and hollow fiber carbon membranes are available at least for laboratory scale experiments. Their merits and demerits are discussed in this chapter.

Chapter 4 is the most comprehensive chapter of the book on the carbon membrane preparation. In this chapter selection of the polymeric precursor membrane, preparation of polymeric membrane, pre-treatment before pyrolysis, pyrolysis and membrane post-treatment are dealt with. Furthermore, the membrane preparation method is described in this chapter as much in detail as possible.

Chapter 5 is for membrane testing. Similar to other inorganic membranes, some of carbon membranes are less flexible than polymeric membranes. This should be taken in consideration in the design of the laboratory scale carbon membrane separator. The construction of the separator is described as much in detail as possible.

Chapter 6 is for carbon membrane characterization. This is also slightly different from the characterization of polymeric membranes, since the degree of carbonization and its effects on physical and morphological properties should be known.

It is expected that the researchers can conduct their own experiments after reading Chaps. 4–6.

Chapter 7 is for the carbon membrane module construction. This is also unique for carbon membranes because of their less favourable mechanical properties compared to the polymeric membranes.

Chapter 8 summarizes the recent progresses in other carbon related membranes. The central issue of this chapter is carbon nano-tubes membranes and mix matrix membranes in which carbon nano-tubes are incorporated.

Chapter 9 will include all aspects of applications of carbon related membranes in separation processes such as reverse osmosis, nanofiltration, pervaporation, gas separation and fuel cell.

In Chap. 10, the cost evaluation of carbon-based membranes is attempted in comparison to the conventional polymeric membranes.

In the last chapter (Chap. 11) the authors attempt to show the future direction in R&D of the carbon-based membranes.

The authors believe that this book is the first book exclusively dedicated to carbon membranes and the carbon-related membranes. The book was written for engineers, scientists, professors, graduate students as well as general readers in universities, research institutions and industry who are engaged in R&D of membranes for separation processes. It is therefore the authors' wish to contribute to the further development of membrane science and technology in general and carbon membranes in particular by showing the future directions in the R&D.

October 5, 2010

A. F. Ismail
D. Rana
T. Matsuura
H. C. Foley

Acknowledgement

Ahmad Fauzi Ismail and Takeshi Matsuura would like to express their sincere thanks to the staff of the Advanced Membrane Technology Research Center (AMTEC), Universiti Teknologi Malaysia, Johor Bahru, Johor, Malaysia for their support, encouragement and their understanding during the period of the book writing. The relationship and friendship between authors and AMTEC members have created a very conducive environment which motivated us and ensured the completion of this book despite many hardships. Among them are Dr. Muhammad Noorul Anam, Mohd. Suhaimi Abdullah, Dr. Lau Woei Jye, Ng Be Cheer, Goh Pei Sean, Dr. Suhaila Sanip, Dayang Salyani, Dr. Nurmin Bolong, Dr. Hatijah Basri, Erna Yuliwati, Dr. Azeman Mustafa, Abdel Latif Hashemifard, Agung Mataram, Juhana Jaafar, Norhaniza Yusof, Farhana Aziz, Mohd Razis, Dr. Amir Mansourizadeh, Dr. Mohd Ali Aroon and Dr. Gholamreza Bakeri just to name a few. T. Matsuura would like to thank Universiti Teknologi Malaysia for his appointment to Distinguished Visiting Professor during the years 2009 and 2010, which enabled him to contribute several chapters to the book. D. Rana and T. Matsuura would also like to thank the staff of the Department of Chemical and Biological Engineering for their incessant support to their research endeavor. The authors would also like to thank Kenneth Howell of Springer US for his encouragement and many invaluable suggestions for the publication of the book.

Contents

Chapter 1
Introduction

1.1 The Development of Porous Inorganic Membranes

The development of porous inorganic membranes dates from before 1945, long before the discovery of today's synthetic organic membranes. However, not much publicity was given to these initial innovations, because the first porous inorganic membranes were made for the separation of uranium isotopes and were mainly used for military purposes or nuclear applications [1]. Indeed, non-nuclear applications of inorganic membranes were only realized at the beginning of the 1980s [2], with their potential becoming apparent once high quality porous ceramic membranes could be produced for industrial applications on a large scale [3]. Since then, they have become important tools for beverage production, water purification and the separation of dairy products [1]. Nowadays, inorganic membranes are used primarily for civilian energy-related applications. Furthermore, they play a significant role in the gas separation processes of the industrial sector.

Hsieh has provided a technical overview of inorganic membranes in his 1990 book [3], in which inorganic membranes are divided into two major categories based on structure; porous inorganic membranes and dense (nonporous) inorganic membranes as shown in Fig. 1.1. Moreover, porous inorganic membranes have two different structures: asymmetric and symmetric. Porous inorganic membranes with pores more than 0.3 μm usually work as sieves for large molecules and particles. Glass, metal, alumina, zirconia, zeolite and carbon membranes are the porous inorganic membranes commercially used. However, other inorganic materials such as cordierite, silicon carbide, silicon nitride, titania, mullite, tin oxide and mica can also be used to produce porous inorganic membranes. These membranes vary greatly in pore size, support material and configuration. Alternatively, dense membranes made of palladium and its alloys, silver, nickel and stabilized zirconia have been used or evaluated mostly for separating gaseous components. Applications of dense membranes are primarily for highly selective separation of hydrogen and oxygen and the transport occurs via charged particles. However, the dense membranes have found to have only very limited usage in the industrial application, primarily due to

A. F. Ismail et al., *Carbon-based Membranes for Separation Processes*,
DOI 10.1007/978-0-387-78991-0_1,

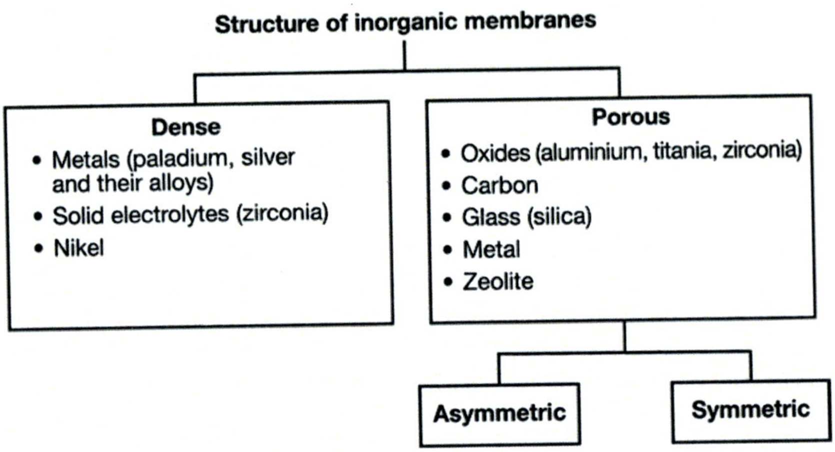

Fig. 1.1 Structure of inorganic membranes

their low permeability compared to that of porous inorganic membranes. Therefore, today's market for the commercial inorganic membranes is dominated by porous membranes [1–3].

At present, interest in the development of porous inorganic membranes that can provide better selectivity, thermal and chemical stability than polymeric membranes is growing. This attention has particularly focused on materials that exhibit molecular sieving properties such as silica, zeolites and carbon [4], which appear to be promising in separation of gas. For example, molecular sieve membranes can exhibit much higher permeabilities and selectivities than the polymeric membranes (Fig. 1.2). Among inorganic membranes, silica-based membranes selectively separate hydrogen from other gases but permselectivity between similar-sized molecules such as oxygen and nitrogen is not sufficient [5]. Zeolites can separate isomers, but it is difficult to obtain a large, crack-free zeolite membrane. Compared with silica and zeolite based membranes, carbon molecular sieve membranes [5, 6] can easily be fabricated and they have very unique features of high selectivity and permeability. Therefore, the purpose of this book review is to give an overview of the development of carbon molecular sieve membranes during the past 30 years. This book also looks towards the future direction of carbon membranes development in the new millennium.

The concept of carbon membrane for gas separation can be found in the early nineteen seventies. Barrer et al. compressed non-porous graphite carbon into a plug, called carbon membrane [8]. Bird and Trimm used poly(furfuryl alcohol) (PFA) to prepare unsupported and supported carbon molecular sieve membranes. During carbonization, they encountered shrinkage problems, which lead to cracking and deformation of the membrane. Hence, they failed to obtain a continuous membrane [9].

Carbon molecular sieves produced from the pyrolysis of polymeric materials have proved to be very effective for gas separation in adsorption applications by Ko-

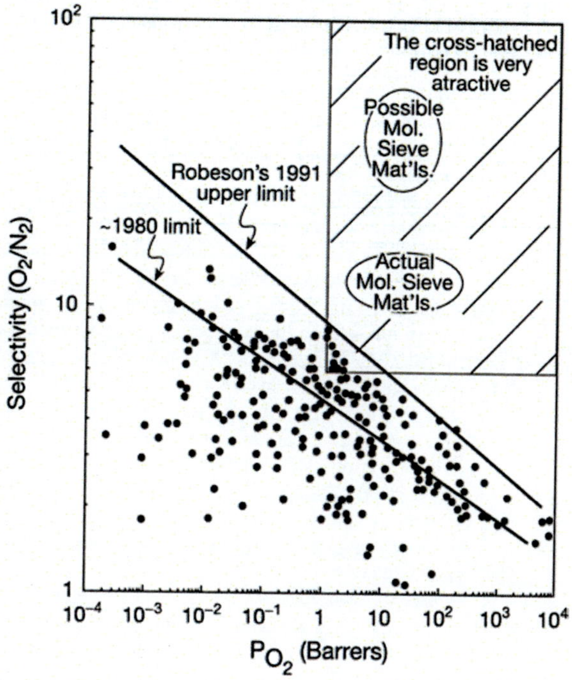

Fig. 1.2 Comparison between properties of polymeric membranes and molecular sieve membranes with the upper bound of performance. (From [7])

resh and Soffer [10–13]. Molecular sieve carbon can be easily obtained by pyrolysis of many thermosetting polymers such as polyacrylonitrile (PAN), poly(vinylidene chloride) (PVDC), PFA, cellulose, cellulose triacetate, saran copolymer, phenol formaldehyde resins and various coals such as coconut shell [10]. They described that the pore dimensions of carbon were dependent on morphology of the organic precursor and the chemistry of pyrolysis [14]. During the research on molecular-sieve carbon adsorbents, they have shown that the molecular sieving effect of non-graphitizing carbons was extremely specific and adjustable by mild activation and sintering steps to the discrimination range 2.8–5.2 Å [15].

Pyrolysis of thermosetting polymers typically cellulosic, phenolic resin, oxidized PAN as well as pitch mesophase have been found to yield an exact mimic of the morphology of the parent material. They do not proceed through melting or softening during any stage of the pyrolysis process. Hence, pyrolysis can produce a carbon molecular sieve membrane from a thermosetting polymer membrane [14]. Due to the research in developing carbon membranes that has occured after the initial work of Koresh and Soffer [14–16], apparently crack-free molecular sieving hollow fiber membranes can be successfully prepared by carbonizing cellulose hollow fibers. They have shown the dependence of permeabilities and selectivities on temperature, pressure and extent of pore for both adsorbing and non-adsorbing permeants [14, 15]. However, those membranes are lacking mechanical strength for practical application.

References

1. Kaizer K, Verweiji H (1996) Progress in inorganic membranes. Chemtech 26 (1): 37-41
2. Soria R (1995) Overview on industrial membranes. Catal Today 25 (3-4): 285-290
3. Hsieh HP (1988) Inorganic membranes. AIChE Symp Ser 84 (261): 1-18
4. Fuertes AB, Centeno TA (1995) Preparation of supported asymmetric carbon molecular sieve membranes. J Membr Sci 144 (1-2): 105-111
5. Hayashi J, Mizuta H, Yamamoto M, Kusakabe K, Morooka S (1997) Pore size control of carbonized BPTA-pp'ODA polyimide membrane by chemical vapor deposition of carbon. J Membr Sci 124 (2): 243-251
6. Chen YD, Yang RT (1994) Preparation of carbon molecular sieve membrane and diffusion of binary mixtures in the membrane. Ind Eng Chem Res 33 (12): 3146-3153
7. Moaddeb M, Koros WJ (1997) Gas transport properties of thin polymeric membranes in the presence of silicon dioxide particles. J Membr Sci 125 (1): 143-163
8. Ash R, Barrer RM, Lowson RT (1973) Transport of single gases and of binary gas mixtures in a microporous carbon membrane. J Chem Soc Faraday Trans I 69 (12): 2166-2178
9. Bird AJ, Trimm DL (1983) Carbon molecular sieves used in gas separation membranes. Carbon 21 (3): 177-163
10. Koresh JE, Soffer A (1980) Study of molecular sieve carbons. Part 1. Pore structure, gradual pore opening, and mechanism of molecular sieving. J Chem Soc Faraday Trans I 76 (12): 2457-2471
11. Koresh J, Soffer A (1980) Study of molecular sieve carbons. Part 2. Estimation of cross-sectional diameters of non-spherical molecules. J Chem Soc Faraday Trans I 76 (12): 2472-2485
12. Koresh J, Soffer A (1980) Molecular sieving range of pore diameters of adsorbents. J Chem Soc Faraday Trans I 76 (12): 2507-2509
13. Koresh J, Soffer A (1981) Molecular sieve carbons. Part 3. Adsorpton kinetics according to a surface barrier model. J Chem Soc Faraday Trans I 77 (12): 3005-3018
14. Koresh JE, Soffer A (1983) Molecular sieve carbon permselective membrane. Part I. Presentation of a new device for gas mixture separation. Sep Sci Technol 18 (8): 723-734
15. Koresh JE, Soffer A (1986) Mechanism of permeation through molecular-sieve carbon membrane. Part 1. The effect of adsorption and the dependence on pressure. J Chem Soc Faraday Trans I 82 (7): 2057-2063
16. Koresh JE, Soffer A (1987) The carbon molecular sieve membranes. General properties and the permeability of CH_4/H_2 mixture. Sep Sci Technol 22 (2-3): 973-982

Chapter 2
Transport Mechanism of Carbon Membranes

2.1 Transport of Gas Through CMSMs

Mass transfer of gas through a porous membrane can involve several processes depending on the pore structure and the solid [1]. There are four different mechanisms for the transport: Poiseuille flow; Knudsen diffusion; partial condensation/capillary diffusion/selective adsorption and molecular sieving [2, 3]. The transport mechanism exhibited by most of carbon membranes is the molecular sieving mechanism as shown in Fig. 2.1. The carbon membranes contain constrictions in the carbon matrix, which approach the molecular dimensions of the absorbing species [4].

In this manner, they are able to separate the gas molecules of similar sizes effectively. According to this mechanism, the separation is caused by passage of smaller molecules of a gas mixture through the pores while the larger molecules are obstructed. It exhibits high selectivity and permeability for the smaller component of a gas mixture [3]. Carbon matrix itself is impervious suggesting that permeation through carbon membranes can be attributed entirely to the pore system which consists of relatively wide openings with narrow constrictions. The openings contribute the major part of the pore volume and are thus responsible for the adsorption capacity, while the constrictions are responsible for the stereoselectivity of pore penetration by host molecules and for the kinetics of penetration [5]. Hence, the diffusivity of gases in carbon molecular sieve may change abruptly depending on the size and the shape of molecules because carbon molecular sieve has pore sizes close to the dimension of gas molecules [6].

Carbon molecular sieve membrane (CMSM) has been identified as a very promising candidate for gas separation, both in terms of separation properties and stability. These molecular sieves are porous solids that contain constrictions of apertures that approach the molecular dimensions of the diffusing gas molecules. At these constrictions, the interaction energy between the molecule and the carbon is comprised of both dispersive and repulsive interactions. When the opening becomes sufficiently small relative to the size of the diffusing molecule, the repulsive forces dominate and the molecule requires activation energy to pass through the constrictions. In this region of activated diffusion, molecules with only slight differences in size can be effectively separated through molecular sieving [7]. Therefore, the mechanism of gas

A. F. Ismail et al., *Carbon-based Membranes for Separation Processes*,
DOI 10.1007/978-0-387-78991-0_2,

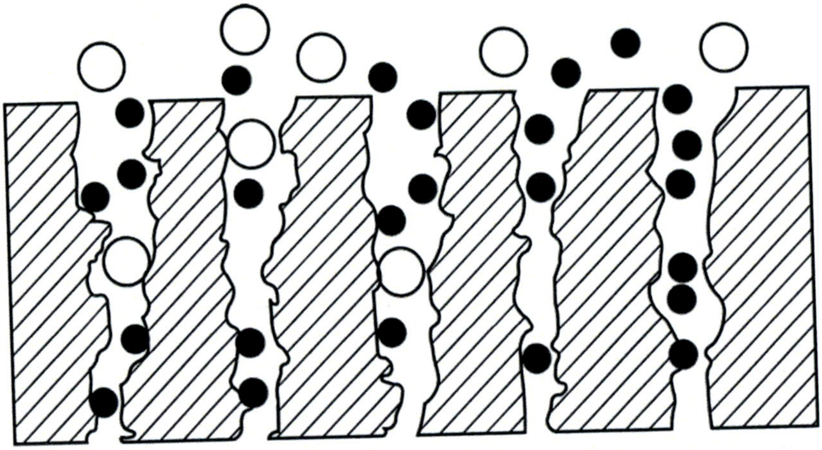

Fig. 2.1 Typical molecular sieving mechanism

permeation and uptake through porous solids is closely related to the internal surface area and, dimensions of the pores and to the surface properties of the solid, rather than to its bulk properties of the solid as in the case with polymers [5].

Carbon molecular sieve membranes suitable for gas separation have been prepared by pyrolyzing thermosetting polymers. CMSMs with pore diameter 3–5 Å have ideal separation factors, ranging from 4 to more than 170 for various gas pairs [2]. The permeation characteristics of the molecular sieve carbon membrane can be varied by changing the high temperature treatment parameters [8].

Another transport mechanism of carbon membrane is selective adsorption-surface diffusion mechanism. Adsorption-selective carbon membranes separate non-adsorbable or weakly adsorbable gases (O_2, N_2, CH_4) from adsorbable gases, such as NH_3, SO_2, H_2S and chlorofluorocarbons (CFCs). The difference between the structure of adsorption-selective carbon membranes (ASCMs) and CMSMs is the size of the micropores. ASCMs have a carbon film with micropores slightly larger than CMSMs, probably in the range of 5–7 Å [9]. It is known that the performance of an asymmetric membrane is governed by the structure of the thin active layer [10]. Meanwhile, the great difference between carbon asymmetric membranes and polymeric asymmetric membranes seems to be in the skin layer as shown in Fig. 2.2. In contrast to polymeric membranes, carbon membranes may be considered as a refractory porous solid where the permeants are non-soluble and merely penetrate through the pore system [8]. It is greatly different from the transport mechanism of polymeric membranes—solution-diffusion mechanism. Figure 2.3 shows the solution-diffusion mechanism in the dense layer of a polymeric membrane. Size (diffusivity) and condensability (solubility) selectivity factors interact with polymer to determine which component passes though the membrane faster [11]. However, carbon membrane requires a very fine control of the pore sizes (diameter <4 Å) and also often requires operation at an elevated temperature in order to provide practically acceptable flux for the membrane thickness may extend to a range of several microns [9]. The influ-

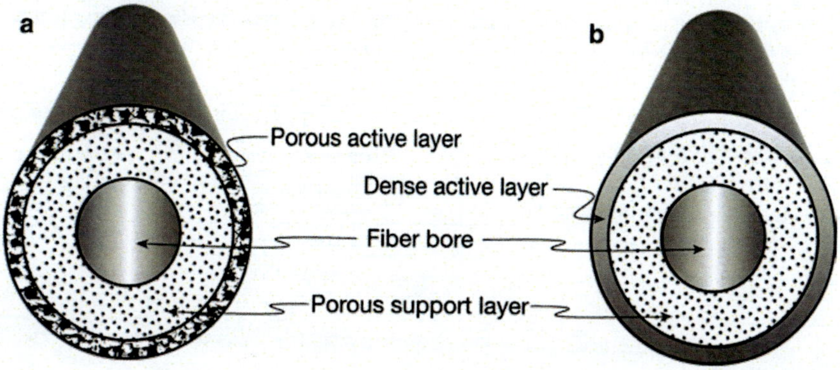

Fig. 2.2 Comparison of (**a**) carbon hollow fiber membrane with (**b**) polymeric hollow fibrer membrane

Fig. 2.3 Solution-diffusion transport mechanism. (From [11])

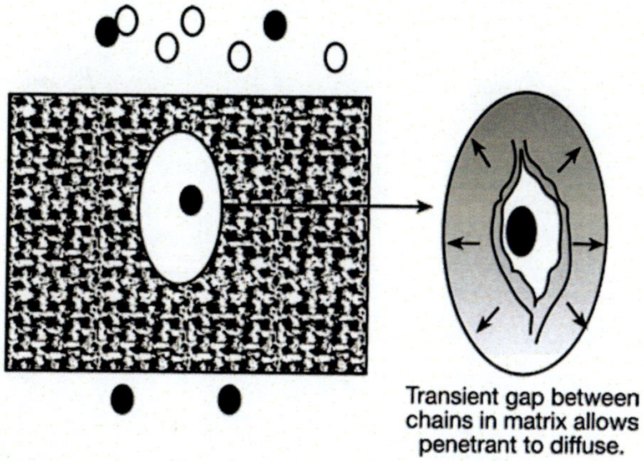

ence of adsorption of permeants on the pore walls has also to be taken into consideration, especially when permeants have relatively high boiling point [8].

2.2 Solution-Diffusion Model for Single Gas Transport

The transport through CMSMs is often studied by using the solution (sorption)-diffusion model. Because of simplicity of the model, a brief description is given below for the transport involved in a single gas system.

According to the Fick's first law

$$J = -D\frac{dc}{dx} \tag{2.1}$$

where J is the gas flux, dc/dx is the concentration gradient to the direction of gas flow and D is diffusivity.

When the solubility of gas in the membrane follows the Henry's law relationship

$$c = Sp \tag{2.2}$$

where c is the concentration in the membrane phase, p is the external pressure and S is the proportionality constant called solubility.

Combining Eqs. (2.1) and (2.2) and integrating from the feed side to the permeate side of the membrane, we obtain

$$J = \frac{SD}{l}\Delta p \tag{2.3}$$

where l is the membrane thickness and Δp is the transmembrane pressure difference.

Permeability is defined as:

$$P = SD \tag{2.4}$$

Hence,

$$J = \frac{P}{l}\Delta p \tag{2.5}$$

often l can not be measured. Then, P/l, called permeance, is measured instead of permeability.

Permeability, solubility and diffusivity can all be represented by Arrhenius type equations; i.e.

$$P = P_0 \exp\left(\frac{-E_p}{RT}\right) \tag{2.6}$$

$$S = S_0 \exp\left(\frac{-\Delta H_s}{RT}\right) \tag{2.7}$$

$$D = D_0 \exp\left(\frac{-E_d}{RT}\right) \tag{2.8}$$

where P_0, S_0 and D_0 are the pre-exponential factor of the respective term. R and T are the universal gas constant and the absolute temperature, respectively. E_p is the activation energy of gas permeation, ΔH_s is the heat of solution (sorption) and E_d is the activation energy for diffusion.

From Eqs. (2.6)–(2.8)

$$E_p = \Delta H_s + E_d \tag{2.9}$$

usually, ΔH_s is negative and E_d is positive. Therefore, E_p may be negative or positive, depending on whether solution (sorption) or diffusion governs the transport process.

Sorption does not necessarily follow Henry's law. For a glassy polymer an assumption is made that there are small cavities in the polymer and the sorption at the cavities follows Langmuir's law. Then, the concentration in the membrane is given as the sum of Henry's law adsorption and Langmuir's law adsorption

$$c = k_p p + \frac{c_h' bp}{1 + bp} \tag{2.10}$$

where k_p is the Henry's law constant, b is the hole affinity constant and c_h' is the saturation constant.

Equation (2.10) is also used to interpret the results from CMSMs. It should be noted that the applicability of solution (sorption)-diffusion model has nothing to do with the presence or absence of the pore.

2.3 Solution-Diffusion Model for the Transport of Binary Gas Mixtures

A study by Chen and Yang is described below in detail to show how the transport model is applied for the CMSM [12]. The nomenclature of the original work is retained even though it is different from the one used in Eqs. (2.1)–(2.10). However, analogy in the model development is obvious.

In the study conducted by Chen and Yang, the diffusivities of binary mixtures were measured using the laboratory fabricated CMSM and the results were compared with the authors' own kinetic theory developed for the prediction of binary diffusivities from pure component diffusivities.

The CMSM was prepared by coating polyfurfuryl alcohol (PFA) on a graphite support followed by pyrolysis. The graphite disk was 4.45 cm in diameter and 0.476 cm in thickness. A thin layer of PFA was coated on one face of the graphite disk. The heating protocol of the coated graphite is as follows: 90°C for 3 h in air; heating to 300°C at 1.5°C/min in N_2 stream and held at 300°C for 2 h; heating to 500°C at the same heating rate in N_2 stream and held at 500°C for 6 h. The membrane was then cooled to room temperature. The coating and pyrolysis procedure were repeated 5 times. The thickness of the carbon molecular sieve (CMS) layer was 15 μm.

Equilibrium isotherms of CH_4 and C_2H_6 were established by using the gravimetric method, following the weight change of the CMS. The CMS (particles) sample was prepared by pyrolysing PFA according to the same heating protocol but without the graphite substrate. After pyrolysis the the carbonized PFA sample was ground and sieved to 50 mesh for the adsorption experiments.

The diffusivity measurement was made by the diffusion cell. The laboratory prepared CMSM was loaded between two chambers of the diffusion cell. Pure helium (He) gas was fed into the lower chamber (permeate side) while He carrying different concentrations of CH_4, C_2H_6 or CH_4/C_2H_6 mixture was fed to the upper chamber (feed side) of the diffusion cell. The pressures on both chambers were kept equal. Hence, the gas transport was solely by the diffusion. From the flow rate and the concentration of the permeant at the outlet of the permeate side stream, the permeant flux can be calculated. The flux measurement was done after waiting for 8 h, at least, to ensure the establishment of the steady state.

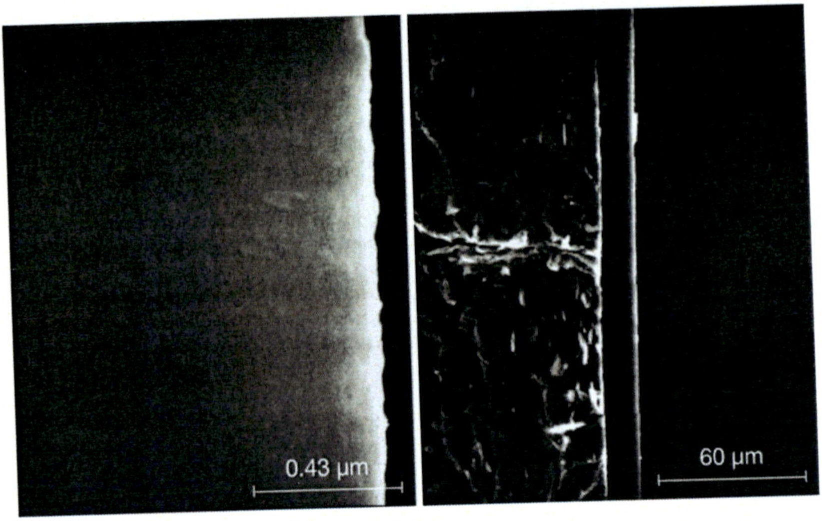

Fig. 2.4 SEM cross-sectional images of (*right*) CMS layer formed on the surface of graphite substrate, and (*left*) CMS layer and its surface. (From [12])

Figures 2.4 and 2.5 show the SEM pictures of the laboratory prepared CMSM and the substrate graphite. Figure 2.4 depicts that the CMSM coated on top of the graphite support is crack-free. The thickness of the CMSM layer is 15 μm. Figure 2.5 shows that the surface of the CMSM and the graphite support. The CMSM layer is much smoother with a roughness within 0.02 μm. The pore size in the graphite support is 5–10 μm.

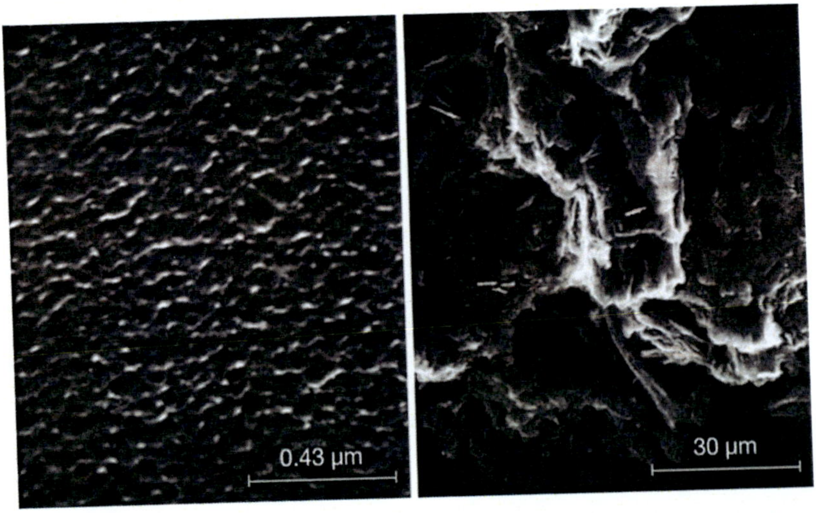

Fig. 2.5 SEM images of (*left*) the surface of CMS layer, and (*right*) the surface of graphite substrate. (From [12])

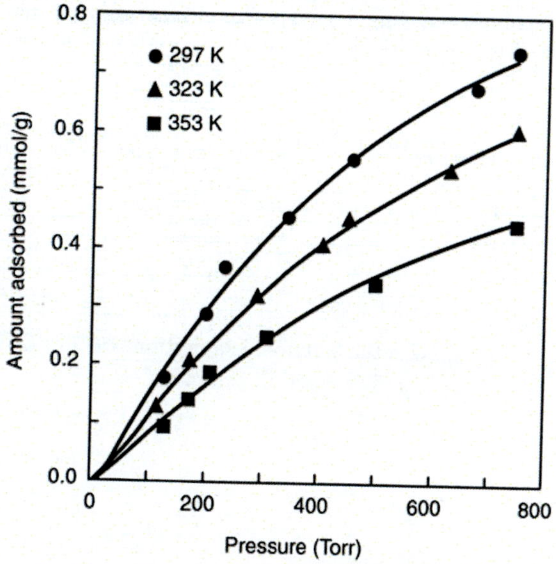

Fig. 2.6 Equilibrium isotherms of CH_4 on CMS. (From [12])

The adsorption isotherms for CH_4 and C_2H_6 are given in Figs. 2.6 and 2.7. The data were fitted to the Sippe type isotherm;

$$\frac{q}{q_s} = \frac{bp^n}{1 + bp^n} \tag{2.11}$$

where p is the gas pressure, q is the amount adsorbed and q_s is the saturated amount adsorbed. b is the Langmuir constant and n is isotherm constant.

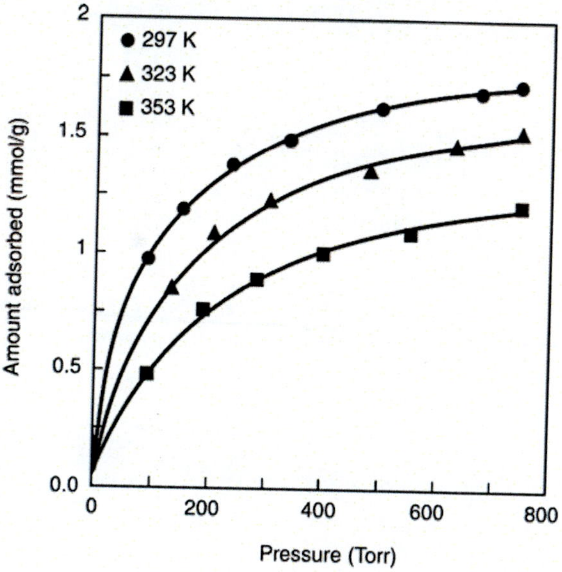

Fig. 2.7 Equilibrium isotherms of C_2H_6 on CMS. (From [12])

Regarding the single component gas diffusion, the flux can be written as:

$$J = -D\frac{\partial q}{\partial x} \tag{2.12}$$

where D is the diffusivity and x is distance.

Diffusion in molecular sieves is strongly concentration dependent, and is given by:

$$D = \frac{D_0}{1-(1-\lambda)\theta} \tag{2.13}$$

where θ is the fractional saturation $\theta = q/q_s$, D_0 is the diffusivity at zero adsorption and λ is and interaction parameter.

At the steady state the gas flux is constant. Then, substituting Eq. (2.13) into Eq. (2.12) and integrating,

$$J = \frac{D_0}{\Delta x}\frac{q_s}{(1-\lambda)}\ln\left[\frac{q_s-(1-\lambda)q_L}{q_s-(1-\lambda)q_H}\right] \tag{2.14}$$

where Δx is the thickness of CMS layer. The subscripts H and L are for the higher (upper) and lower chamber. Since $q_H >> q_L$, the equation is further reduced to

$$J = -\frac{D_0}{\Delta x}\frac{q_s}{(1-\lambda)}\ln\left[1-(1-\lambda)\frac{bp^n}{1+bp^n}\right] \tag{2.15}$$

Table 2.1 shows the parameters involved in Eqs. (2.11) and (2.13) at different temperatures.

When an assumption is made that the diffusivity is not concentration dependent, the flux becomes:

$$J = \frac{D}{\Delta x}\frac{q_s bp^n}{1+bp^n} \tag{2.15a}$$

The experimental CH_4 and C_2H_6 fluxes are shown in Figs. 2.8 and 2.9. The symbols are the experimental data. The solid lines are calculated results from Eq. (2.15) using the parameters given in Table 2.1. The broken lines are the best fit data based on

Table 2.1 Single-component equilibrium and diffusivity parameters

Gas	Temperature (K)	q_s (mmol/g)	$b \times 10^4$ (1/Torr)	n	$D_0 \times 10^8$ (cm²/s)	λ	$D \times 10^8$ (cm²/s)
CH_4	297	1.145	4.91	1.24	1.695	0.056	2.379
	323	1.051	4.51	1.21	2.429	0.196	3.047
	353	0.797	4.21	1.22	3.428	0.280	4.445
C_2H_6	297	2.072	20.0	0.84	0.146	0	0.293
	323	1.821	53.0	1.03	0.188	0	0.343
	353	1.549	49.4	0.99	0.315	0	0.524

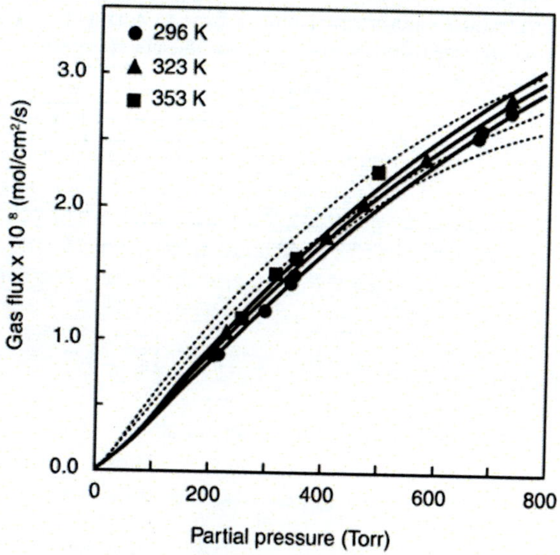

Fig. 2.8 Steady-state flux of CH_4 through CMSM: *solid curves* are fitted with conc.-dependent diffusivity and *dashed curves* are with constant diffusivity. (From [12])

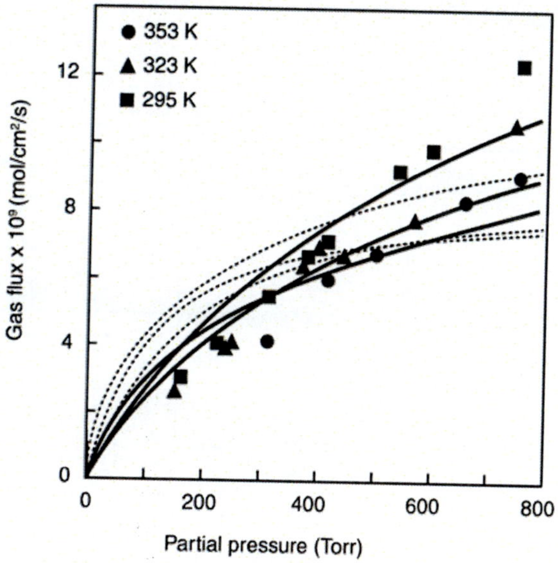

Fig. 2.9 Steady-state flux of C_2H_6 through CMSM: *solid curves* are fitted with conc.-dependent diffusivity and *dashed curves* are with constant diffusivity. (From [12])

Eq. (2.15a). Equation (2.15) gives better results in fitting than Eq. (2.15a), meaning the effect of concentration on the diffusivity can not be ignored.

The parameter λ defined for a specific molecule A is given as:

$$\lambda_A = e^{-(\varepsilon_{AV} - \varepsilon_{AA})/RT} \tag{2.16}$$

ε_{AV} and ε_{AA} mean the bond energy of the molecule A and the vacant site and the bond energy of molecule A and the molecule A that already occupied the vacant site, respectively.

Table 2.2 Energetic parameters and pre-exponential factors

Gas	E_a (kcal/mol)	$D_0^* \times 10^8$ (cm^2/s)
CH_4	2.51	1.22
C_2H_6	2.97	0.21

In Table 2.1, λ for C_2H_6 is equal to zero at all temperatures, meaning that ε_{AV} is much greater than ε_{AA}, while λ for CH_4 decreases with decreasing temperature, meaning that the A–V bonding increases as the temperature decreases.

D_0 obtained at three temperatures can be given in the form of Arrhenius equation as:

$$D_0 = D_0^* e^{-E_a/RT} \tag{2.17}$$

where D_0^* is the pre-exponential factor and E_a is the activation energy of diffusion.

The values for D_0^* and E_a are shown in Table 2.2.

For a binary system of components A and B, the fluxes are:

$$J_A = -D_{AA}\frac{\partial q_A}{\partial x} - D_{AB}\frac{\partial q_B}{\partial x} \tag{2.18}$$

$$J_B = -D_{BA}\frac{\partial q_A}{\partial x} - D_{BB}\frac{\partial q_B}{\partial x} \tag{2.19}$$

And the concentration dependent Fickian diffusivities are:

$$D_{AA} = D_{A0}\left[\frac{1-(1-\lambda_{AB})\theta_B}{1-(1-\lambda_A)\theta_A-(1-\lambda_{AB})\theta_B}\right] \tag{2.20}$$

$$D_{AB} = D_{A0}\left[\frac{1-(1-\lambda_{AB})\theta_A}{1-(1-\lambda_A)\theta_A-(1-\lambda_{AB})\theta_B}\right] \tag{2.21}$$

$$D_{BA} = D_{B0}\left[\frac{1-(1-\lambda_{BA})\theta_B}{1-(1-\lambda_B)\theta_B-(1-\lambda_{BA})\theta_A}\right] \tag{2.22}$$

$$D_{BB} = D_{B0}\left[\frac{1-(1-\lambda_{BA})\theta_A}{1-(1-\lambda_B)\theta_B-(1-\lambda_{BA})\theta_A}\right] \tag{2.23}$$

The cross-term λ_{AB} is defined as:

$$\lambda_{AB} = e^{-(\varepsilon_{AV}-\varepsilon_{AB})/RT} \tag{2.24}$$

A similar expression is applicable for λ_{BA}.

The cross-term activation energies are given by

$$\varepsilon_{AB} = \varepsilon_{BA} = (\varepsilon_{AA}\varepsilon_{BB})^{1/2} \tag{2.25}$$

By substituting the diffusivity Eqs. (2.20)–(2.23) into the flux Eqs. (2.18) and (2.19) and integrating the flux equations over q_A by keeping the other component at a constant average $\bar{q}_B$, the following equations are obtained.

$$J_i = -\frac{1}{\Delta x}\sum_{j=1}^{2}\left[\bar{D}_{ij}(\bar{q}_{jout} - q_{jin})\right] \quad \mathrm{i = A\ and\ B} \tag{2.26}$$

where

$$\bar{D}_{AA} = \frac{1}{q_{Aout} - q_{Ain}}\int_{q_{Ain}}^{q_{Aout}} D_{AA}(q_A\bar{q}_B)dq_A \tag{2.27}$$

$$\bar{D}_{AB} = \frac{1}{q_{Bout} - q_{Bin}}\int_{q_{Bin}}^{q_{Bout}} D_{AB}(\bar{q}_A q_B)dq_B \tag{2.28}$$

Similarly, $\bar{D}_{BA}$ and $\bar{D}_{BB}$ can be obtained.

To calculate the q_s for the mixed system adsorbed phase averaging is used

$$\frac{1}{q_s} = \frac{X_A}{q_{sA}} + \frac{X_B}{q_{sB}} \tag{2.29}$$

and

$$\theta_A = \frac{q_A}{q_s}, \theta_B = \frac{q_A}{q_s}, \theta_A + \theta_B \le 1 \tag{2.30}$$

where X is the adsorbate mole fraction at the equilibrium. To know q_A and q_B for the binary mixture adsorption

$$q_A = \frac{q_{sA} b_A P_A^{nA}}{1 + b_A P_A^{nA} + b_B P_B^{nB}} \tag{2.31}$$

is used and similarly for q_B.

Finally, for the prediction based on single component system, the following equations are used.

$$J_A = -D_{AA}\frac{dq_A}{dx} \tag{2.32}$$

$$J_B = -D_{BB}\frac{dq_B}{dx} \tag{2.33}$$

$$D_{AA} = \frac{D_{A0}}{1 - (1 - \lambda_{AA})\theta_A} \tag{2.34}$$

Table 2.3 Comparison of prediction based on the binary system and the single component system

Feed gas mole fraction		Experimental flux × 10^9 (mol/cm^2 s)		Flux predicted, binary theory × 10^9 (mol/cm^2 s)		Flux predicted, single component × 10^9 (mol/cm^2 s)	
CH_4	C_2H_6	CH_4	C_2H_6	CH_4	C_2H_6	CH_4	C_2H_6
0.611	0.021	18.8	0	17.5	Small	15.9	Small
0.138	0.580	3.18	11.6	1.67	6.07	0.989	5.94
0.093	0.430	1.99	9.60	1.19	5.36	0.737	5.27
0.465	0.317	13.6	5.06	8.53	3.97	5.66	3.58
0.349	0.625	8.04	10.4	4.78	5.85	2.82	5.47
0.218	0.125	5.49	2.26	4.87	2.56	3.67	2.42

$$D_{BB} = \frac{D_{B0}}{1-(1-\lambda_{BB})\theta_B} \tag{2.35}$$

Table 2.3 shows experimental flux data for the binary system together with predictions based on binary system and single gas component system.

It is obvious from Table 2.3 that prediction based on the binary system produces much better results.

References

1. Bird AJ, Trim DL (1983) Carbon molecular sieves used in gas separation membranes. Carbon 21 (3): 177-180
2. Hsieh HP (1988) Inorganic membranes. AIChE Symp Ser 84 (261): 1-18
3. Rao MB, Sircar S (1993) Nanoporous carbon membranes for separation of gas mixtures by selective surface flow. J Membr Sci 85 (3): 253-264
4. Jones CW, Koros WJ (1995) Characterization of ultramicroporous carbon membranes with humidified feeds. Ind Eng Chem Res 34 (1): 158-163
5. Koresh JE, Soffer A (1987) The carbon molecular sieve membranes. General properties and the permeability of CH_4/H_2 mixtures. Sep Sci Technol 22 (2-3): 973-982
6. Centeno TA, Fuertes AB (1999) Supported carbon molecular sieve membranes based on phenolic resin. J Membr Sci 160 (2): 201-211
7. Jones CW, Koros WJ (1994) Carbon molecular sieve gas separation membranes-1. Preparation and characterization based on polyimide precursors. Carbon 32 (8): 1419-1425
8. Koresh JE, Soffer A (1986) Mechanism of permeation through molecular-sieve carbon membrane. Part 1. The effect of adsorption and the dependence on pressure. J Chem Soc Faraday Trans I 82 (7): 2057-2063
9. Fuertes AB (2000) Adsorption-selective carbon membrane for gas separation. J Membr Sci 177 (1-2): 9-16
10. Ismail AF, Shilton SJ, Dunkin IR, Gallivan SL (1997) Direct measurement of rheologically induced molecular orientation in gas separation hollow fiber membranes and effects on selectivity. J Membr Sci 126 (1): 133-137
11. Singh A, Koros WJ (1996) Significance of entropic selectivity for advanced gas separation membranes. Ind Eng Chem Res 35 (4): 1231-1234
12. Chen YD, Yang RT (1994) Preparation of carbon molecular sieve membrane and diffusion of binary mixtures in the membrane. Ind Eng Chem Res 33 (12): 3146-3153

Chapter 3
Configurations of Carbon Membranes

3.1 Flat (Supported and Unsupported) Carbon Membranes

Carbon membranes can be divided into two categories: unsupported and supported carbon membranes [1]. Unsupported membranes have three different configurations: flat (film), hollow fibre and capillary, while supported membranes consist of two configurations: flat and tube. Figure 3.1 shows the configurations of carbon membranes.

Porous carbon films have been prepared from Kapton-type polyimide (PI) to produce supported and unsupported carbon membranes by Hattori et al. [2–4]. They reported that the carbon molecular sieve (CMS) film used for gas separation should be as thin as possible in order to enhance the separation efficiency. However, the thin film should be supported by a porous plate for handling convenience. The flat homogeneous carbon films prepared by pyrolysis at 800°C yielded O_2/N_2 selectivities of 4.2 [4].

Rao and Sircar [5–7] introduced nanoporous supported carbon membranes which were prepared by pyrolysis of PVDC layer coated on a macroporous graphite disk support. The diameter of the macropores of the dried polymer film was reduced to the order of nanometer as a result of a heat treatment at 1,000°C for 3 h. These membranes with mesopores could be used to separate hydrogen–hydrocarbon mixtures by the surface diffusion mechanism, in which gas molecules were selectively adsorbed on the pore wall. This transport mechanism is different from the molecular sieving mechanism. Therefore, these membranes were named as selective surface flow (SSF^{TM}) membranes. It consists of a thin (2–5 μm) layer of nanoporous carbon (effective pore diameter in the range of 5–6 Å) supported on a mesoporous inert support such as graphite or alumina (effective pore diameter in the range of 0.3–1.0 μm). The procedures for making the selective surface flow membranes were described in [5, 7]. In particular, the requirements to produce a surface diffusion membrane were shown clearly in [7].

A solution to overcome reproducibility problems of nanoporous carbon (NPC) membranes has been introduced by Acharya and Foley [8]. They have used spray

A. F. Ismail et al., *Carbon-based Membranes for Separation Processes*,
DOI 10.1007/978-0-387-78991-0_3, © Springer Science+Business Media, LLC 2011

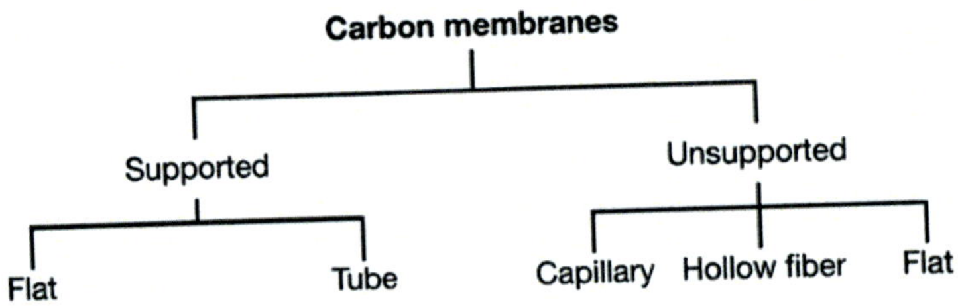

Fig. 3.1 Configurations of carbon membranes

coating system for the production of thin layers of NPC on the surface of a porous stainless steel support. A solution of PFA in acetone was sprayed onto the support in the form of a fine mist using an external mix airbrush with nitrogen gas. That was the first reported case of the technique being used for supported carbon membrane synthesis. The advantages of this preparation method are reproducibility, simplicity, giving good performance for oxygen—nitrogen separation. The resulting membranes were found to have oxygen over nitrogen selectivities up to 4 and oxygen fluxes of the order of 10^{-9} mol/m^2 s Pa.

Chen and Yang [9] prepared a large, crack-free carbon molecular sieve membrane (CMSM) supported on a macroporous substrate by coating a layer of PFA followed by controlled pyrolysis. Diffusion of binary mixtures was measured and the results were compared with the kinetic theory for predicting binary diffusivities from pure component diffusivities. Good agreement was obtained between theoretical predictions and experimental data for binary diffusion, as shown in Chap. 2.

Suda and Haraya [10] were successful in preparation of flat, asymmetric carbon molecular sieve membranes, which exhibited the highest gas permselectivities among those fabricated in the past research by pyrolysis of a Kapton type PI derived from pyromellitic dianhydride (PMDA) and 4,4′-oxydianiline (ODA). They used the permeation measurements and X-ray powder diffraction to relate the relationship between the gas permselectivity and microstucture of the CMSM. They proposed that the decrease of the interplanar spacing, amorphous portion and pores upon heating might be the origin of the "molecular sieving effect".

Suda and Haraya [11] also clarified the factors that determined the micro-structure and the permeation properties of CMS dense membranes derived from Kapton PI film [11]. They have gained insight into the permeation mechanism through the study of permeability versus kinetic diameter in connection with diffusivity and sorptivity. The authors suggested that the factors determining the micro-stucture and the gas permeation properties of CMSMs are not completely governed by the precursor because the permeation properties are significantly affected by several factors: the choice of polymer precursor, the membrane formation method and the pyrolysis condition.

Shusen et al. [12] used one-step preparation method of asymmetric supported carbon molecular sieve membranes, consisting of the formation of phenol formaldehyde film followed by pyrolysis and unequal oxidation steps. Micro-pores were formed as a result of small gaseous molecules channeling their way out of the solid matrix of the polymer during pyrolysis. The micropore structure was further widened by oxidation, which removed carbon chains in the pores. The pore structure was narrowed by high temperature sintering. All of the preparation conditions which lead to the shrinkage of the pore of the carbon membrane would be beneficial for improvement of selectivities while the conditions for widening the pore size would be favorable for increasing permeabilities [13]. They proposed that the key point to create a carbon membrane with asymmetric structure was to keep different oxidation atmospheres on two sides of the membranes in the activation process, for example, a relatively strong activation condition on one side and a relatively weak activation condition on the opposite side [12, 13].

Kita et al. [14] synthesized an unsupported polypyrrolone film by means of a casting method. The authors found that the membranes which had been carbonized at 700°C for 1 h gave the highest performance. The membranes exhibited excellent stability up to 500°C without weight loss.

The flat carbon membranes were produced for gas separation from coal tar pitch by Liang et al. [15]. The result showed that the separation power of carbon membranes prepared from coal tar pitch was generally higher by at least three orders of magnitude compared with polymeric membranes.

The preparation method of flat supported carbon molecular sieve membranes has been investigated by using different polymeric materials by Fuertes and Centeno. They used 3,3′4,4′-biphenyltetracarboxylic dianhydride (BPDA)—4,4′-phenylene diamine (pPDA) [1, 16], phenolic resin [17] as precursor to make flat CMSMs supported on a macroporous carbon substrate. In a later study, they chose polyetherimide (PEI) as a precursor to prepare flat supported CMSMs [18]. PEI was chosen because it was one of PI based materials which can be used economically. On the other hand, these PEI carbon membranes showed performance similar to the CMSMs prepared by Hayashi et al. [19], which was obtained from a laboratory-synthesized PI (BPDA-ODA).

Furthermore, they also used two other commercially available PI type polymers with the trade names Matrimid and Kapton to prepare supported carbon composite membranes in a single casting step [20]. They reported that the different structures of carbon membranes could be obtained depending on the polymeric precursors, the casting solution and the preparation condition. However, preparation conditions had a great effect on the structure and separation properties of the Matrimid-based carbon membranes. Recently, they investigated the preparation of supported CMSM formed by a microporous carbon layer, obtained by carbonization of a poly(vinylidene chloride-co-vinyl chloride) (PVDC-co-PVC) film [21]. They discovered that the preoxidation of PVDC-co-PVC samples did not significantly affect the micropore volume of the carbonized materials. However, the pretreatment in air at 200°C for 6 h led to a less permeable carbon membrane than the untreated membrane but with higher selectivity. The selectivity of CO_2/N_2 was increased from 7.7 to 13.8 after the pretreatment.

There are researchers involved in the study of flat sheet homogeneous membrane particularly focused on the development of entropic contributions to diffusivity selectivity as the polymer matrix evolved to a rigid carbon matrix [22]. Polymer precursor membranes pyrolyzed at intermediate steps in the pyrolysis process and finally pyrolyzed membranes were tested for the purpose to study the development of gas separation properties as the material progresses from a polymer to a completely carbonized membrane.

3.2 Carbon Membranes Supported on Tube

The CMSMs have been produced by dip-coating of BPDA-ODA solution on an α-Al_2O_3 porous support tube followed by pyrolysis at 500–900°C in an inert atmosphere by Hayashi et al. [19]. The sorptivity and diffusivity of penetrants in the carbonized membrane were greatly improved by carbonization because of the increased micropore volume (free space) and segmental stiffness. The carbonization procedure was optimized and excellent permselectivities of penetrants were obtained.

The researchers modified the resulting CMSMs by chemical vapor deposition (CVD) using propylene as the carbon source at 650°C [23]. The study showed that the CVD modification was effective to increase the CO_2/N_2 and O_2/N_2 selectivity because the pore structure was further controlled and the micropores were narrowed.

Hayashi et al. [24, 25] also found that a carbonized membrane prepared with a BPDA-pp'ODA PI gave higher C_3H_6/C_3H_8 and C_2H_4/C_2H_6 permselectivities than those of polymeric membranes. This was in accord with the fact that carbonized membranes possessed a micropore structure, which was capable of recognizing size differences of alkane and alkene molecules. Additionally, Hayashi et al. [25] evaluated the stability of a membrane based on BPDA-ODA PI and carbonized at 700°C by exposing it to air at 100°C for 1 month. It was suggested that the CMSMs were usable for a prolonged period in an atmosphere which contained low levels of oxidants. Their study also showed that the permeation properties of carbon membrane could be improved by treating carbon membrane under an oxidizing atmosphere [25]; i.e., they oxidized BPDA-ODA carbon membrane with a mixture of O_2–N_2 at 300°C or with CO_2 at 800–900°C and found the excessive oxidation fractured the carbon membrane. The researchers concluded that carbonization under the optimum condition shifted the trade-off relationship of the BPDA-pp'ODA PI membrane toward the direction of higher selectivity and permeability [24].

Microporous carbon membranes have been prepared [26] by carbonization and activation of an asymmetric phenolic resin structure comprised of a dense resol layer, and supported on a highly permeable macroporous novolak resin tube.

BPDA-ODA/2,4-diaminotoluene (DAT) copolyimide, which contains methyl groups, was used as a precursor for CMSM coated on a support tube by Yamamoto et al. [27]. Methyl groups were expected to be decomposed during the post-treatment under an oxidative atmosphere and result in expanded micropores. They reported that the permeation properties of the resulting membranes were depending on the composition of the precursor films, carbonization temperature and oxidation condition. In spite of the permeance that increased with increasing permeation temperature, the separation coefficients were not greatly influenced by the oxidation and carbonization treatments. They suggested that the oxidation in air by increasing temperature up to 400°C with a 1-h hold and carbonization up to 700°C was most suitable for increasing permeance with no adverse effect on separation coefficient. The trade-off line for the BPDA-ODA carbon membrane for O_2/N_2 system was threefold higher in the direction of separation coefficient than that for PI membrane reported by Stern. Researchers have concluded that optimization of the treatment procedure was more important than changes in diamine portion of the co-polyimide.

A further study regarding CMSMs have been made by Kusakabe et al. [28] in which CMSMs were formed by carbonizing BPDA-pp′ODA PI membranes at 700°C and then oxidized with either an O_2–N_2 mixture or pure O_2 at 100–300°C under controlled conditions. The thin polymer layer was formed on the outer surface of α-alumina tube by dip-coating. The study showed that the oxidation increased permeance without greatly damaging the permselectivities. This was because the oxidation at 300°C for 3 h significantly increased the micropore volume but the pore size distribution was not broadened. The result was similar to the previous research [25] regarding the effect of oxidation on gas permeation of CMSMs based on BPDA-pp′ODA polyimide.

They also formed the condensed polynuclear aromatic (COPNA) resin film on a porous α-alumina support tube. Next, a pinhole-free CMSM was produced by carbonization at 400–1,000°C [29]. The mesopores of the COPNA-based carbon membranes did not penetrate through the total thickness of each membrane and served as channels which increased permeances by linking the micropores. CMSMs produced using COPNA and BPDA-pp′ODA polyimdes showed similar permeation properties even though they had different pore structures. This suggests that the micropores are responsible for the permselectivities of the carbonized membrane. Besides that, Fuertes [30] used phenolic resin in conjunction with the dip coating technique to prepare adsorption-selective carbon membrane supported on ceramic tubular membranes.

There are other different coating methods on porous stainless steel support media in the production of carbon membranes supported on tube including: brush coating; spray-coating and ultrasonic deposition of the polymer resin. For example, Shiflett and Foley reported various approaches to prepare carbon molecular sieve layers on the stainless steel support by ultrasonic deposition [31].

Alternatively, Wang et al. [32] used a gas phase coating technique, vapor deposition polymerization (VDP) to prepare supported carbon membranes from furfuryl alcohol. They reported that the membranes prepared by VDP had comparable CO_2/CH_4 selectivities to but lower CO_2 permeabilities than certain PFA-based membranes prepared by dip-coating techniques.

3.3 Carbon Capillary Membranes

Asymmetric capillary CMSMs were prepared using Kapton PI and their gas permeation properties reported by Haraya et al. [33]. Capillary CMSM must have a controlled asymmetric structure, consisting of a dense surface layer with molecular sieving properties and a porous support layer in order to attain both high permselectivity and permeance. However, it is not easy to control the structure of the capillary CMSM. The researchers described that the structure of membrane was constructed in the gelation step of polyamic acid (PAA) and was also maintained in the imidization step. However, the membrane was shrunk by about 30% at the pyrolysis steps. They observed that the surface layer became thinner and the pore dimension became larger with acceleration in the exchange rate of solvent with coagulant. Slow gelation process would result in a thicker dense surface layer.

Petersen et al. prepared CMSMs (capillary tubes) by using a precursor derived from Kapton [34]. An integral asymmetric capillary carbon membrane was prepared by coagulation of a PAA solution which was imidized to a Kapton capillary and finally pyrolyzed to a capillary carbon membrane.

3.4 Carbon Hollow Fiber Membranes

A number of special techniques were summarized by Linkov et al. [35], which had been developed to obtain narrow pore-size distribution in carbon membranes. Those techniques consist of introduction of monomers with low carbon residual into polyacrylonitrile (PAN), irradiation of polymer films with high-energy ions, in-situ polymerization on the surface of dip-coated polymeric precursors, treatment with concentrated hydrazine solution and the dispersion of a finely divided inorganic material in the casting solution of PAN. They reported that the carbonization of highly asymmetric PAN precursors, produced by the use of various combinations of solvent and non-solvents in precipitation media, resulted in the formation of a range of flexible hollow fiber carbon membranes with high porosity and good mechanical properties. Morphology of the membranes as well as the possibility of altering the pore structure was studied. It was suggested that precursor preparation (solution formulation and fabrication procedure) and stabilization as well as carbonization conditions had possibility to alter the pore sizes of carbon membranes.

The VDP method was used to coat hollow fiber carbon membranes by Linkov et al. [36]. Then the coated membranes were heated in a nitrogen atmosphere to produce composite carbon-PI membranes. The composite membranes had small wall and active skin thickness with good mechanical properties. They have resistance against high pressures and have high flexibility.

Polyimide derived from a reaction of 2,4,6-trimethyl-1,3-phenylene diamine, 5,5-[2,2,2-trifluoro-1-(trifluoromethyl)ethylidene]-1,3-isobenzofurandione and 3,3′,4,4′-biphenyl tetra carboxylic acid dianhydride was used by Jones and Koros to prepare carbon molecular sieve asymmetric hollow fiber membranes [37]. These membranes were developed and optimized for air separation applications. However, they were also effective for the separation of other gas mixtures such as CO_2/N_2, CO_2/CH_4 and H_2/CH_4. The selectivities obtained were much higher than those found for conventional polymeric materials without sacrificing productivity.

Jones and Koros found potential problems or weaknesses of carbon membranes in their studies [38, 39]. Carbon membranes generally have nonpolar surfaces. As a result, they are organophilic. Therefore, ultra-microporous carbon membranes would be very vulnerable to adverse effects from exposure to organic contaminants due to its adsorption characteristics of organics. Membrane performance will be lost severely if feed streams have as low as 0.1 ppm organics. As adsorption of organic compounds proceeds, capacity for other compounds is diminished and losses in membrane performance occur rapidly. Once a monolayer has been established, resistance to other permeating species will be prohibitive. However, a unique regeneration technique developed by Jones and Koros [39], seems to be very promising for removing a number of organic contaminants. Pure propylene at unit or near-unit activity was found to be suitable for the regeneration process. The propylene most likely acted as a cleaning agent, removing other sorbed compounds from the carbon surface. Propylene exposure resulted in a small "opening up" of the pore structure and membrane performance was recovered.

Jones and Koros [38] also found that the micropores of carbon membranes would gradually be plugged with water at room temperature, resulting in decrease of permeabilities of non-polar gases and selectivities. The reason is the surface of membrane carbonized at relatively low temperature is affected by oxygen remaining in the inert purge gas during pyrolysis [40]. The surface is partially covered with oxygen containing functional groups, thus giving the membrane a hydrophilic character [28]. The resulting oxygen-containing surface complexes will act as primary sites for water sorption. Sorbed water molecules then attract additional water molecules through hydrogen bonding, leading to the formation of clusters. The cluster grows and coalesces, leading to bulk pore filling. As the amount of sorbed water in microporous carbon adsorbents increases, it will greatly diminish the diffusion rate of other permeating species [38].

The problem can be overcome by coating the membrane with a highly hydrophobic film, which does not prohibitively reduce the flux of other permeating species. Therefore, the resulting carbon composite membranes demonstrate a greater resistance to the adverse effects from water vapor while retaining very good separation properties [41]. Thus, Kusakabe et al. [28] reported that the modification of the surface properties of CMSM is a key technology for the selective gas separation.

The effect of PI pyrolysis conditions on CMSM properties was studied by Geiszler and Koros [40]. They compared the carbon hollow fibre membrane performances prepared by vacuum pyrolysis and inert purge pyrolysis. In addition, they also studied other pyrolysis variables such as the processing temperature, purge gas flow rate and residual oxygen concentration in the purge gas. They observed that pyrolysis atmospheres and flow rates of purge gas strongly influenced H_2/N_2 and O_2/N_2 selectivities of CMSMs. It is noteworthy that pyrolysis condition has significant influence on the carbon membranes performance.

Kusuki et al. [42] made the asymmetric carbon membranes by carbonization of asymmetric PI hollow fiber membranes. The effects of different experimental conditions on the membrane performance were investigated. They reported that those carbon membranes showed high permselectivities.

Asymmetric carbon membranes were made by carbonization of asymmetric PI hollow fiber membranes by Tanihara et al. [43]. In their study, it was found that the permeation properties of carbon membranes were hardly affected by feed pressure and exposure to toluene vapor. Furthermore, there was only little change in the permeation properties of the carbon membrane with the passage of time.

Ogawa and Nakano [44] investigated the effect of gelation conditions on the properties of the carbonized hollow fiber membrane. The carbonized hollow fiber membrane was formed by gelation of PAA film, imidization and carbonization. The micro-structure of the carbonized membrane was evaluated by the micropore volumes, which depended on gelation temperature and pH of coagulant. Hence, the gelation process was important to control micro-structure, permeance and permselectivity of the carbonized membrane. They observed that the gelation time was not a predominant factor to control the micropore volume, the permeance and CO_2/CH_4 permselectivity. However, they found that the gelation temperature would influence the permeation properties of the carbon membranes.

Table 3.1 Configurations of CMSM that appeared in the literature until year 2000

Researcher(s)	Configuration	Period (year)	Reference(s)
Ash et al.	–	1973	[46]
Bird and Trimm	Flat	1983	[47]
Koresh and Soffer	Hollow fiber	1980–1987	[48–51]
Bauer et al.	Tube	1991	[52]
Hattori et al.	Flat	1992	[2–4]
Rao and Sircar	Flat	1993, 1996	[5–7]
Chen and Yang	Flat	1994	[9]
Linkov et al.	Hollow fiber	1994	[35, 36]
Jones and Koros	Hollow fiber	1994–1996	[37–39, 41]
Hayashi et al.	Tube	1995–1997	[19, 23–25]
Suda and Haraya	Flat, capillary	1995–1997	[10, 11, 33, 53]
Shusen et al.	Flat	1996	[12, 13]
Kita et al.	Flat	1997	[14]
Petersen et al.	Capillary	1997	[34]
Kusuki et al.	Hollow fiber	1997	[42]
Katsaros et al.	Tube	1997	[26]
Yamamoto et al.	Tube	1997	[27]
Kusakabe et al.	Tube	1998	[28, 29]
Geiszler and Koros	Hollow fiber	1996	[40]
Tanihara et al.	Hollow fiber	1999	[43]
Okamoto et al.	Hollow fiber	1999	[54]
Liang et al.	Flat	1999	[15]
Acharya and Foley	Flat	1999	[8]
Fuertes and Centeno	Flat	1998–2000	[1, 16–18, 20, 21, 30]
Ogawa and Nakano	Hollow fiber	1999–2000	[44, 45]
Ghosal and Koros	Flat	2000	[22]
Shiflett and Foley	Tube	2000	[31]
Wang et al.	Tube	2000	[32]

Moreover, they reported that the permeance decreased with the increase of pH value of the coagulant. The micropore volume also decreased remarkably in the alkali region, resulting in reduction of micropore size. However, the permselectivity increased with the increase of pH. The researchers proposed that the most important factor to achieve both high permeance and high permselectivity in the carbonized membrane was pH control of the coagulant (water). They realized that the high permeance of CO_2 and the high CO_2/CH_4 permselectivity were obtained under the specific conditions of gelation: time 6 h, temperature 275 K and pH 9.4. They concluded that the transport of CO_2 was mainly governed by the adsorption effect and transport of CH_4 was restricted by the molecular sieving effect, yielding high CO_2/CH_4 permselectivity [44].

The researchers also investigate the difference in permeation behaviors of CO_2 and CH_4 between the single component and the multi-component system from the view point of the microporous structure, which was created through the formation process of the carbonized membrane [45].

Table 3.1 summarizes the configurations of carbon membranes found in the literature. It is noticed that most of the carbon membranes produced from the 1980s to early the 1990s are flat disk or flat sheet membranes. Only in the middle of the 1990s, carbon membranes supported on tubes were fabricated, followed by carbon capillary membranes and carbon hollow fiber membranes. Flat sheet carbon membranes are more suitable for laboratory or research applications while carbon membranes supported on tube, carbon capillary membranes and carbon hollow fiber membranes are more practical and suitable to apply in industry.

References

1. Fuertes AB, Centeno TA (1998) Preparation of supported asymmetric carbon molecular sieve membranes. J Membr Sci 144 (1-2): 105-111
2. Hatori H, Yamada Y, Shiraishi M (1992) Preparation of macroporous carbon films from polyimide by phase inversion method. Carbon 30 (2): 303-306
3. Hatori H, Yamada Y, Shiraishi M, Nakata H, Yoshitomi S (1992) Carbon molecular sieve films from polyimide. Carbon 30 (2): 305-306
4. Hatori H, Yamada Y, Shiraishi H, Nakata H, Yoshitomi S (1992) Carbon molecular sieve films from polyimide. Carbon 30 (4): 719-720
5. Rao MB, Sircar S (1993) Nanoporous carbon membranes for separation of gas mixtures by selective surface flow. J Membr Sci 85 (3): 253-264
6. Rao MB, Sircar S (1996) Performance and pore characterization of nanoporous carbon membrane for gas separation. J Membr Sci 110 (1): 109-118
7. Rao MB, Sircar S (1993) Nanoporous carbon membrane for gas separation. Gas Sep Purif 7 (4): 279-284
8. Acharya M, Foley HC (1999) Spray-coating of nanoporous carbon membranes for air separation. J Membr Sci 161 (1-2): 1-5
9. Chen YD, Yang RT (1994) Preparation of carbon molecular sieve membrane and diffusion of binary mixtures in the membrane. Ind Eng Chem Res 33 (12): 3146-3153
10. Suda H, Haraya K (1995) Molecular sieving effect of carbonized kapton polyimide membrane. J Chem Soc Chem Commun: 1179-1180

11. Suda H, Haraya K (1997) Gas permeation through micropores of carbon molecular sieve membranes derived from kapton polyimide. J Phys Chem B 101 (20): 3988-3994
12. Shusen W, Meiyun Z, Zhizong W (1996) Asymmetric molecular sieve carbon membranes. J Membr Sci 109 (2): 267-270
13. Shusen W, Meiyun Z, Zhizong W (1996) Carbon membrane for gas separation. Sep Sci Technol 31 (16): 2299-2306
14. Kita H, Yoshino M, Tanaka K, Okamato K (1997) Gas permselectivity of carbonized polypyrrolone membrane. J Chem Soc Chem Commun: 1051-1052
15. Liang CH, Sha GY, Guo SC (1999) Carbon membrane for gas separation derived from coal tar pitch. Carbon 37 (9): 1391-1397
16. Fuertes AB, Centeno TA (1999) Preparation of supported carbon molecular sieve membrane. Carbon 37 (4): 679-684
17. Centeno TA, Fuertes AB (1999) Supported carbon molecular sieve membranes based on phenolic resin. J Membr Sci 160 (2): 201-211
18. Fuertes AB, Centeno TA (1998) Carbon molecular sieve membranes from polyetherimide. Microporous Mesoporous Mater 26 (1-3): 23-26
19. Hayashi J, Yamamoto M, Kusakabe K, Morooka S (1995) Simultaneous improvement of permeance and permselectivity of 3,3′4,4′-biphenyltetracarboxylic dianhydride-4,4′-oxydianiline polyimide membrane by carbonization. Ind Eng Chem Res 34 (12): 4364-4370
20. Fuertes AB, Nevskaia DM, Centeno TA (1999) Carbon composite membranes from matrimid and kapton polyimide. Microporous Mesoporous Mater 33 (1-3): 115-125
21. Centeno TA, Fuertes AB (2000) Carbon molecular sieve gas separation membranes based on poly(vinylidene chloride-co-vinyl chloride). Carbon 38 (7): 1067-1073
22. Singh-Ghosal A, Koros WJ (2000) Air separation properties of flat sheet homogeneous pyrolytic carbon membranes. J Membr Sci 174 (2): 177-188
23. Hayashi J, Mizuta H, Yamamoto M, Kusakabe K, Morooka S (1997) Pore size control of carbonized BPTA-pp′ODA polyimide membrane by chemical vapor deposition of carbon. J Membr Sci 124 (2): 243-251
24. Hayashi J, Mizuta H, Yamamoto M, Kusakabe K, Morooka S (1996) Separation of ethane/ethylene and propane/propylene systems with a carbonized BPDA-pp′ODA polyimide membrane. Ind Eng Chem Res 35 (11): 4176-4181
25. Hayashi J, Yamamoto M, Kusakabe K, Morooka S (1997) Effect of oxidation on gas permeation of carbon molecular sieve membranes based on BPDA-pp′ODA polyimide. Ind Eng Chem Res 36 (6): 2134-2140
26. Katsaros FK, Steriotis TA, Stubos AK, Mitropoulos A, Kanellopoulos NK, Tennison S (1997) High pressure gas permeability of microporous carbon membranes. Microporous Mater 8 (3): 171-176
27. Yamamoto M, Kusakabe K, Hayashi J, Morooka S (1997) Carbon molecular sieve membrane formed by oxidative carbonization of a copolyimide film coated on a porous support tube. J Membr Sci 133 (2): 195-205
28. Kusakabe K, Yamamoto M, Morooka S (1998) Gas permeation and micropore structure of carbon molecular sieving membranes modified by oxidation. J Membr Sci 149 (1): 59-67
29. Kusakabe K, Gohgi S, Morooka S (1998) Carbon molecular sieving membranes derived from condensed polynuclear aromatic (COPNA) resins for gas separation. Ind Eng Chem Res 37 (11): 4262-4266
30. Fuertes AB (2000) Adsorption-selective carbon membrane for gas separation. J Membr Sci 177 (1-2): 9-16
31. Shiflett MB, Foley HC (2000) On the preparation of supported nanoporous carbon membranes. J Membr Sci 179 (1-2): 275-282
32. Wang H, Zhang L, Gavalas GR (2000) Preparation of supported carbon membranes from furfuryl alcohol by vapor deposition polymerization. J Membr Sci 177 (1-2): 25-31
33. Haraya K, Suda H, Yanagishita H, Matsuda S (1995) Asymmetric capillary membrane of a carbon molecular sieve. J Chem Soc Chem Commun: 1781-1782

34. Petersen J, Matsuda M, Haraya K (1997) Capillary carbon molecular sieve membranes derived from kapton for high temperature gas separation. J Membr Sci 131 (1-2): 85-94
35. Linkov VM, Sanderson RD, Jacobs EP (1994) Highly asymmetrical carbon membranes. J Membr Sci 95 (1): 93-99
36. Linkov VM, Sanderson RD, Rychkov BA (1994) Composite carbon-polyimide membranes. Mater Lett 20 (1-2): 43-46
37. Jones CW, Koros WJ (1994) Carbon molecular sieve gas separation membranes- I. Preparation and characterization based on polyimide precursors. Carbon 32 (8): 1419-1425
38. Jones CW, Koros WJ (1995) Characterization of ultramicroporous carbon membranes with humidified feeds. Ind Eng Chem Res 34 (1): 158-163
39. Jones CW, Koros WJ (1994) Carbon molecular sieve gas separation membranes – II. Regeneration following organic exposure. Carbon 32 (8): 1427-1432
40. Geiszler VC, Koros WJ (1996) Effect of polyimide pyrolysis conditions on carbon molecular sieve membrane properties. Ind Eng Chem Res 35 (9): 2999-3003
41. Jones CW, Koros WJ (1995) Carbon composite membranes: A solution to adverse humidity effects. Ind Eng Chem Res 34 (1): 164-167
42. Kusuki Y, Shimazaki H, Tanihara N, Nakanishi S, Yoshinaga T (1997) Gas permeation properties and characterization of asymmetric carbon membranes prepared by pyrolyzing asymmetric polyimide hollow fiber membrane. J Membr Sci 134 (2): 245-253
43. Tanihara N, Shimazaki H, Hirayama Y, Nakanishi S, Yoshinaga T, Kusuki Y (1999) Gas permeation properties of asymmetric carbon hollow fiber membranes prepared from asymmetric hollow fiber. J Membr Sci 160 (2): 179-186
44. Ogawa M, Nakano Y (1999) Gas permeation through carbonized hollow fiber membranes prepared by gel modification of polyamic acid. J Membr Sci 162 (1-2): 189-198
45. Ogawa M, Nakano Y (2000) Separation of CO_2/CH_4 mixture through carbonized membrane prepared by gel modification. J Membr Sci 173 (1): 123-132
46. Ash R, Barrer RM, Lowson RT (1973) Transport of single gases and of binary gas mixtures in a microporous carbon membrane. J Chem Soc Faraday Trans I 69 (12): 2166-2178
47. Bird AJ, Trimm DL (1983) Carbon molecular sieves used in gas separation membranes. Carbon 21 (3): 177-180
48. Koresh JE, Soffer A (1987) The carbon molecular sieve membranes. General properties and the permeability of CH_4/H_2 mixture. Sep Sci Technol 22 (2-3): 973-982
49. Koresh JE, Soffer A (1986) Mechanism of permeation through molecular - sieve carbon membrane. Part 1. The effect of adsorption and the dependence on pressure. J Chem Soc Faraday Trans I 82 (7): 2057-2063
50. Koresh JE, Soffer A (1980) Study of molecular sieve carbons. Part 1. Pore structure, gradual pore opening and mechanism of molecular sieving. J Chem Soc Faraday Trans I 76 (12): 2457-2471
51. Koresh JE, Soffer A (1983) Molecular sieve carbon permselectivities membrane Part I. Presentation of a new device for gas mixture separation. Sep Sci Technol 18 (8): 723-734
52. Bauer JM, Elyassini J, Moncorge G, Nodari T, Totino E (1991) New developments and application of carbon membranes. Key Eng Mater 61-62: 207-212
53. Suda H, Haraya K (1997) Alkene/alkane permselectivities of a carbon molecular sieve membrane. J Chem Soc Chem Commun: 93-94
54. Okamoto K, Kawamura S, Yoshino M, Kita H, Hirayama Y, Tanihara N, Kusuki Y (1999) Olefin/paraffin separation through carbonized membranes derived from an asymmetric polyimide hollow fiber membrane. Ind Eng Chem Res 38 (11): 4424-4432

Chapter 4
Preparation of Carbon Membranes

4.1 Precursor Selection

The properties of the polymeric precursor are probably one of the most important factors to be considered in order to produce a CMSM of high quality. Because of the variety of polymeric precursors available for carbon membrane preparation, the most suitable characteristics for carbon membrane must be first identified for preparing successful carbon membranes as the pyrolysis of different precursors may result in different types of carbon membrane. From practical experience and knowledge gathered during carbon membrane preparation exercises, two most dominant characteristics have been identified; thermo-resistance and molecular arrangement at molecular levels. These characteristics allow further fine-tuning of membrane morphology as well as separation performance optimization. In addition to high temperature resistance, thermosetting polymers neither liquefy nor soften at any stage of pyrolysis [1]. Also, suitable precursor materials for carbon membrane preparation will not cause any pin-holes or cracks to appear after the pyrolysis step [2]. Another important reason why polymer precursors are preferred is that the resulting membranes are invariably prepared with fewer impurities than those derived from activated carbon precursor [3].

At present, different polymeric materials such as polyimides (PIs) and the polyimide (PI)-like polymers [4–12], phenolic resins (PRs) [13–15], polyfurfuryl alcohol (PFA) [16], and phenol formaldehyde resins (PFRs) [17, 18], cellulose [19] and others are utilized for polymeric precursors. These polymeric precursors have thermosetting properties, i.e. they do not melt when heated and thus can retain structure and shape during heating and pyrolysis. Among the possible polymeric precursors, PIs have been studied most intensively for carbon membrane preparation, due to their excellent physical properties and the tunable chemical compositions utilizing different molecular structures composed of dianhydride and diamine monomers. Figure 4.1 shows some examples of polyimides. The most common commercial PI used for preparing carbon membrane is Kapton (DuPont trademark) PI. CMSM with homogeneous fine pores and without cracks or large pores normally can be achieved by carbonizing Kapton film at 800°C [20].

A. F. Ismail et al., *Carbon-based Membranes for Separation Processes*,
DOI 10.1007/978-0-387-78991-0_4,

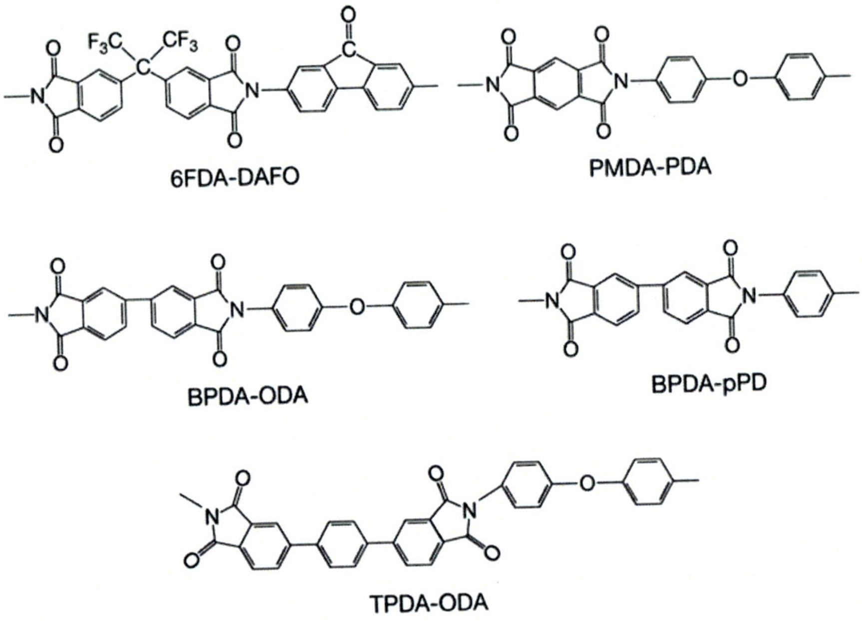

Fig. 4.1 Chemical structures of some polyimide (PI) precursors

In the following sections some of the most typical polymeric precursors are discussed.

4.1.1 Polyacrylonitrile (PAN)

In the area of carbon fiber production, PAN fibers have been recognized as one of the most important and promising precursors for producing high performance carbon fibres. It dominates nearly 90% of world wide sales of carbon fiber [21]. There are numerous advantages of PAN fibers, including a high degree of molecular orientation, higher melting point (PAN fiber tends to decompose below its melting point, T_m of 317–330°C) and a greater yield of the carbon fiber [22].

PAN fibers form a thermally stable, highly oriented molecular structure when subjected to a low temperature heat treatment, which is not significantly disturbed during the carbonization treatment at higher temperatures. This means that the resulting carbon fibers have good mechanical properties [22]. One of the main problems of unsupported carbon membrane (i.e. hollow fiber) is brittleness. Therefore, by using PAN which is widely employed in the production of high strength carbon fiber, this problem can be minimized [23].

Schindler and Maier [24] were the first researchers involved in the preparation of the PAN carbon membrane based on their US patent issued in 1990. They prepared porous capillary carbon membranes by nitrogen gas pyrolysis system at pyrolysis temperatures ranging from 800 to 1,600°C. Before preoxidation and carbonization, the precursor membranes were subjected to pretreatment with hydrazine solution. Later in 1992, Yoneyama and Nishihara [25] from Mitsubishi Rayon, Japan, patented their invention on the preparation of porous hollow fiber carbon membranes.

During the period of 1994–1995, Linkov et al. conducted research on the properties of PAN based carbon membranes by blending acrylonitrile-methyl methacrylate copolymer [26] with polyethylene glycol (PEG) [27], and polyvinylpyrrolidone (PVP) [27] in order to alter the pore size distribution of the carbon membranes. The membrane precursors that include PAN, blend of PAN-PEG 600, PAN-PEG 20,000 and PAN-PVP 30,000 were spun into hollow fibers and the hollow fibers were subjected to differential scanning calorimetry (DSC). As Fig. 4.2 shows, in the case of PVP containing precursor the temperature of cyclization of nitrile groups increased from 309 to 330°C. The infrared study also could confirm that there is stronger association between the copolymer and PVP than between the copolymer and PEG. Figure 4.3 shows the pore size distribution of the carbonized membranes measured by the mercury intrusion method. The pore sizes are close to 100 Å and the pore size distribution was narrower when the precursor contained PVP. With PEG the pore size distribution became broader with an increase in molecular weight.

Furthermore, PAN hollow fiber carbon membranes were coated with zeolites [28], PI [29] and perfluoro-sulfonated ionomer (PSI) [30] to produce composite membranes. After this period, there was not much research involving the use of PAN as a precursor for carbon membranes until late 2003, when Davis and Ismail [31] prepared their carbon hollow fiber membrane from PAN precursor.

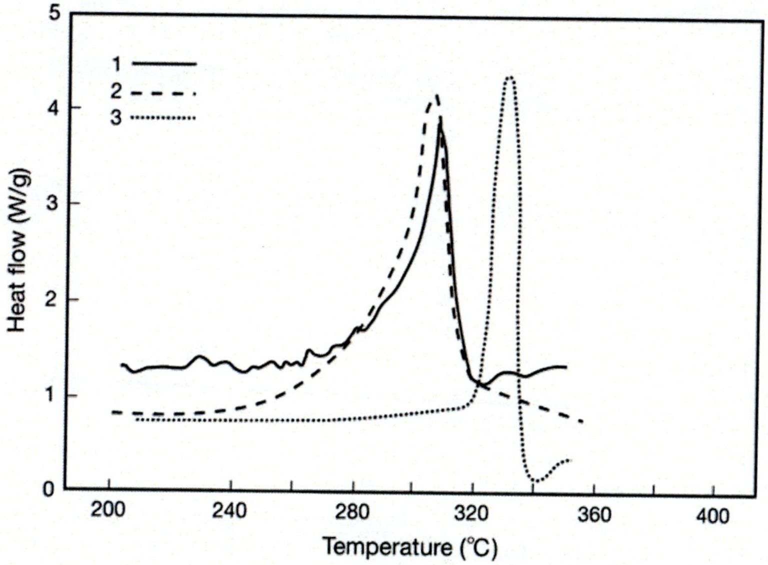

Fig. 4.2 DSC thermograms of precursors for carbon membranes; (*1*) pure PAN, (*2*) PAN-PEG 20,000, and (*3*) PAN-PVP 30,000. (From [27])

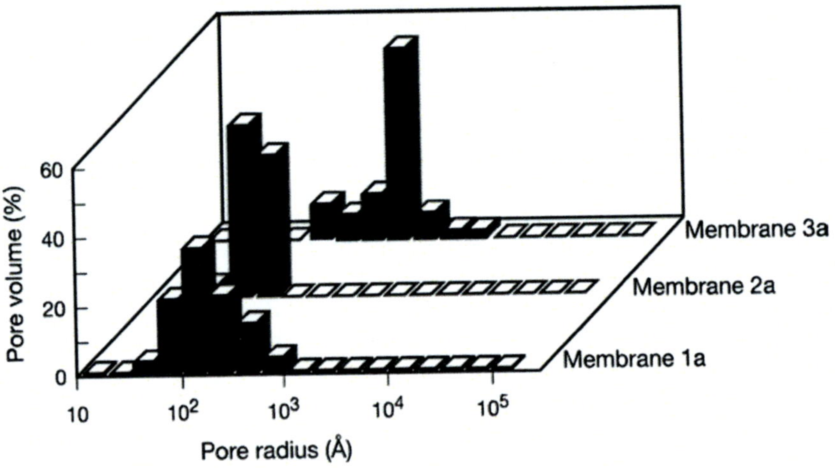

Fig. 4.3 Mercury intrusion diagrams of carbon membranes produced from A-type precursors containing low carbon residual polymers: (*1a*) PEG 600, (*2a*) PVP 30,000, and (*3a*) PEG 20,000. (From [27])

4.1.2 Polyimide and Derivatives

Polyimides (PIs) are among the most thermally stable classes of polymers and can typically be heated above 300°C decomposing before their melting points are reached. Because PIs do not go through a melting phase transition and lose their shape, PIs are good precursors for glassy carbon [31]. Many researchers have used PI as their precursor for making carbon membranes (Table 4.1). For example, Jones and Koros [32] reported that the best carbon membranes, in terms of both separation and mechanical properties, were produced from the pyrolysis of aromatic PIs. Moreover, Hattori et al. [33] found that PIs blended with PEG were also good precursors for porous carbon membranes.

PIs are rigid, high melting point, high glass transition temperature (T_g), thermally stable polymers synthesized by the condensation reactions of dianhydrides with diamines (Fig. 4.4) [34]. Normally, PIs are synthesized in lab scale with different types of dianhydrides and diamines in order to tailor its separation properties when being used as membrane material (Fig. 4.4).

A typical synthesis of PI might involve separately suspending powder 3,3′,4,4′-biphenyltetracarboxylic dianhydride (BPDA) and 4,4′-oxydiamine (ODA) in a solvent such as N,N′-dimethylacetamide (DMAC) and mixing each suspension together drop-wise under an inert environment to produce a homogeneous polyamic acid (PAA) solution [36]. Then, the PAA is subjected to an imidization process, which transforms PAA into PI polymer. Further details about the preparation of PIs can be found in the following papers [35, 37–39]. One of common commercial PIs used for preparing carbon membrane is Kapton (DuPont trade name) PI. Hattori et al. [40] prepared successfully CMSMs with homogeneous fine pores and without cracks or

Table 4.1 Representative examples of carbon membrane precusors and pyrolysis conditions

Precursor	Configuration	Pyrolysis conditions (temperature[a]; heating rate[b]; soak time[c]; atmosphere)	Reference
Acrylonitrile	Hollow fiber	800–1,600; –; 5–60; N_2	[24]
Acrylonitrile	Hollow fiber	600–1,200; –; 10; N_2	[25, 104]
PAN	Hollow fiber	250–800; 9; 10–180; N_2	[31, 91]
PAN	Hollow fiber	600–950; 1; –; N_2	[26]
PAN	Hollow fiber	900; 5; –; N_2	[27]
Cellulose	Hollow fiber or supported film	400–900; 1–10; –; Ar	[2]
Cellulose	Hollow fiber or supported film	500–800; –; 12 h; Ar	[95]
Cellulose	Hollow fiber	120–400; 0.1–0.6; –; Ar	[81, 105]
Coal and pitch	Plate	600–900; 1; –; N_2	[80]
Condensed poly-nuclear aromatic (COPNA)	Supported film	400–1,000; 5; 0; N_2	[64]
Kapton and matrimid	Supported film	450–700; 0.5; 1 h; vacuum	[42]
Kapton and matrimid	Film	500–800; 1 or 4; 2 or 8 h; vacuum	[87]
Kapton polyimide	Film	800; 3; –; –	[40, 106]
Kapton polyimide	Film	1,000; 10; 2 h; vacuum	[100, 107]
Kapton polyimide	Film	600–1,000; 10; 2 h; vacuum or Ar	[89]
Kapton polyimide	Capillary	950; 5; 120; –	[41]
Phenol formaldehyde	Flat	800–950; –; 120–180; N_2	[108]
Phenol formaldehyde	Tubular	800; 50; –; N_2	[109]
Phenol formaldehyde	Supported film	800; –; –; vacuum	[110]
Phenol formaldehyde	Supported film	900; 0.5; 60; Ar	[15]
Phenol resin	Supported film	700–850; –; 3 h; N_2 or CO_2	[111]
Phenolic resin	Supported film	900; 25°C/h; 1 h; N_2	[112]
Phenolic resin	Supported film	500–1,000; 5; –; vacuum	[12]
Phenolic resin	Supported film	700; 0.5; –; vacuum	[78]
Phenolic resin	Supported film	700; –; –; vacuum	[47, 48]
Phenolic resin	Supported film	700; –; –; vacuum	[14]
Phenolic resin	Supported film	800; 50; –; N_2	[113]
Phenolic resin	Supported film	700; –; –; vacuum	[46]
Poly(dimethyl silane)	Supported film	300–950; –; –; Ar	[114]
Poly (paraphenylene pyromellitimide) blend with PEG	Film	600; 3; 1 h; Ar	[33]
PVDC-AC	Supported film	600; 1; 3 h; N_2	[115]
PVDC-AC	Supported film	1,000; 1; 3 h; N_2	[116]
PVDC-AC	Supported film	1,000; 1; 3 h; N_2	[117]
PVDC-AC	Supported film	600; 1; 3 h; N_2	[118]
PVDC-AC	Supported film	500–1,000; 1; –; vacuum	[79]
PEI	Supported film	800; 0.5; 1 h; vacuum	[43]
PEI	Supported film	350; 1; 30; Ar and followed by 600°C for 4 h	[44, 119]
PFA	Supported film	500–800; 10; 2 h; He or N_2	[51]

Table 4.1 (continued)

Precursor	Configuration	Pyrolysis conditions (temperature[a]; heating rate[b]; soak time[c]; atmosphere)	Reference
PFA	Supported film	600; 5; 1–2 h; He	[54]
PFA	Supported film	300; 1.5; 2 h; N_2 and followed by 500°C for 6 h	[59]
PFA	Supported film	150–600; 5; 0–2 h; He or N_2	[55]
PFA	Supported film	450; 5; 120; He	[56]
PFA	Supported film	150–600; 5; 0–120; He	[52]
PFA	Supported film	200–600; 5; 120; He	[53]
PFA	Supported film	600; 5; 1–2 h; He	[57]
PFA	Supported film	450; 5; 120; He	[58]
PFA	Supported film	450–600; 1; 60; –	[60]
Polyimide	Supported film	550–700; 0.5; –; vacuum	[70]
Polyimide	Supported film	550; 0.5; 1 h; vacuum	[37]
Polyimide	Supported film	800; 0.25; 2 h; vacuum	[92]
Polyimide	Supported film	500–900; 5; 0; N_2	[39, 120]
Polyimide	Supported film	700–800; 5; 0; Ar	[60, 121]
Polyimide	Supported film	600–900; 5; 0; N_2	[99]
Polyimide	Supported film	500–700; 5; 0; N_2	[35]
Polyimide	Hollow fiber	500–550; 0.25–13.3; 2 h; vacuum	[32, 88, 102]
Polyimide	Hollow fiber	500–550; 0.25–13.3; 2 h; vacuum and inert gas	[31, 85]
Polyimide	Hollow fiber	550 and 800; 0.25; 2 h; vacuum or He	[34, 122]
Polyimide	Hollow fiber	600–900; 1; 1 h; N_2	[93]
Polyimide	Hollow fiber	750; –; 60; vacuum	[77]
Polyimide	Hollow fiber	600–1,000; –; 3.6; N_2	[74]
Polyimide	Hollow fiber	500–900; –; 0.5s–20; N_2	[123]
Polyimide	Hollow fiber	750; 2.6; 3 h; N_2 or 850; 2.6; 3 h; vacuum	[38, 124]
Polyimide	Hollow fiber	500–700; 5; –; N_2	[76]
Polyimide	Hollow fiber	500–900; –; 0.1–30; N_2 or He or Ar	[75]
Polypyrrolone	Film	300–800; 5–8; 1–10 h; N_2	[125]
Sulfonated phenolic resin	Supported film	250–800; 5; 1.5 h; N_2	[49]

[a] °C unless otherwise stated
[b] °C/min unless otherwise stated
[c] Minute (min) unless otherwise stated

large pores, by carbonizing Kapton film at 800°C. Kapton PI was also successfully carbonized in the form of capillary membrane which can retain high gas permeability even at high temperatures [41]. The membrane exhibited a permselectivity of 2,000 for He/N_2 at 0°C and 170 at 250°C, respectively. Recent study on Kapton PI-based carbon membrane showed that the permselectivities for O_2/N_2, CO_2/CH_4 and CO_2/N_2 are 4, 16 and 9, respectively, at 25°C [42].

Polymer	Chemical structure
Matrimid® 5218 (BTDA-DAPI)	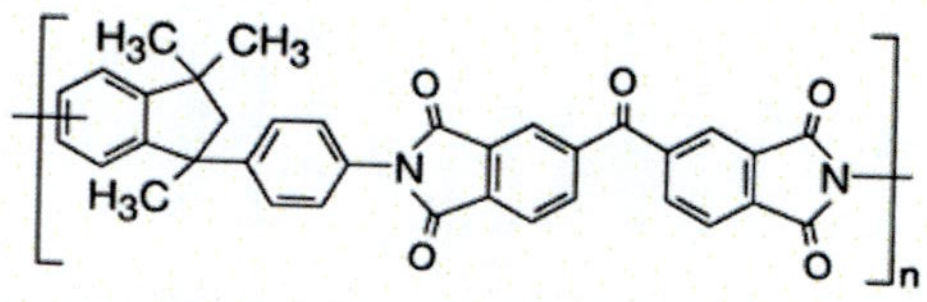 3,3′,4,4′-benzophenone tetracarboxylic dianhydride (BTDA) and 5(6)-amino-1-(4′-aminophenyl)-1,3-trimethylindane (DAPI)
6FDA/BPDA-DAM	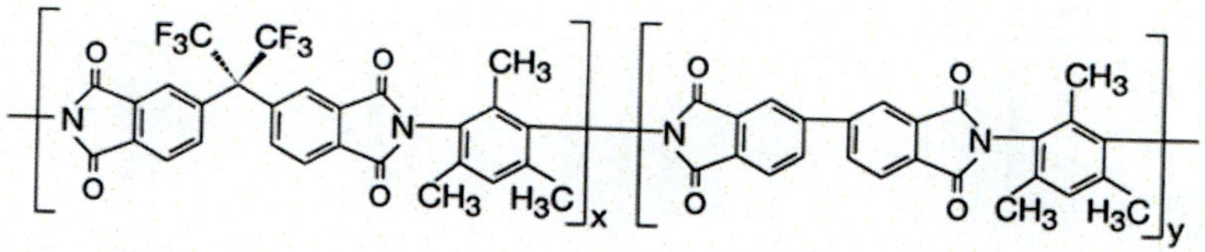 4,4′-(hexafluoroisopropylidene) diphthalic anhydride (6FDA), 3,3′-4,4′-biphenyl tetracarboxylic acid dianhydride (BPDA) and 2,4,6,-trimethyl-1,3-phenylene diamine (DAM)

Fig. 4.4 Basic molecular structures of polyimide precursors. (From [35])

Another economical PI used to form carbon membrane is polyetherimide (PEI), which is manufactured by General Electric under the trade name of Ultem® 1,000. This material has been used to prepare polymeric membranes due to its superior strength and chemical resistance [43, 44]. An early work regarding the use of PEI as a carbon membrane precursor was disclosed by Fuertes and Centeno [43]. They cast a small quantity of the polymer dope by spin coating on a finely polished surface of a porous carbon disk obtained by agglomeration of fine graphite particles with a phenolic resin, before it was gelled in isopropanol coagulation bath. The gelled polymeric layer was dried in air at room temperature and carbonized under vacuum by heating up to 1,073 K at the heating rate of 0.5 K/min and kept at 1,073 K for 1 h. SEM micrographs of precursor film and the carbonized membrane are shown in Fig. 4.5. Figure 4.5a shows a symmetric sponge like structure of the precursor film. Figure 4.5b shows clear distinction of the dense thin carbon film of about 3 μm thickness and the porous carbon support. Comparison of Fig. 4.5a and b suggests that the PEI passes through a plastic stage during carbonization. As a result, the structure of the porous polymeric membrane is completely changed into a dense and uniform carbon film after the carbonization. The fabrication of defect-free dense carbon layer as distinct from the porous sublayer was enabled by two factors. (1) The rapid coagulation of polymer solution in the gelation bath, which prevents the penetration of the polymer solution into the porous carbon substrate and (2) the plastic matter formed during the carbonization has a high viscosity and does not penetrate into the porous membrane support. Figure 4.6 shows the permeability of the carbon membrane for various gases. The permeability increases as the kinetic diameter changes from 3.8 Å (methane) to 2.6 Å (helium). The activation energy seems to increase as the molecular size increases. Both are the typical features of the sieve mechanism. The selectivity is higher at lower temperatures, as shown in Table 4.2.

Sedigh et al. [44] dip-coated a PEI film inside a ceramic tube and then carbonized it at 600°C under flowing argon for 4 h. The authors investigated the transport characteristics of their PEI carbon membranes using single gas (CH_4, H_2, CO_2), binary gas mixtures (CO_2/CH_4, H_2/CH_4, CO_2/H_2) and also ternary mixtures of $CO_2/H_2/CH_4$. CO_2/CH_4 separation factors as high as 145 for the equi-molar binary and 155 for the ternary mixture were obtained with this PEI-based CMSM. More recently, Countinho et al. [45] prepared PEI carbon hollow fiber membrane that can be served as a catalyst support in a membrane reactor. They investigated pyrolysis process parameters as well as stabilization parameters to find an optimum condition for preparing PEI based carbon membranes.

4.1.3 *Phenolic Resin*

Phenolic resins (PRs) are very popular and inexpensive polymers, which are employed in a wide range of applications, spanning from commodity construction to high technology materials [12]. Phenolic resins present desirable features when used as carbon membrane precursors:

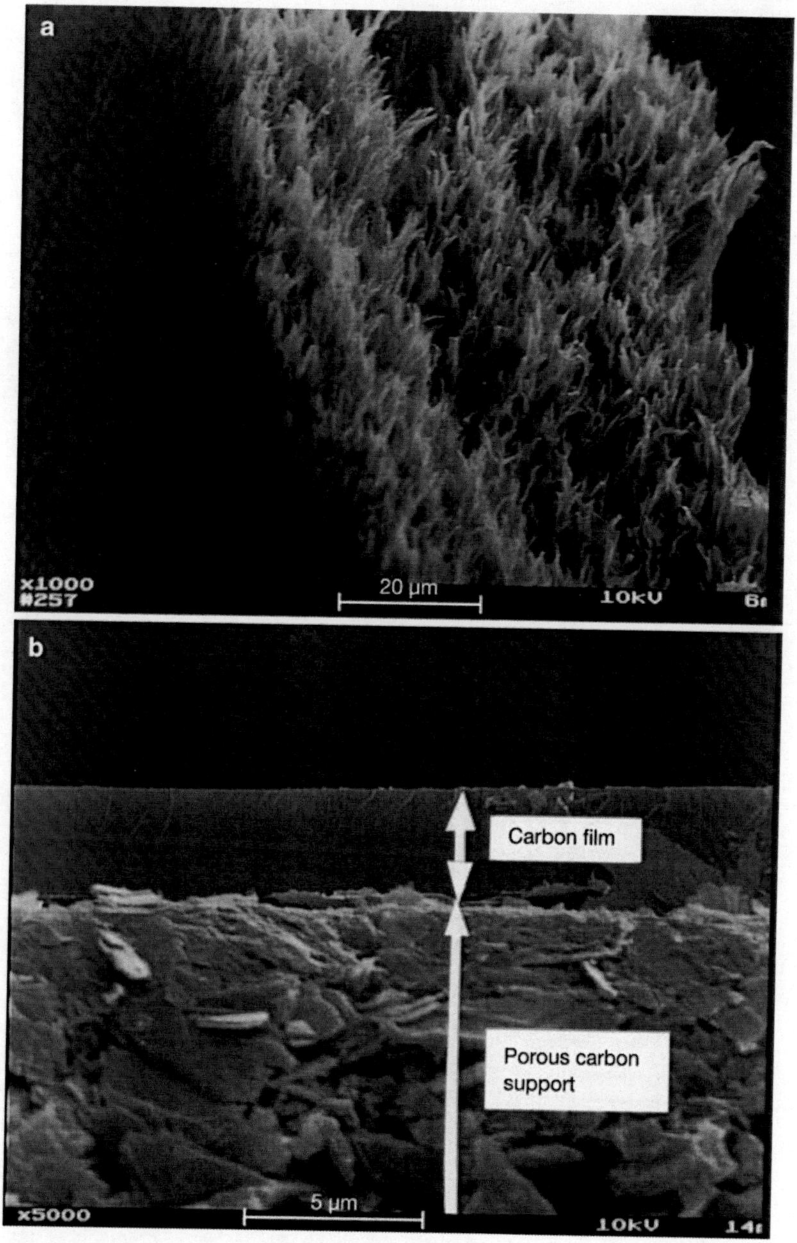

Fig. 4.5 Scanning electron microphotographs of (**a**) a polymeric film and (**b**) the supported CMSM. (From [43])

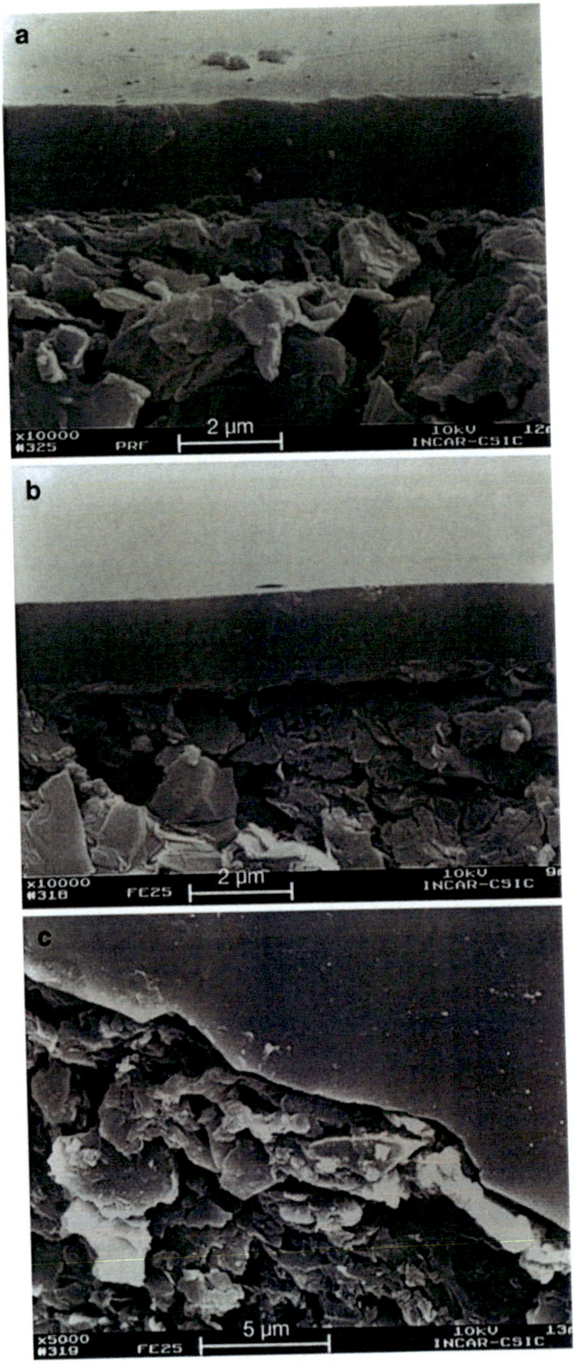

Fig. 4.6 SEM micrographs of membranes: (**a**) cross-section of phenolic resin film, (**b**) cross-section of carbon membrane and (**c**) top view of carbon membrane. (From [12])

Table 4.2 Permselectivity values of various gas pairs at different temperatures

Temperature (°C)	O_2/N_2	He/N_2	CO_2/CH_4	CO_2/N_2
25	7.4	121	25	15
100	6.7	42	31	14
150	5.1	20	20	9.1

1. The pyrolysis of PRs provides carbon films with molecular sieve properties [78].
2. PRs provide a high carbon yield after carbonization [12].
3. PRs are low cost polymers with numerous applications in different fields [78].

Fuertes and co-workers [12, 46–48, 78] investigated intensively the preparation of carbon membranes from PR precursors. According to Centeno and Fuertes [12], they coated a small quantity of liquid phenolic resin (Novolak type) on the finely polished surface of a porous carbon disk by means of a spin coating technique. The supported phenolic resin film was cured in air at 150°C for 2 h, and then carbonization was carried out in a vertical tubular furnace (Carbolite) at different temperatures (between 500 and 1,000°C) under vacuum. Figure 4.6 shows the SEM micrograph of the membranes. The polymeric film (Fig. 4.6a) coated on top of the porous substrate is dense and has a thickness of around 2 μm. The thickness of the carbon membrane shown in Fig. 4.6b is also about 2 μm. Figure 4.6c shows the top view of the fractured membrane. The top surface is very smooth. Helium gas permeance of membranes carbonized at different temperatures is shown in Fig. 4.7.

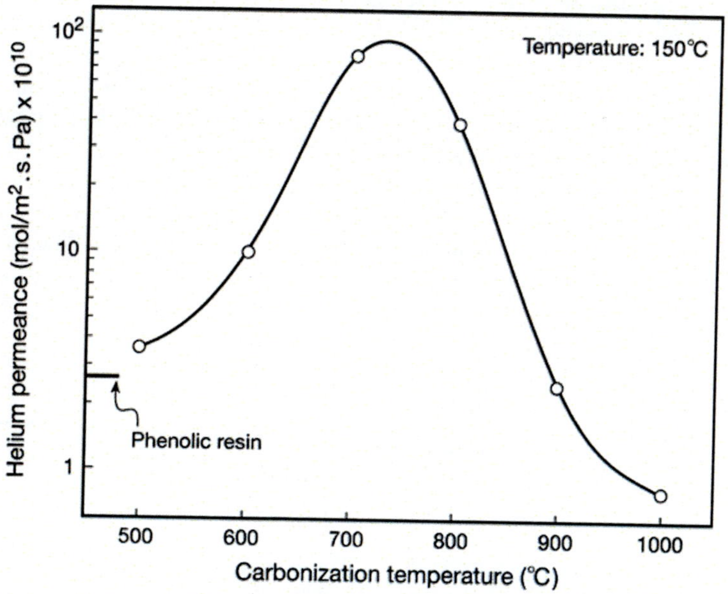

Fig. 4.7 Effect of carbonization temperature on helium permeance. (From [12])

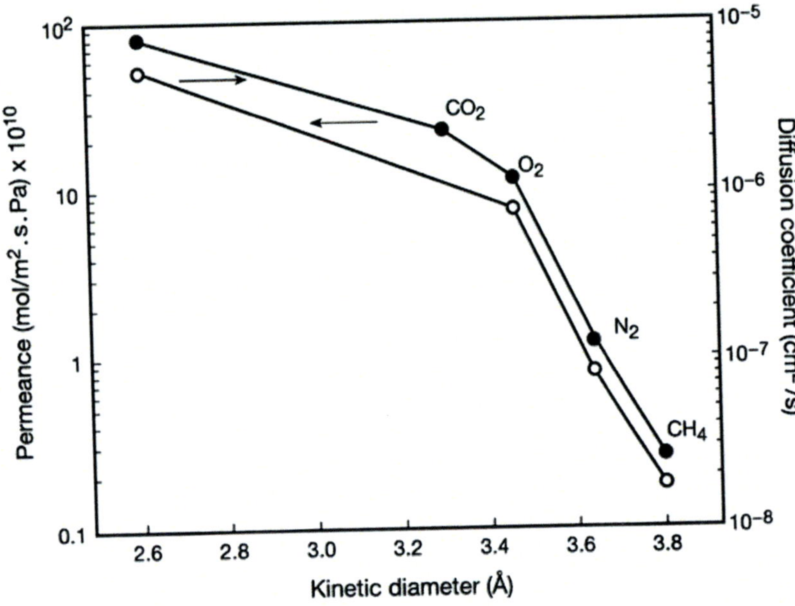

Fig. 4.8 Variation of permeance and diffusivity with the kinetic diameter of gases (carbonization temperature 700°C). (From [12])

The permeance shows maximum at 700°C. At this temperature He permeance was 80×10^{-10} mol/m^2 s Pa, which is 30 times as high as that of the cured phenolic resin film (2.7×10^{-10} mol/m^2 s Pa). The He permeance decreased at above 700°C. Heating to temperatures higher than 700°C caused the membrane pores to shrink and finally to disappear. Figure 4.8 shows the permeance of various gases. The permeance showed a sharp decrease for gases with a molecular size (kinetic diameter) in the range 3–4 Å. From the permeance and diffusivity data of CH_4, the pore size was calculated to be 4.2 Å for the membrane made by carbonization at 700°C. As for the selectivity of the membrane, Fig. 4.9 shows the permeance ratio for the CO_2/N_2 gas pair at various temperatures. Interestingly, remarkable increase in the separation factor was observed particularly when the temperature was as low as 25°C. Table 4.3 shows that there is little change in CO_2 permeance between the pure gas and mixed gas experiments. However N_2 permeance was less than half. This is the reason why the permeance ratio from the mixed gas experiments was much higher than the pure gas experiments. This fact suggests that the presence of CO_2 suppresses the N_2 gas transport. The same is true for CO_2 and CH_4, as the data indicated in Fig. 4.10 and Table 4.3 indicate.

Thus, the PR carbonized at 700°C showed an O_2/N_2 selectivity of around 10, and separation factors of 160 and 45 for CO_2/CH_4 and CO_2/N_2, respectively, in gas mixture separation experiments [12].

The PR was also used to prepare adsorption selective carbon membranes (ASCM) [47]. The main feature of this work is that the carbonized membrane is subjected to air oxidation at temperatures above 100°C for 0.5–6 h to increase the pore size

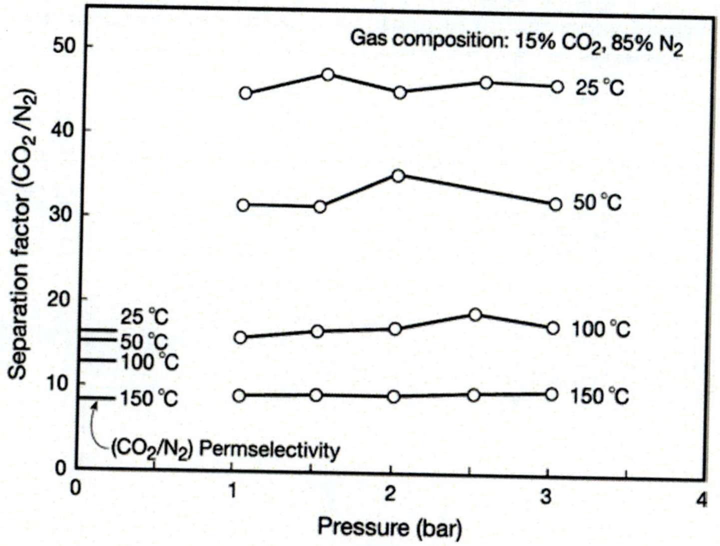

Fig. 4.9 Effect of operating pressure and temperature on the ideal separation factor of CO_2/N_2 mixtures (carbonization temperature 700°C). (From [12])

Table 4.3 Comparison of permeances from the pure gas and mixed gas experiments

	Permeance × 10^{10} (mol/m^2 s Pa)		
	CO_2	N_2	CH_4
Pure gas	20.0	1.1	0.23
15% CO_2 + 85% N_2	19.5	0.44	–
10% CO_2 + 90% CH_4	24.5	–	0.15

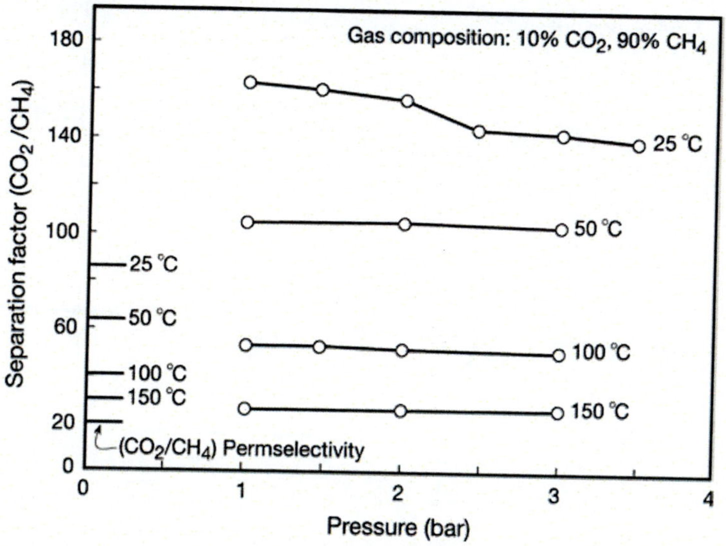

Fig. 4.10 Effect of operating pressure and temperature on the ideal separation factor of CO_2/CH_4 mixtures (carbonization temperature 700°C). (From [12])

of the CMSMs. By this way, the membrane is transformed from a molecular sieve to an adsorption selective membrane. Figure 4.11a shows the permeance versus kinetic diameter plot for the membrane carbonized at 700°C. The hydrocarbons larger than methane have kinetic diameters nearly equal to or larger than 4 Å and their

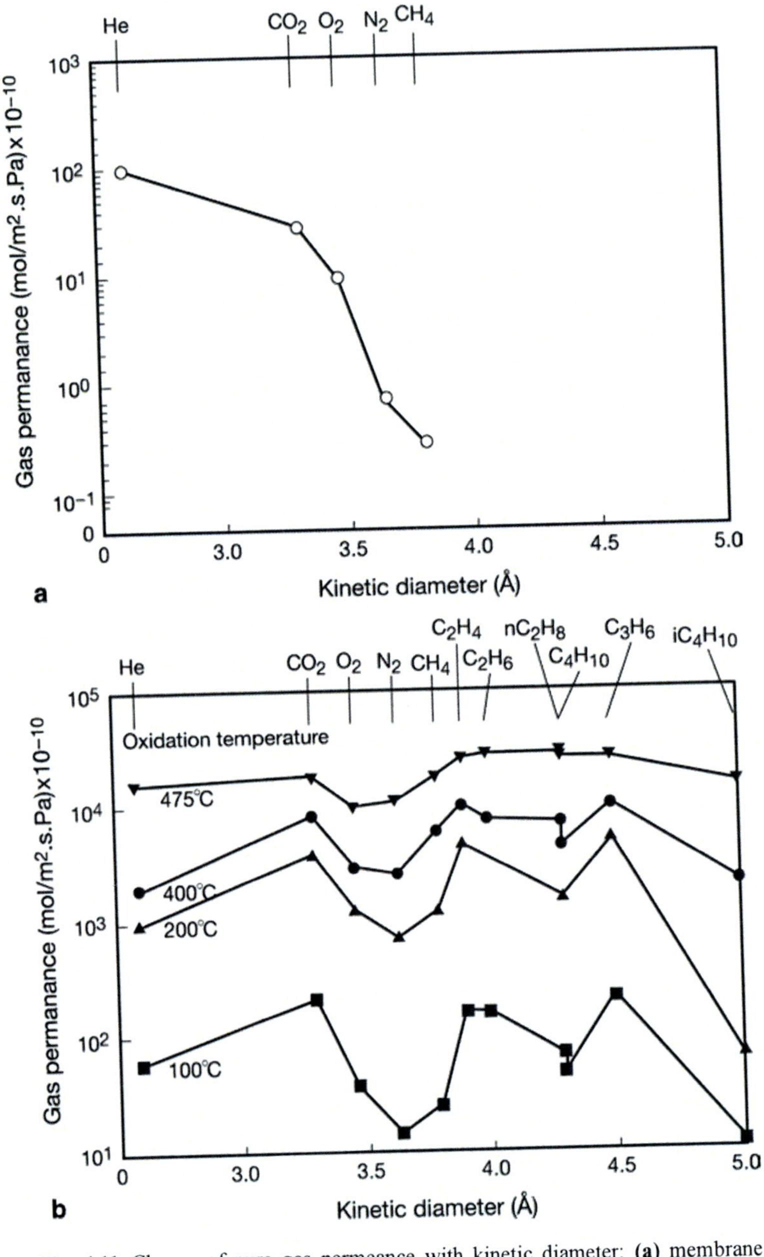

Fig. 4.11 Change of pure gas permeance with kinetic diameter; **(a)** membrane carbonized at 700°C, **(b)** carbon membranes oxidized at different temperatures. (From [47])

permeances are practically equal to zero. The curve shows that the gas transport is a typical sieve mechanism. Figure 4.11b shows the effect of the oxidation temperature on the pattern observed in permeance versus kinetic diameter curves. It can be observed that as a consequence of oxidation, the membranes are now permeable to hydrocarbons with kinetic diameters of more than 4 Å and hydrocarbons permeate at higher or comparable permeation rates than those of gases. A minimum in permeance is observed when the permeating gas is nitrogen, and the minimum is the deepest when the oxidation temperature is 100°C. This is because of the high adsorption of the condensable hydrocarbon vapor on the pore wall and its surface flow. This membrane seems particularly useful for the separation of non-adsorbable or weakly adsorbable gases (i.e. He, H_2, air, O_2, CH_4, CO_2, etc) from adsorbable gases such as hydrocarbons (C^{2+}), NH_3, SO_2, H_2S, etc. For example, a carbon membrane oxidized at 250°C for 0.5 h shows the following permeance (in mol/m^2 s Pa $\times$ 10^{-10}) and selectivity (α=hydrocarbon/N_2) values for the separation of a complex gas mixture consisting of 16.3% CH_4, 16.1% C_2H_6, 16.2% C_3H_8, 20% n-C_4H_{10}, 31.4% N_2. CH_4 172 (α=3), C_2H_6 700 (α=12), C_3H_8 1730 (α=30), n-C_4H_{10} 4270 (α=74).

Zhou et al. [49] introduced sulfonic acid groups in a thermosetting phenolic resin to make a carbon membrane precursor. The sulfonic acid groups evolve upon heating as small molecular gases or fragments such as sulfur dioxide and water, and leave void spaces in the thermoset matrix during the pyrolysis step. The decomposition of sulfonic acid groups occurred before substantial carbonization took place. A detailed description of the preparation method was provided by Zhou and co-workers [49].

4.1.4 Polyfurfuryl Alcohol

Poly(furfuryl alcohol) (PFA) is an amorphous polymer with a non-graphitizable structure, and potentially a good precursor to prepare CMSM [50]. Because of its simple molecular structure and formation mechanism, it is an appropriate material for fundamental experimental and atomistic simulation studies. Hence, Acharya [51] chose PFA based on the knowledge that PFA derived carbon could have desirable properties such as narrow pore size distribution and chemical stability. However, PFA did not appear to have the best mechanical and elastic properties required for forming a thin film on a rigid support [51]. Since PFA is in liquid state at room temperature, it can be used in supported film only.

PFA was used extensively as a precursor during the preparation of nanoporous carbon (NPC) membrane for gas separation by Foley and co-workers [52–58]. An early work had uncovered the use of a spray-coating technique in order to produce a thin layer of NPC in the porous surface of a stainless steel disk support [51, 54]. Selectivity for O_2/N_2 separation of up to 4 was achieved for the supported membrane carbonized at 600°C in flowing helium [54]. Later Foley and co-workers improved their preparation technique by using an ultrasonic deposition system which could uniformly distribute the polymeric solution on to the support [52, 53, 57]. They have fabricated a number of thin film supported nonporous carbon membranes

Table 4.4 Membrane preparation conditions and carbon coating results

SNPCM	PFA (wt%)	Temperature (K)	Time (min)	Coats	Mass (mg)	Thickness (μm)	Additives
5	25	873	120	6	33.9	20.9	
11	25	723	120	3	17.3	10.7	
17	25	573	120	3	18.6	11.4	
27	25	723	60	4	11.9	7.3	
28	25	723	0	5	15.0	9.3	
32	25, 10	723	120	3	14.8	9.1	
43	25	43, 723	60	4	7.7	4.7	
73	30	723	120	4	23.9	14.7	1 wt% SSZ-13
74	30	723	120	3	14.5	8.9	1 wt% SSZ-13
76	30	800, 723, 650	120	3	20.8	12.8	
77	30	800, 723, 650	120	3	18.4	11.4	1 wt% SSZ-13
83	30	723	120	3	17.2	10.6	10 wt% TiO_2
85	30	723	120	2	10.2	6.3	25 wt% PEG1500
86	30	723, 673	120	2	9.8	6.0	25 wt% PEG900
95	30	723	120	4	22.2	13.7	
96[a]	30	723	120	6	29.4	18.1	
97	30	723	120	3	25.6	15.8	15 wt% SSZ-13
511	30	673, 723	120, 60	5	349.0	18.2	

[a] Pyrolysed in hydrogen

(SNPCM) by their approach. The membrane preparation conditions are summarized in Table 4.4.

Table 4.5 summarizes the performance of the membranes. It is clear that the gas permeation is not by Knudsen diffusion even for the membranes of very high permeance.

Figure 4.12 shows the O_2/N_2 selectivity versus membrane mass per surface area. It can be noticed that the membranes with the highest selectivities are clustered about a carbon mass/surface area of 3.4 and 3.6 mg/cm^2. This suggests that there is a critical thickness of the membrane for high selectivity, which is 22 ± 1 μm based on the solid density of 1.6 g/cm^3. Below the critical value, the amount of defects is very large, providing Poiseuille flow. On the other hand, if the thickness is higher than the critical value, the film begins to crack, reducing the separation factor.

They successfully prepared reproducible NPC membranes with an improved O_2/N_2 selectivity which is within a range of 2–30. Even though SNPCM-43 membrane does not necessarily exhibit the highest selectivity, its separation might be high enough for some applications. In fact sufficiently large separation factors for H_2/CH_4, CO_2/CH_4, N_2O/N_2 and H_2/CO_2 of 600, 45, 17 and 14, respectively, could be achieved [52].

Other studies involving the use of PFA precursor for carbon membrane synthesis were done by Chen and Yang [59], Wang et al. [60], and Sedigh et al. [50].

Table 4.5 Pure gas permeances for various gases and separation factors

SNPCM	Permeance (mol/m^2 s Pa × 10^{10})					Separation factor			
	N_2	O_2	He	H_2	N_2O	O_2/N_2	He/N_2	H_2/N_2	N_2O/N_2
5	65.0	136	213	409		2.1	3.3	6.3	
11	0.0183	0.557	3.26	6.05		30.4	178	331	
17	0.0079	0.106	1.25	1.38		13.5	160	175	
27	0.372	2.33	6.28	13.0	5.23	6.3	16.9	35.1	14.1
28	0.143	0.984	3.22	8.14	2.48	6.9	22.4	56.8	17.3
32	0.254	2.04	6.82	14.0	3.43	8.0	26.9	55.0	13.5
43	1.24	6.20	9.02	25.6		5.0	7.3	20.6	
73	0.486	3.52	5.76	15.6		7.2	11.9	32.1	
74	0.320	2.87				9.0			
76	0.038	1.08	2.75	7.21		28.3	72.3	189	
77	0.043	0.433				10.2			
83	0.050	1.02	2.96	7.95		20.4	59.4	159	
85	17.6	77.7				4.4			
96	34.6	117		332		3.4		9.6	
95	0.280	2.78	5.83	17.5		9.9	20.8	62.4	
96	22.5	92.1				4.1			
97	0.28	1.75	4.37	13.3		6.3	15.6	47.6	
511	2.21	11.1				5.0			
SPCM 2,3,4	2.11	6.27				3.0			
Knudsen						0.94	2.65	3.73	0.80

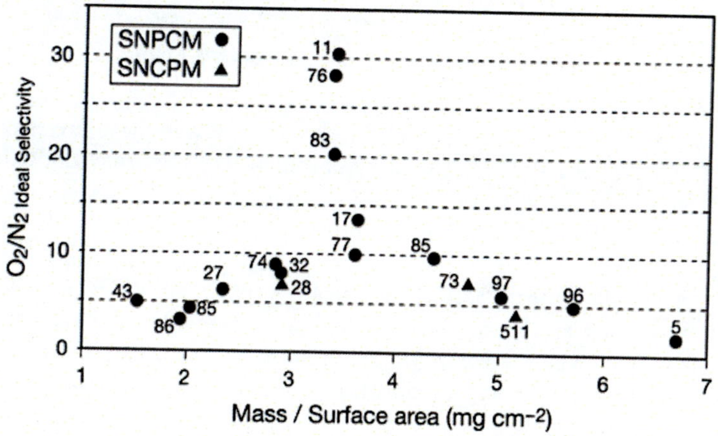

Fig. 4.12 O_2/N_2 selectivity versus membrane mass/surface area. (From [52])

4.1.5 *Recent Works on the CMSM Precursors*

Attempts are still being made to improve the performance of CMSMs. Several unique works were selected from the recent literature to be described more in detail as below.

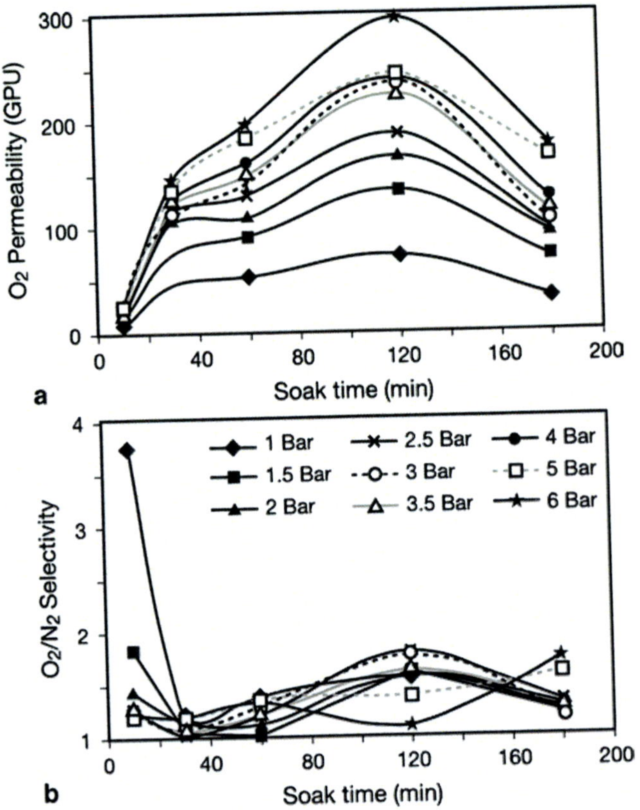

Fig. 4.13 Influence of soak time on (**a**) oxygen permeability, and (**b**) membrane selectivity for different feed pressures. (From [31])

Polyacrylonitrile hollow fibers were spun by the dry/wet phase inversion process by David and Ismail [31]. Thermostabilization was performed by oxidative stabilization process at 250°C for 30 min under pure oxygen or compressed air. The carbonization was done at 500°C under nitrogen for a certain amount of time, which is called "soak time".

Figure 4.13a shows the effect of the "soaking time" on the O_2 permeance. A "soak time" as short as 10 min gives the permeance smaller than PAN membrane of zero soak time since the carbonization is not completed. Partially carbonized membranes possess impermeable bulk structure and the formation of the porous structure is yet incomplete. The gas transport is difficult to occur either by solution diffusion or by sieving mechanism. The maximum permeance is achieved at 120 min. A maximum in the selectivity is also achieved at this soak time. At a longer soak time, the pore size is reduced and the permeance starts to decrease.

Combination of cross-linking and carbonization was applied to fabricate novel PI membrane to achieve high gas separation performance and better physical/chemical stability [61]. This was done by preparing PI from 4,4′-diaminodiphenylacetylene (p-intA) and 2,2′-bis (3,4-dicarboxyphenyl) hexafluoropropane dianhydride (6FDA). In order to avoid excessive cross-linking a co-PI containing 2,3,5,6-tetra-

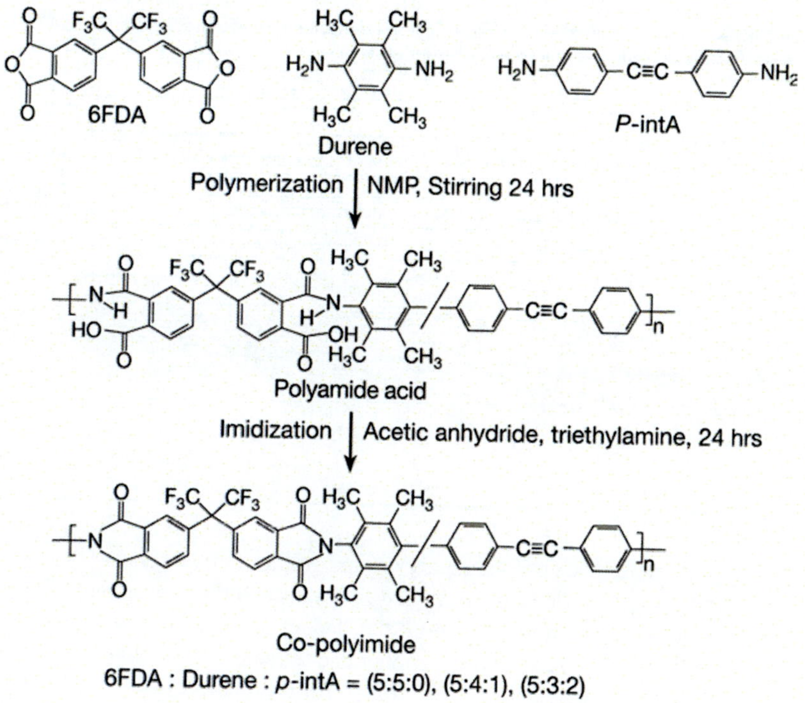

Fig. 4.14 The schemes of co-polyimde synthesis and structure. (From [61])

methyl-1,4-phenylenediamine (Durene) was synthesized. The reaction scheme of the co-PI is shown in Fig. 4.14.

The PIs were coded according to the Durene:p-intA ratio, such as 6FDA-D (no p-intA), 6FDA-D4A1 (Durene:p-intA=4:1) and 6FDA-D3A2 (Durene:p-intA =3:2). The thermal cross-linking was carried out at 400°C, while carbonization was up to 800°C. Depending on the heat cure temperature, the membranes were coded such as 6FDA-D-400 and 6FDA-D-800. Gas permeability and selectivity of these membranes are summarized in Table 4.6.

With respect to the membranes without thermal treatment, the permeability decreases while the selectivity increases as the amount of p-intA increases in the copolymer. As for the membranes cross-linked at 400°C, permeability increases and selectivity decreases slightly by heat treatment of 6FDA-Durene membrane without p-intA (Comparison of 6FDA-D and 6FDA-D-400). But when p-intA is incorporated, permeability decreases and selectivity increases by the heat treatment. The effect of heat treatment is enhanced with an increase in the amount of p-intA in the copolymer. This is due to the cross-linking that occurs at 400°C. After carbonization at 800°C, both permeability and selectivity increase noticeably (Comparison of 6FDA-D and 6FDA-D-800). By incorporating p-intA, permeability decreases slightly while the selectivity increases. The large increase in selectivity indicates the change of gas transport mechanism from solution-diffusion to molecular sieving.

Table 4.6 Gas permeabilties and permselectivities of polyimide precursors, cross-linked polyimides and cross-linked and carbonized polyimides

	Permeability (Barrer)				Selectivity	
	O_2	N_2	CH_4	CO_2	CO_2/CH_4	O_2/N_2
Polyimide						
6FDA-D	186	55	45	612	14	3.4
6FDA-D4A1	80	21	15	317	21	3.8
6FDA-D3A2	45	11	6.8	180	27	4.3
Cross-linked polyimide						
6FDA-D-400	218	67	57	728	13	3.3
6FDA-D4A1-400	52	13	7.5	186	25	4.1
6FDA-D3A2-400	28	5.7	3.9	113	29	4.9
Cross-linked and carbonized polyimide						
6FDA-D-800	399	61	44	2700	61	6.5
6FDA-D4A1-800	293	40	35	2430	69	7.3
6FDA-D3A2-800	257	28	16	1200	77	9.0

The polymer blends of PI and polyvinylpyrrolidone (PVP) was used as precursor, assuming that the partially intermingled PVP phase in the continuous PI matrix will produce porous carbon structures by the decomposition of PVP during pyrolysis [62]. PI was synthesized from benzophenone tetracarboxylic dianhydride (BTDA) and 4,4′-oxydianiline (ODA). Membranes were prepared by casting the polymer dope, either PI or PI/PVP blend, on a glass plate. The cast polymer film was imidized by stepwise heating up to 250°C and then carbonized at either 550 or 700°C.

Characterization was done by DSC, TGA and WAXD.

Figure 4.15 depicts the permeance of the prepared membranes for various gases. The trends in the displacing data are reflected exactly in the permeance data, i.e. permeance increases as PVP content in the precursor increases and the permeance decreases as the carbonization temperature increases. The selectivities for the gas pairs of He/N_2, CO_2/N_2 and O_2/N_2, on the other hand, show only slight decreases as the PVP content is increased in the precursor, but show remarkable increases as the carbonization temperature is increased.

For the fabrication of CMS membranes from commercially available relatively inexpensive polymeric material, Centeno and Fuertes chose poly(vinylidene chloride-co-vinyl chloride) (PVDC-co-PVC) copolymer commercially available under the trade name of Saran [63]. The polymer solution was spin-coated on a finely polished surface of porous carbon support. In some cases the polymer film was preoxidized in air at 150 or 200°C. The carbonization was carried out at either 500 or 1,000°C.

Figure 4.16 presents the effect of carbonization temperature on He permeance. The figure shows that the membrane carbonized at 700°C gives the maximum flux.

Interestingly, when the oxidative pretreatment of the membrane is carried out at 150°C for 2.5 days, there is no notable permselectivity. However, when the pretreatment is conducted at 200°C for 6 h, the permeance decreases but the selectivity increases (Fig. 4.17).

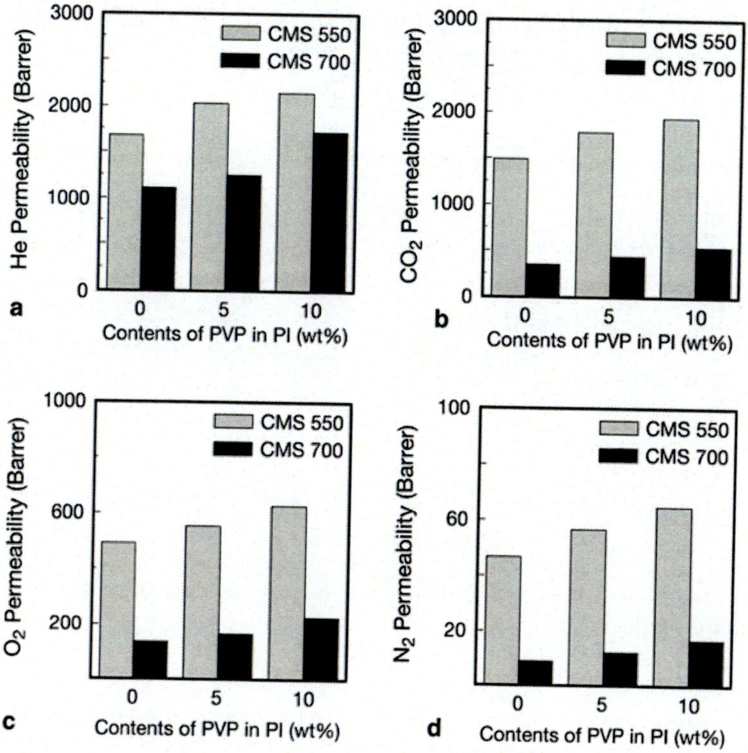

Fig. 4.15 Gas permeabilities as functions of the PVP content and the carbonization temperature. (From [62])

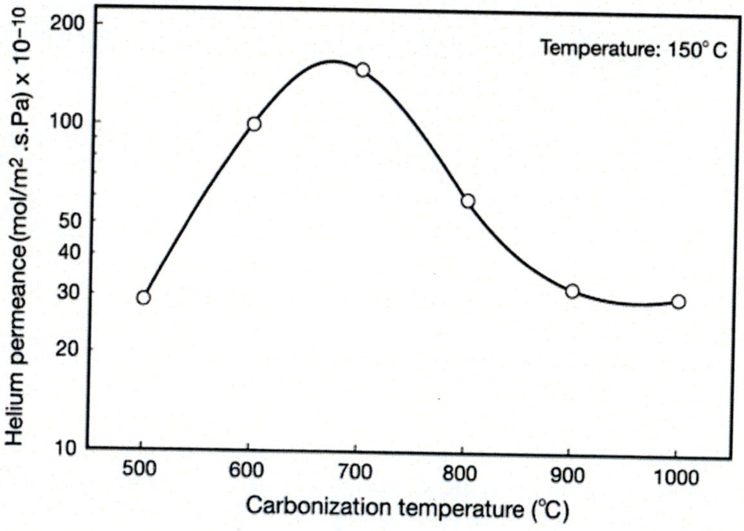

Fig. 4.16 Effect of carbonization temperature on helium permeance. (From [63])

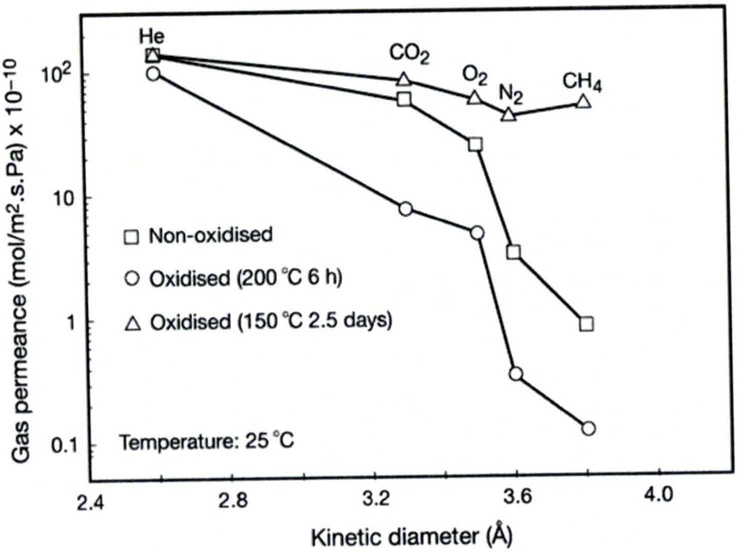

Fig. 4.17 Gas permeance versus kinetic diameter of gas molecules for different preoxidation temperatures. (From [63])

Condensed polynuclear aromatic (COPNA) resins were used as precursor for carbonization by Kusakabe and co-workers [64]. Since COPNA resin is a thermoplastic resin with three-dimensional structure, the pore structure of the carbonized COPNA membrane is expected to be different from that of the membrane fabricated by carbonization of PI which is linear. COPNA compounds were synthesized from polycyclic aromatic compound (PCA) such as pyrene, phenanthrene and 1,4-benzenedimethanol (BDM) using p-toluene sulfonic acid as the catalyst by the procedure given in Fig. 4.18.

In PCA the molar ratio of pyrene to phenanthrene was fixed to 7:3. The degree of cross-linking was changed by changing the BDM/PCA ratio. The thin layer of COPNA was prepared by dip-coating on the outer surface of a porous α-alumina tube. The film was then carbonized in a deoxygenated nitrogen stream by heating at a rate of 5 K/min to 400–1,000°C. The maximum permeance was obtained at 600°C of carbonization temperature (see Fig. 4.19). The permeance increased as the kinetic diameter of the gas decreased, indicating that the permeation mechanism was the sieving mechanism.

Trimethylsilyl (TMS) substituent was introduced to polyphenylene oxide (PPO) with n-BuLi is a catalyst. Homogeneous hollow fibers were spun by the dry-wet spinning method [65].

Figure 4.20 shows that the maximum permeability as well as selectivity was obtained at the carbonization temperature of 923 K for the gas pairs of CO_2/CH_4 and O_2/N_2 and Fig. 4.21 depicts that permeability increased but the selectivity decreased as the TMS content in PPO increased. The permeability decreased as kinematic

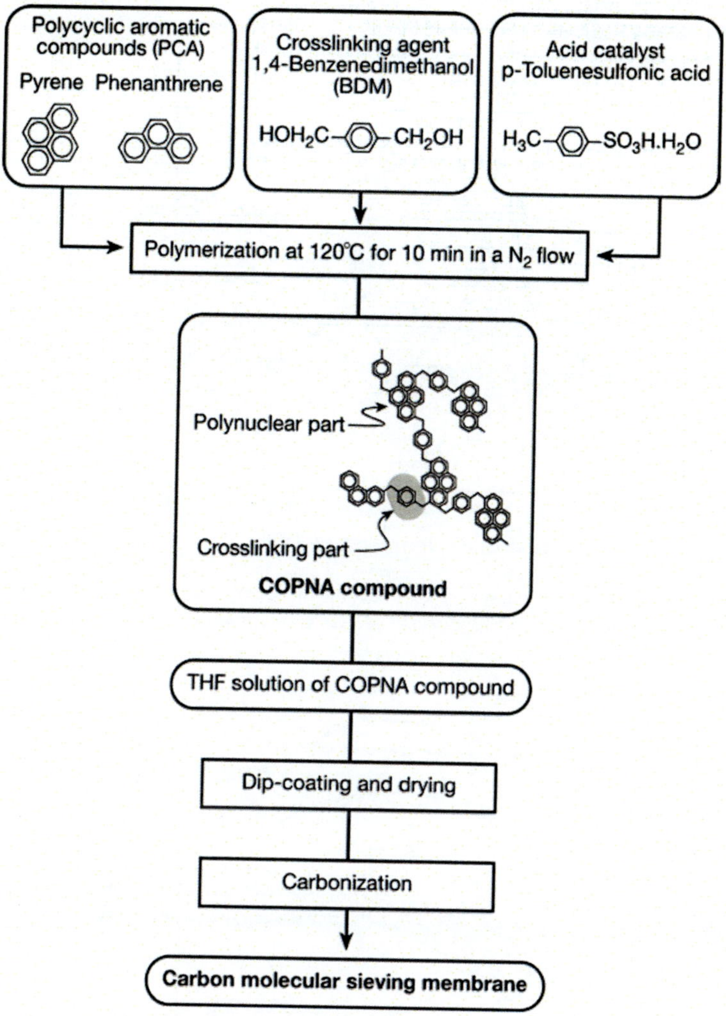

Fig. 4.18 Procedure for the preparation of CMSM from COPNA. (From [64])

diameter of the gas increased, indicating that the transport mechanism is sieving mechanism.

The method to prepare carbon membranes was described from wood pulp hydrolyzed to different extents by Lie and Hägg [66]. Different heating protocols were also shown. A simple, energy effective and rapid regeneration method for membranes that are conductors or semi-conductors has been developed by applying low voltage, direct current on an iron-doped carbon.

Wood pulp (kraft pulp) from a mixture of spruce and pine was supplied from Södra Cell Tofte, Norway. It is composed of cellulose and hemicellulose. The pulp was dissolved in trifluoroacetic acid (TFA) to a concentration of 1 wt%. The disso-

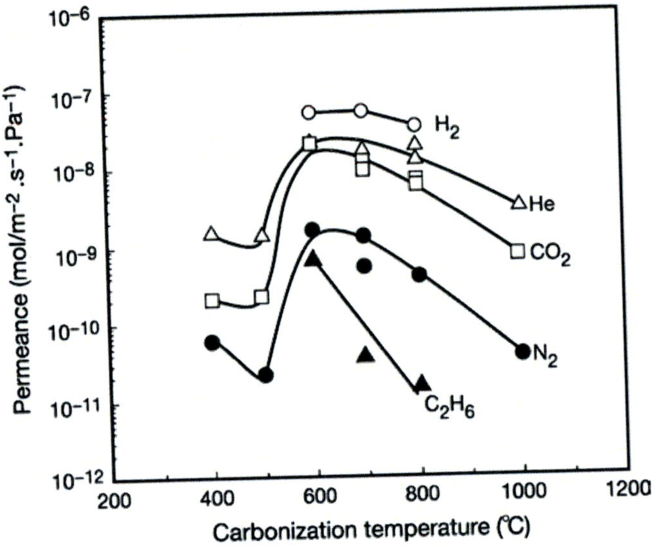

Fig. 4.19 Effect of carbonization temperature on the permeance of COPNA based CMSMs (BDM/PCA = 1.25; permeation temperature, 100°C). (From [64])

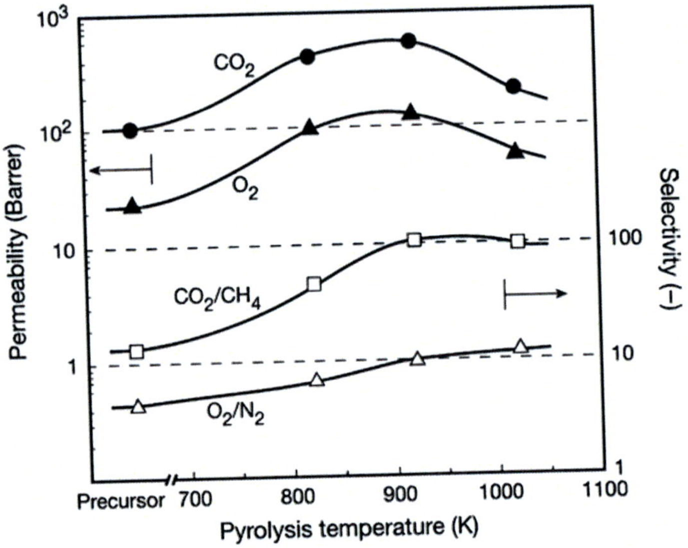

Fig. 4.20 Effect of pyrolysis temperature on the gas transport properties of PPO based CMSM (gas permeation at 298 K). (From [65])

lution of pulp in TFA lasted 6, 74 days or 2–5 weeks before iron nitrate was added in various concentrations to some of the solutions, which were then stirred overnight and ultrasonicated for 2 min. Next, a film was cast on a Teflon™ dish at room temperature. The cast film precursor was left at room temperature and after a certain

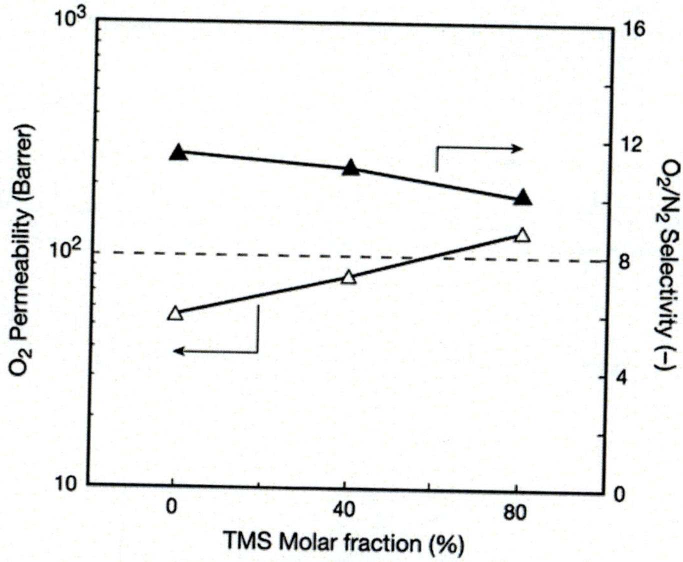

Fig. 4.21 Effect of the molar content of TMS on O_2 permeability and O_2/N_2 selectivity of PPO based CMSMs (membrane pyrolyzed at 923 K). (From [65])

period, called "hydrolysis time", dried in a vacuum oven at 105°C for about 18 h. Films were carbonized under mild vacuum (0.5 mbar) in a tubular furnace of special design. The basic heating protocol was a final temperature of 550–850°C to be kept at the temperature for 2 h. The temperature was raised at a heating rate of 1°C/min with several dwells. The first two dwells were necessary to remove traces of water and solvent. On the other hand, the other dwells were necessary to allow carbon matrix to rearrange and form micropores in between turbostratically arranged layers of graphene sheets. The system was allowed to cool from the final temperature naturally to a temperature less than 50°C, before the furnace was purged with ambient air and the films were removed.

Ferric nitrate may possibly react with carbon matrix releasing CO_2. The release of gas should make the ion-nitrate containing membranes more open and permeable than the membranes without addition of ferric nitrate. In the gas permeation test, single gas permeation was measured at 30°C and 2 bar using the constant pressure system. The order of testing was always N_2, H_2, CH_4, O_2, CO_2 and SF_6 and finally N_2 again.

Electrochemical regeneration of the membrane was attempted after the membrane was used. When an increasing current was applied on a test membrane, the carbon starts to burn at 100 mA current. Hence, the electric current far below this failure limit was used. To apply the current, the single core cables were attached to the feed side membrane surface with conductive glue that was heat cured at 65–70°C or 10 min. The permeation test results are given in Table 4.7 and Fig. 4.22.

Hydrolysis time of 6 days resulted in films showing no selectivity for the gases. After 2 weeks a critical threshold in the hydrolysis occurred and CO_2/CH_4 sepa-

Table 4.7 Results of gas permeation tests[a] of the membranes hydrolyzed for 6, 14 and 74 days

Film	Thickness	Permeability (Barrer)						Aging[b]
		N_2	H_2	CH_4	O_2	CO_2	SF_6	
C-6	86	29×10^3	55×10^3	32×10^3	17×10^3	28×10^3	21×10^3	−18
C-14	40	4.2	9.4×10^2	4.6	54	1.9×10^2	0.63	+14
C-74	46	5.7	8.6×10^2	1.3	48	1.5×10^2	–	−24

[a] At 2 bar and 30°C
[b] Percent change in N_2 permeability

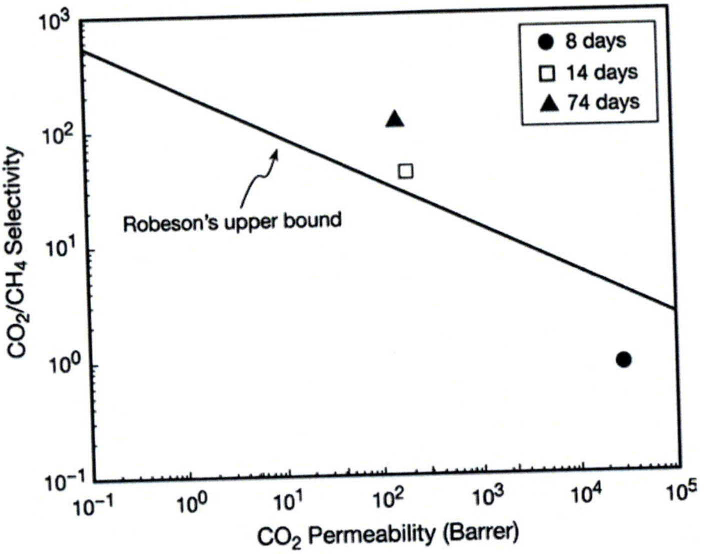

Fig. 4.22 Separation performance with single gases at 2 bar and 30°C for carbon membranes (550°C, 2 h soak period) with different hydrolysis times. (From [66])

ration increased with the hydrolysis time according to Fig. 4.22. Apparently, the selectivity is affected by the hydrolysis time but the permeability is not. Increased hydrolysis time probably causes a change in the carbon matrix, where the average pore width is shifted towards the sieving diameter of CO_2. The narrowing of the pore largely prevents the methane permeation.

The permeation test results for the carbon membranes of different final temperatures and soak periods are presented in Table 4.8, Figs. 4.23 and 4.24.

Looking into the zero soaking time, the permeability increases as the final temperature increases, which is in accordance with the gradual pore opening with increasing temperature. However, at 850°C the permeability decreases due to the sintering of the carbon matrix and the volume loss was found to be greater than the weight loss. Alternatively at 650°C they were equal. However, at temperatures lower than 650°C, the volume loss was lower than the weight loss.

Table 4.8 Permeation results from pure pulp carbonization

Protocol	Thickness (μm)	Permeability (Barrer)						Aging[a]
		N_2	H_2	CH_4	CO_2	O_2	SF_6	
500°C, no soak	23	6.6	4.6×10^2	2.8	2.2×10^2	49	0.050	–0.76
550°C, no soak	52	8.6	11×10^2	2.8	3.1×10^2	90	0.23	–1.0
550°C, 2 h soak	40	4.2	9.4×10^2	4.6	1.9×10^2	54	0.63	+14
650°C, no soak	66	15	13×10^2	4.4	4.8×10^2	1.3×10^2	0.61	–6.1
850°C, no soak	33	3.7	55	2.0	11	5.8	–	–48

At 2 bars and 30°C
[a] Percent change N_2 permeability

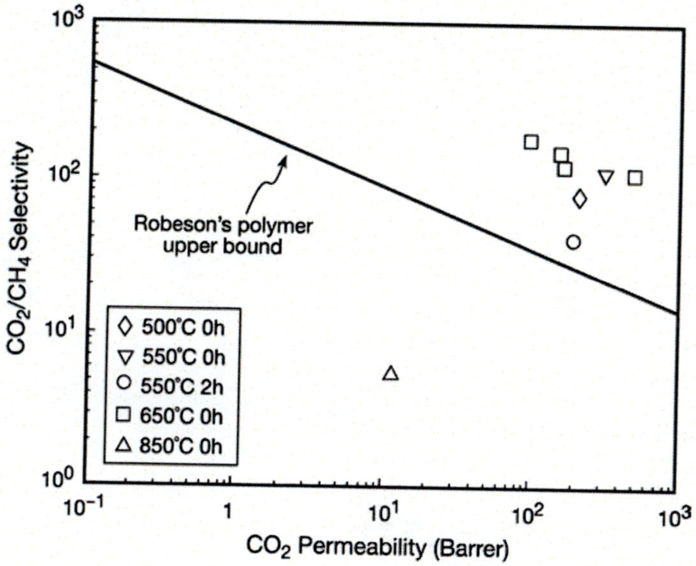

Fig. 4.23 CO_2/CH_4 separation performance of carbon membranes prepared with different heating protocols (single gas experiments at 2 bar and 30°C, hydrolysis time, 2–5 weeks). (From [66])

For the experiments of electrochemical regeneration, carbon membranes prepared from a dope containing 4.4 wt% of ferric nitrate was chosen. Figure 4.25 shows that the gas permeability increased when a low voltage direct current (17.5 V, 10 mV) was applied to the metal doped carbon membrane during the permeation test. By switching on and off the current, the gas permeability increased or decreased. Relative gas permeability, defined as (the permeability with the electric current)/(permeability without electric current) is plotted in Fig. 4.26. It shows that there is a clear correlation between the relative gas permeability and critical temper-

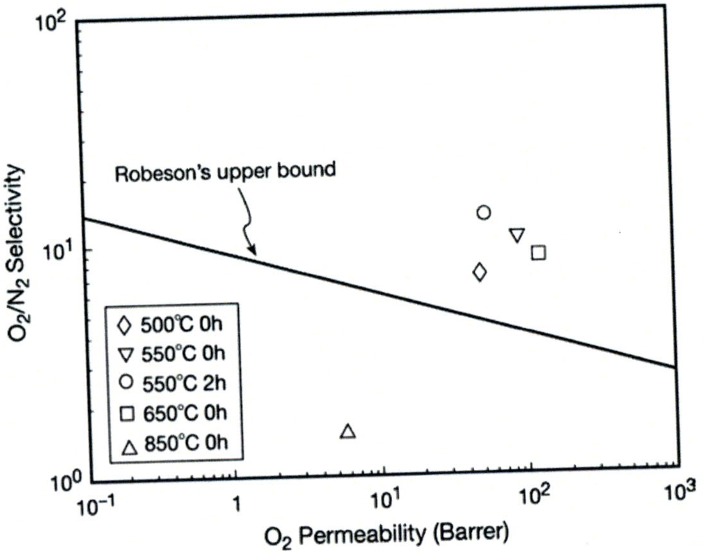

Fig. 4.24 O_2/N_2 separation performance of carbon membranes prepared with different heating protocols (single gas experiments at 2 bar and 30°C, hydrolysis time, 2–5 weeks). (From [66])

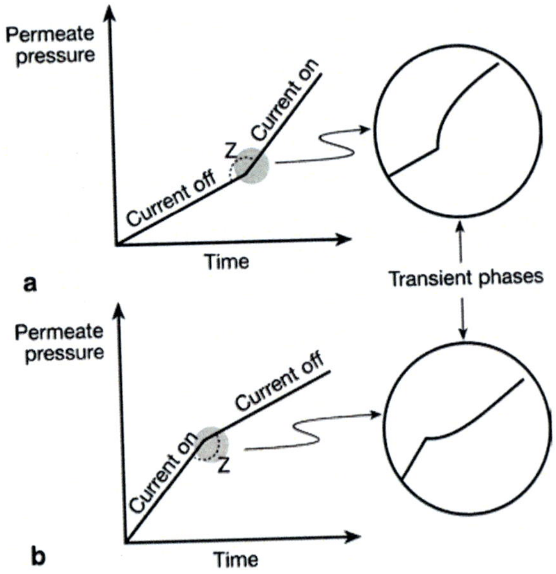

Fig. 4.25 Procedure for the permeation test and the shapes of the transient phases. (From [66])

ature of the gas. From these results, it is speculated that the enhanced permeability with electric current is due to (1) the generation of heat by the current application. The increased temperature will increase the desorption rate of the gas. If the desorption step is the rate limiting step for the gas transport through the membrane, current

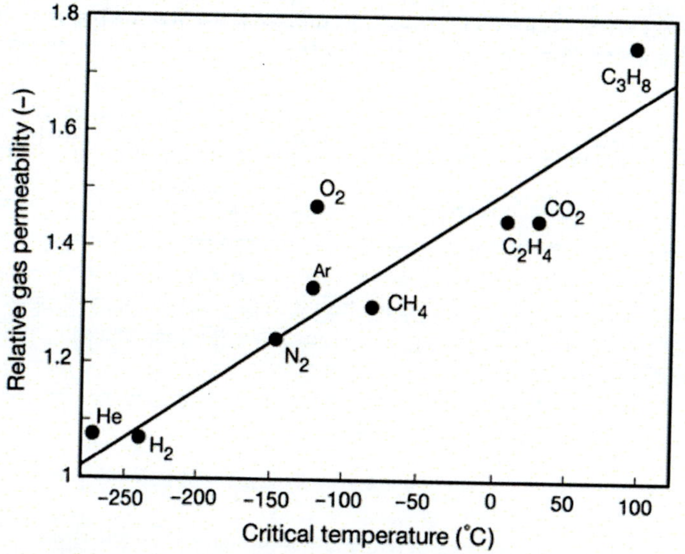

Fig. 4.26 Relative gas permeability as a function of the critical temperature of the gas (10 mA, 30°C, 2 bar feed). (From [66])

application will enhance the gas permeation. (2) the generation of magnetic field by current application. This will cause the orientation of diffusing polar gas molecules, leading to the entropy reduction, which will increase the chance of overcoming the pore entrance barrier.

Long term experiments were performed with the same membrane for hydrogen and methane permeation under the conditions, 30–50°C and 0.5–2.0 bar for 2 weeks. The current (5–15 mA) was applied for 2 h on average each day. After the series of the test, the current was switched off and the hydrogen and methane permeability tests were performed. Hydrogen permeability increased by 14% while the methane permeability increased 7%. In other words, both permeability and selectivity increased by applying the electric current.

4.2 Polymeric Membrane Preparation

Polymeric membranes or precursors for fabricating carbon membranes must be prepared at optimum conditions in order to produce a carbon membrane of high quality. If the polymeric membrane is of poor quality, a pyrolysis process cannot be expected to satisfactorily produce a carbon membrane with superior properties. Therefore, precursor membranes must be prepared in defect free form in order to minimize problems in subsequent processing during the manufacture of carbon membranes.

A polymeric membrane can be produced in two main configurations as a precursor for carbon membranes, namely unsupported membranes and supported membranes. Unsupported membranes have three different configurations: flat (film); hollow fiber; and capillary. Supported membranes can adopt one of two configurations: flat and tubular. Detailed descriptions of these two categories can be found in Ismail and David's review [67]. In most cases, supported polymeric membranes are prepared because of the poor mechanical stability (i.e. brittleness) of unsupported carbon membranes

For making the supported carbon membranes, various options are available for coating the supports with thin polymeric films, such as ultrasonic deposition [52, 53], dip coating [68], vapour deposition [60], spin coating [43], and spray coating [54]. Ultrasonic deposition (UD) provides nearly zero spray velocity, droplet sizes that are often narrowly distributed in sizes from 10 to 10^2 μm, accurate compared to spray and misting [52].

Acharya in his thesis listed the requirements for a method to be acceptable when coating with a polymeric solution [51] namely:

1. Coating should be able to control the amount of material being deposited on the support;
2. Coating should produce a uniform distribution of material on the support;
3. Coating should work for a number of different supports; and
4. Coating should not destroy or alter the support in any way.

Because of the shrinkage of the polymeric material during pyrolysis, the coating procedure has to be repeated until a defect free carbon molecular sieve is obtained [69]. Details of the surface of the substrate may be translated into the carbon membrane film, thereby originating small pinholes that destroy the molecular sieve properties required for gas separation. The coating of the support surface with an intermediate layer reduces the number of defects existing on the original substrate [70]. Moreover, the rapid coagulation of polymer prevents the infiltration of solution into the porous carbon substrate, and an excellent polymeric film is achieved [43]. Chen and Yan [59] noted that further work was needed to understand the effect of different supports on the quality of the membrane produced. The choice of a support material is based on:

- Economics (i.e. cost, availability, morphology);
- Durability;
- Heat transfer characteristics;
- Chemical reactivity; and
- Compatibility with the carbon.

For unsupported polymeric membranes (i.e. hollow fibers), spinning parameters are crucial factors that must be controlled during hollow fiber preparation. These parameters include the amount and types of polymer, solvents, and additives mixed into the spinning dope, the dope and bore fluid flow rates, the fibre take-up velocity, the air gap distance (unless wet spinning is used), and the coagulant bath temperature [31, 71]. The main challenges in hollow fiber spinning, as proposed by Puri [72], are:

- Synthesizing polymeric melts and solutions having appropriate thermodynamic and rheological properties.

- Determining the kinetic and thermodynamic conditions involved in transforming the solution/melt to a solid.
- Annealing and crystallizing (and orienting) the formed bodies.
- Post-treating to alter surface properties and stabilize the products.

As a conclusion, during the preparation of precursor membranes, the process parameters involved must be optimized to form a good precursor for pyrolysis. Moreover, the choices of supported and unsupported carbon membranes depend on the application for which the carbon membranes will be used. Normally, most of today's gas separation membranes are formed into hollow fiber modules [73]. Hollow fiber membranes offer a greater area per specific module volume than flat film membranes, by a factor of 1–40 [31].

4.3 Pretreatment of Precursor

Pretreatment stabilizes the structure of the precursors, acts to maintain the molecular structure of the carbon chains, and/or enhance the uniformity of pore formation during the pyrolysis process. Current pretreatment includes oxidation, chemical treatment, physical method such as stretching. Oxidation or thermostabilization is the most popular and commonly used method to pretreat the polymeric precursors. This preteatment stabilizes the structure of the precursors so that they can withstand the high temperatures in several pyrolysis steps. Thermostabilization can maximize the carbon yields of resultant membranes by preventing excessive volatilization of elemental carbon during pyrolysis. Oxidation has been carried out by Kusuki et al. [74], who thermally treated the precursors in atmospheric air at 400°C for 30 min before pyrolysis. Tanihara and Kusuki [75], Okamoto and co-workers [76], and David and Ismail [31] have also applied thermostabilization.

Pretreatment of the membrane with certain chemicals can provide enhanced uniformity of the pore system formed during pyrolysis. Among the chemicals commonly used for chemical pretreatment are hydrazine, dimethylformamide (DMF), hydrochloric acid (HCl) and ammonium chloride (NH_4Cl) [77]. During the chemical pretreatment, the membrane is fully immersed in the appropriate solution, followed by washing and drying before it is subjected to heat treatment. In certain cases, it has also been proven that it may be advantageous to evacuate the pores of the membrane by applying a low air pressure, and subsequently fill them with nitrogen gas at normal pressure prior to pretreatment with a chemical reagent. In this way, the membranes with higher carbon contents can be obtained [24]. Schindler and Maier [24] proposed chemical pretreatment using chemical reagent, where the capillary PAN membranes were pretreated in aqueous hydrazine solution before carbonization.

Stretching is a pretreatment that is usually applied to hollow fiber precursors. This technique is adapted from the fabrication of carbon fibers and is sometimes referred to as post-spinning. An ideal post-spinning modification scheme would allow the removal of surface defects, attenuation of variations in filament diameter, and enhanced retention of molecular orientation prior to the heat treatment. Chen modified PAN precursors via physical stretching in DMF [20]. More detailed information for each pretreatment technique is given below.

4.3.1 Oxidation Pretreatment

Kusuki and co-workers [74] reported that certain precursors, when not preoxidized, underwent softening during the pyrolysis step, and the resulting carbon membranes had poor membrane performance. The stabilization or oxidation step is also intended to prevent the melting or fusion of the fibers (or membranes), and to avoid excessive volatilization of elemental carbon in the subsequent carbonization step, and thereby maximizing the final carbon yield from precursor. Thus, the oxidation treatment is considered very important and can have a substantial influence on the resulting performance of a carbon membrane. Oxidation pretreatments can be applied at very different ranges of thermal soak times, depending largely on the precursor used. In all cases, the aim is the same, that is, to contribute to the stabilization of the asymmetric structure of the precursor and provide sufficient dimensional stability to withstand the high temperatures of the pyrolysis steps. Table 4.9 lists selected oxidation pretreatments applied to different precursors by various researchers.

Kusuki et al. studied the gas separation properties of the carbon membranes that were prepared with oxidation pre-treatment and called the treatment thermostabili-

Table 4.9 Preoxidation conditions for various precursors

Precursor	Configuration	Temperature; time; atmosphere[a]	References
Acrylonitrile	Hollow fiber	200–300; 3 h;	[104]
Acrylonitrile	Hollow fiber	180–350; 1–20;	[24]
Polyacrylonitrile	Hollow fiber	250; 30; or O_2	[31, 91]
Polyacrylonitrile	Hollow fiber	265; 30; or N_2	[26]
Polyacrylonitrile	Hollow fiber	270; 30;	[27]
Coal tar pitch	Plate	200–260; 2 h; O_2	[80]
Phenol resin	Supported film	300; 1 h;	[111]
Phenolic resin	Supported film	150–300; 2 h;	[78]
Phenolic resin	Supported film	150; 2 h;	[12, 14, 46–48]
Polydimethylsilane	Supported film	200; 1 h;	[114]
PVDC-co-PVC	Supported film	150–200; 6 h;	[79]
Polyfurfuryl alcohol	Supported film	90; 3 h;	[59]
Polyimide	Hollow fiber	400; 30;	[76]
Polyimide	Hollow fiber	300; 1 h;	[93]
Polyimide	Hollow fiber	400; 30;	[77]
Polyimide	Hollow fiber	400; 30;	[74]
Polyimide	Hollow fiber	400; 30;	[90]
Polyimide	Hollow fiber	250–495; 0.1–100 h;	[75]
Polyimide	Hollow fiber	250–495; 0.1–100 h;	[123]
Polyimide	Supported film	400–500; 0–1 h;	[35]

[a] Temperature (°C); time (min); atmosphere (air) unless otherwise stated

zation [74]. The hollow fibre membranes were prepared from PI consisting of the monomers BPDA and aromatic diamines. The major component (85 wt%) of the diamines was dimethyl-3,7-diaminodiphenyl-thiophene-5,5-dioxide (DDBT) and the minor components were 3,5-diaminobenzoic acid and a CF_3-containing diamine. The hollow fibers were spun from the p-chlorophenol solution of the PI via dry-jet spinning method and dried under the nitrogen at 270°C. The hollow fibers of the precursor were further treated at 400°C for 30 min in atmospheric air before the pyrolysis. When the precursor hollow fiber was not subjected to the oxidative pretreatment, the hollow fiber membrane softened during the pyrolysis and the resulting carbon membrane exhibited only poor performance. Hence, the carbon hollow fiber fabrication should consist of the following three steps: (1) precursor drying at 270°C, (2) oxidative pretreatment at 400°C, and (3) pyrolysis at temperatures above 600°C.

The effect of the oxidative pretreatment as well as the effect of the temperature of pyrolysis for the separation experiments with the 50/50 H_2/CH_4 gas mixture at 80°C is depicted in Fig. 4.27.

The H_2 permeation rate, P′ (H_2), and the permeation rate ratio P′ (H_2)/P′ (CH_4) are also given in the figure. The membrane subjected to the oxidative treatment showed lower P′ (H_2) and higher P′ (H_2)/P′ (CH_4) ratio than the untreated precursor membrane. As for the pyrolyzed membranes, P′ (H_2) kept decreasing with an increase in pyrolysis temperature, while the P′ (H_2)/P′ (CH_4) ratio showed a maximum at about 800°C. More specifically, P′ (H_2)=1.0×10^{-3} cm^3(STP)/(cm^2 s cmHg) and P′ (H_2)/P′ (CH_4) ratio = 132 at the pyrolysis temperature of 700°C while they

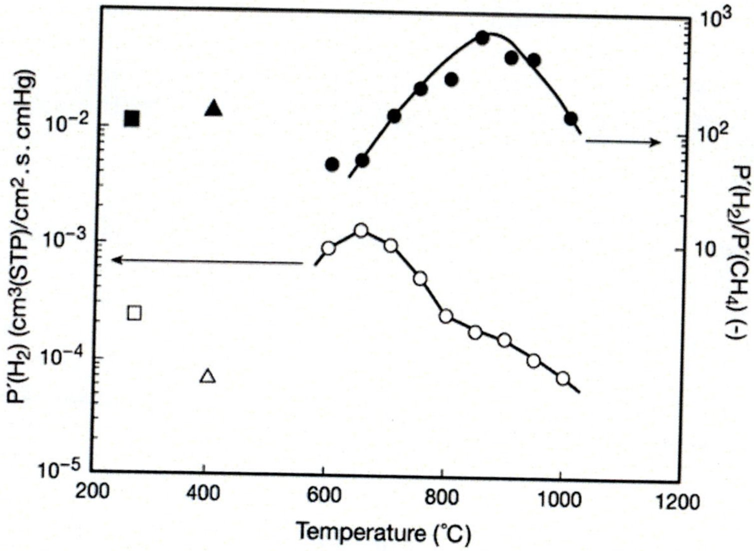

Fig. 4.27 Plots of *P*′ (H_2) and *P*′ (H_2)/*P*′ (CH_4) for feed gas mixture of 50% H_2 in CH_4 at 80°C against heat treatment temperature. *Open* and *closed plots* show *P*′ (H_2) and *P*′ (H_2)/*P*′ (CH_4), respectively. (From [74])

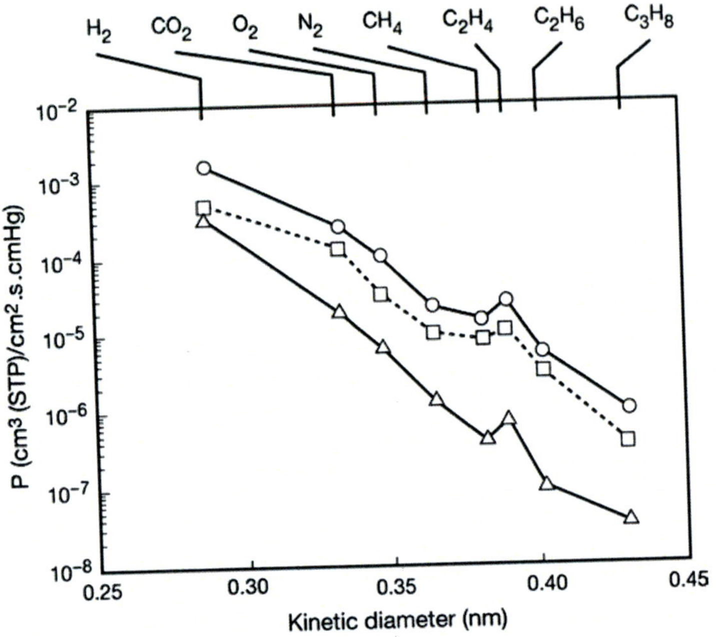

Fig. 4.28 Plots of permeation rates at 120°C against kinetic diameter for pure gases. (□) Asymmetric polyimide hollow fiber membrane, (○) carbon membrane pyrolyzed at 700°C, and (Δ) carbon membrane pyrolyzed at 850°C. (From [74])

were 1.8×10^{-4} cm^3(STP)/(cm^2 s cmHg) and 631 at 800°C, respectively. Figure 4.28 shows the permeance P′ versus the kinematic diameter of the permeant gases. P′ decreases almost linearly with the kinetic diameter of the gas, indicating the gas permeation occurs by the sieve mechanism. The slope of the linear relationship becomes steeper as the pyrolysis temperature increases, indicating the permeance becomes more size dependent. Figure 4.29 shows the temperature dependence of the pure gas permeance for the precursor membrane and the carbon membranes pyrolyzed at 700 and 850°C. Table 4.10 depicts the activation energies obtained from the data shown in Fig. 4.29. It is found that, particularly for the large sized gases the activation energies are smaller for the carbon membranes. This is because gas diffusion occurs through the free spaces created by the thermal motion of the macromolecules. For larger gas molecules, larger energies are required to create sufficiently large spaces. On the other hand, for carbon membranes, the gas permeation occurs through the pores that already exist. No energy is required for the creation of the space. Because of these differences in activation energies, permselectivities decrease largely for the gas pairs such as H_2/hydrocarbon, CO_2/hydrocarbon and O_2/N_2 with an increase in permeation temperature. On the other hand, the change is much less for the carbon membranes.

Centeno and Fuertes [78] observed that air oxidation of a phenolic resin (PR) film, prior to the carbonization step, at temperatures ranging from 150 to 300°C

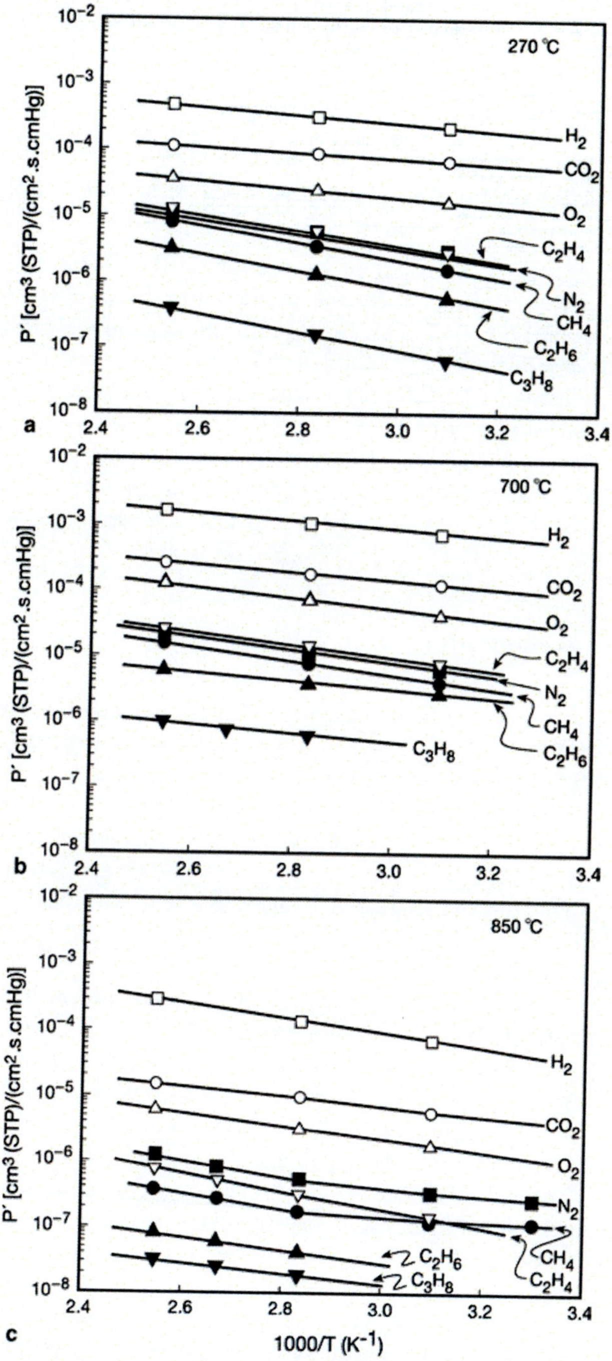

Fig. 4.29 Temperature dependence of permeation rates for pure gases on (**a**) asymmetric polyimide hollow fiber membrane, (**b**) carbon membrane pyrolyzed at 700°C, and (**c**) carbon membrane pyrolyzed at 850°C, respectively. (From [74])

Table 4.10 Activation energy of permeance for different gases

Membrane	Activation energy (kJ/mol)							
	H_2	CO_2	O_2	C_2H_4	N_2	CH_4	C_2H_4	C_2H_6
Polyimide precursor	13	7.6	13	22	22	25	25	26
Carbon membrane pyrolyzed at 700°C	11	11	15	17	19	20	12	11
Carbon membrane pyrolyzed at 850°C	21	14	20	26	(21)	(19)	21	14

for 2 h improved the gas permeation rate of the resulting carbon membrane significantly. This observation implied that the oxidation prior to carbonization led to an enlargement of carbon film pores [78]. However, in the case of CMSM made from poly(vinylidene chloride-co-vinyl chloride)(PVDC-co-PVC), they found that oxidative pretreatment in air at 200°C for 6 h produced a less permeable membrane than the untreated membrane, but a much more selective membrane (i.e. selectivity of O_2/N_2 increases from 7 to 13.8) [79]. Therefore, it is necessary to optimize the preoxidation conditions to gain the best cost-effectiveness due to the permeability and selectivity of carbon membrane that are inversely related to each other [80].

4.3.2 *Chemical Treatment*

Pretreatment of a membrane with certain chemicals can provide enhanced uniformity of the pore system formed during pyrolysis. Among the chemicals commonly used for chemical pretreatment are hydrazine, dimethyl formamide (DMF), HCl and NH_4Cl [77].

During the manufacture of porous carbon membranes by Schindler and Maier [24], an acrylic precursor was subjected to a pretreatment with an aqueous solution of hydrazine. It was found that the pretreatment improved the dimensional stability of the membrane during the subsequent processing steps and, in particular, that the formation and clogging of the pores could be avoided during these steps. Very good results were obtained from the acrylic precursor using a pretreatment with 80% hydrazine hydrate for 30 min at a solution temperature of 90°C.

During the chemical pretreatment, the membrane is fully immersed in the appropriate solution. After that, the membrane is washed and dried before it is fed to the first treatment station. In certain cases, it has also been proven that it may be advantageous to evacuate the pores of the membrane by applying a low air pressure, and subsequently fill them with nitrogen gas at normal pressure prior to pretreatment with an aqueous solution. In this way, one can obtain membranes of carbon contents higher than those made from the precursors whose pores remained full of air during the pretreatment [24].

Another type of chemical pretreatment involves the use of catalysts including mineral acids and acidic salts, such as phosphoric acid and diammonium hydrogen phosphates, before carbonization. However, the preparation of carbon hollow fiber membranes with carbonization catalyst causes problems. In carbon hollow fibers, carbonization must take place uniformly both inside and outside of the fiber, and pitting must be avoided because the selectivity of the membrane depends strongly on the uniformity of the pores produced during carbonization. Pitting occurs immediately if the catalyst is not uniformly distributed throughout the fiber, due to locally catalyzed oxidation on the surface [81].

Another method of catalytic carbonization involves application of a gaseous catalyst such as HCl, or NH_4Cl in a stream of inert gas. This method leads to two useful results [27].

1. The catalyst can be more uniformly distributed throughout the fiber bundle, thereby avoiding hot spots.
2. The inert gas acts as a purging gas removing tars which are formed during pyrolysis and could otherwise impair membrane properties by causing undesirable occlusions therein.

However, all the water must be removed before exposing the filler to the gaseous catalyst in order to prevent local dissociation of the fiber surface, which may lead to the fusion or cementation of adjacent fibers. This may in turn lead to the fracture of some of the cemented fibers due to non-uniform contraction during pyrolysis, and due to local pitting in the area attacked by the water-soluble catalyst. Accordingly, it is critical to apply the catalyst to the fibers only after the last traces of water have been removed [81].

4.3.3 *Stretching*

Stretching or drawing is categorized as a physical pretreatment and usually involves the hollow fiber precursor only. This technique is adapted from the fabrication of carbon fiber and is sometimes referred to as post-spinning treatment. An ideal post-spinning modification scheme would allow the removal of surface defects, attenuation of variations in filament diameter, and an enhanced retention of molecular orientation prior to heat treatment in order to obtain fibers with a good balance of stiffness and strength [20].

Draw ratios are preferably as high as possible within the range in which the structure of fiber is not ruptured. Typical upper limits of the total draw ratio are around 80% of the draw ratio at which fiber break occurs [25]. Drawing can also take place during the spinning process not more than a few times of the original length of the fibers. This offers an advantage that the pore systems obtained after the spinning process can be preserved to a large extent in the carbon membrane and thus ensures greater dimensional stability [24].

Stretching PAN precursor fibers in the preoxidation stabilization stage results in an increase in the Young's modulus of the final carbon fiber. Since the dipole–dipole interactions among the nitrile groups obstruct the molecular chain from becoming fully oriented during stretching, a reduction of these interactions can make the drawing process more effective. In particular, the introduction of solvent molecules or heat has been cited as an effective means of decreasing these interactions in PAN molecules.

4.3.4 Other Pretreatment

The pretreatment was made by soaking the membrane (dense membrane) into liquid alcohols by Tin and co-workers [82]. Dense membranes were prepared from two kinds of polyimide (PI), i.e. Matrimid 5218 (3,3′4,4′-benzophenone tetracarboxilic dianhydride and 5(6)-amino-1-(4′-aminophenyl)-1,3-trimethylindane), and P84 (copolymer of 3,3′,4,4′-benzophenone tetracarboxylic anhydride and 80% methylphenylene-diamine + 20% methylenediamine). The membranes were then immersed into nonsolvent for 1 day at room temperature, followed by drying naturally for 24 h. Then the membrane was subjected to the heating scheme given in Fig. 4.30 for carbonization.

Membrane characterization was made by TGA, WAXD, DSC, and PALS.

The effect of the immersion time was investigated. It was found that the increase in immersion time resulted in lower flux and higher selectivity. Immersion time of 1 day was chosen for the further study because it has given the highest selectivity. Tables 4.11 and 4.12 summarize the results of pretreatment experiments using various alcohols as nonsolvent. Among others ethanol gave the highest selectivity values.

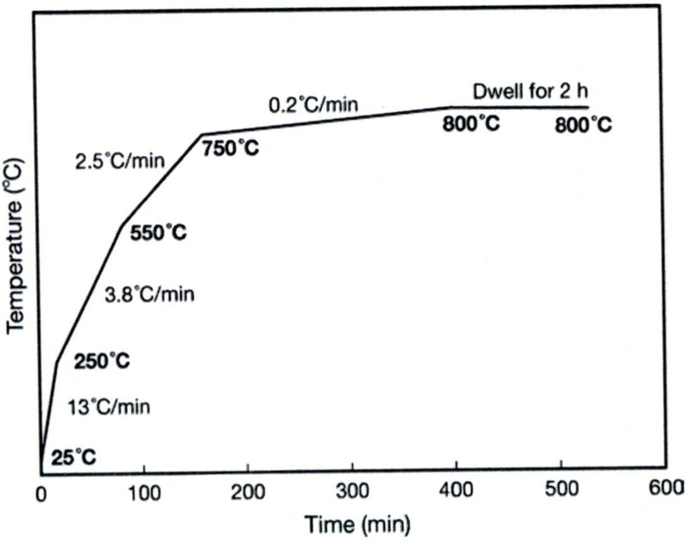

Fig. 4.30 Steps involved in the pyrolysis process for the final temperature of 800°C. (From [82])

Table 4.11 Gas permeation properties of CMSMs derived from Matrimid

	Permeability (Barrer)				Selectivity		
	O_2	N_2	CO_2	CH_4	O_2/N_2	CO_2/CH_4	CO_2/N_2
CM-MAT-800	227	30.3	611	10.0	7.5	61	20
CM-MAT-MeOH-800	138	15.8	423	4.8	8.8	88	27
CM-MAT-EtOH-800	75.4	6.3	191	1.1	12	169	30
CM-MAT-PpOH-800	204	24.1	565	6.7	8.5	84	23
CM-MAT-BtOH-800	186	21	547	7.0	8.9	78	26

Membrane code: CM-polyimide-nonsolvent-pyrolysis temperature

Table 4.12 Gas permeation properties of membranes derived from P84

	Permeability (Barrer)				Selectivity		
	O_2	N_2	CO_2	CH_4	O_2/N_2	CO_2/CH_4	CO_2/N_2
CM-P84-800	158	17.8	499	5.6	8.9	89	28
CM-P84-MeOH-800	132	13.6	402	3.7	9.7	109	30
CM-P84-EtOH-800	101	9.0	278	2.0	11.2	139	31
CM-P84-PpOH-800	144	14.6	428	3.9	9.9	110	29

Membrane code: CM-polyimide-nonsolvent-pyrolysis temperature

4.4 Pyrolysis Process

One of the distinctive features of (CMSMs) is the pyrolysis conditions of a polymeric precursor. When a given precursor and its pyrolysis conditions are properly selected, a desirable micropore size and shape can be tailored. Tailoring of correct pore size is important for the fabrication of high-performance carbon membranes with optimal balance between selectivity and permeability that are suitable for various gas separations. It demands a good understanding of the optimum conduction for the membrane production particularly with respect to the pyrolysis condition. Through the rigorous control of the pyrolysis conditions, i.e. temperature, heating rate, atmosphere and thermal soaking time, gas flow rate, pressure and concentration, and pre/post-treatment conditions, the pore aperture can be nearly continuously tuned, so that a membrane module can be specifically designed for gas separation and reaction [83]. Carbon membranes are prepared by the pyrolysis of polymeric precursors. Pyrolysis is usually carried out at temperatures of 500–1,000°C under vacuum or inert atmosphere. This process removes most of the heteroatoms originally present in the macromolecules, leaving behind a stiff and cross-linked carbon matrix with an amorphous porous structure created by the evolution of gaseous products and the rearrangement of the molecular structure of the starting polymeric precursor during the pyrolysis. Figures 4.31 and 4.32 show the typical hollow fiber membranes with the substructure and the skin layer morphology produced under different pyrolysis temperature.

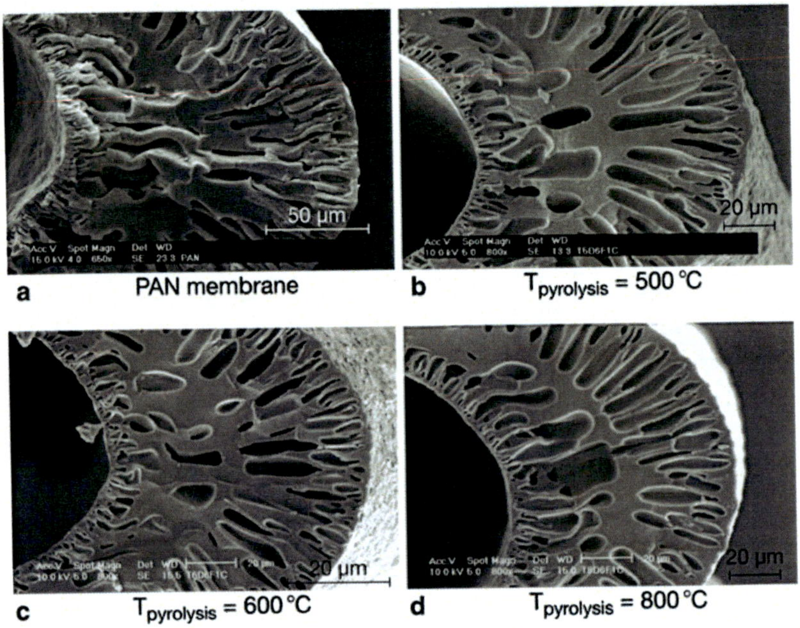

Fig. 4.31 Structural change of the typical hollow fiber membrane substructure at different pyrolysis temperatures. (From [84])

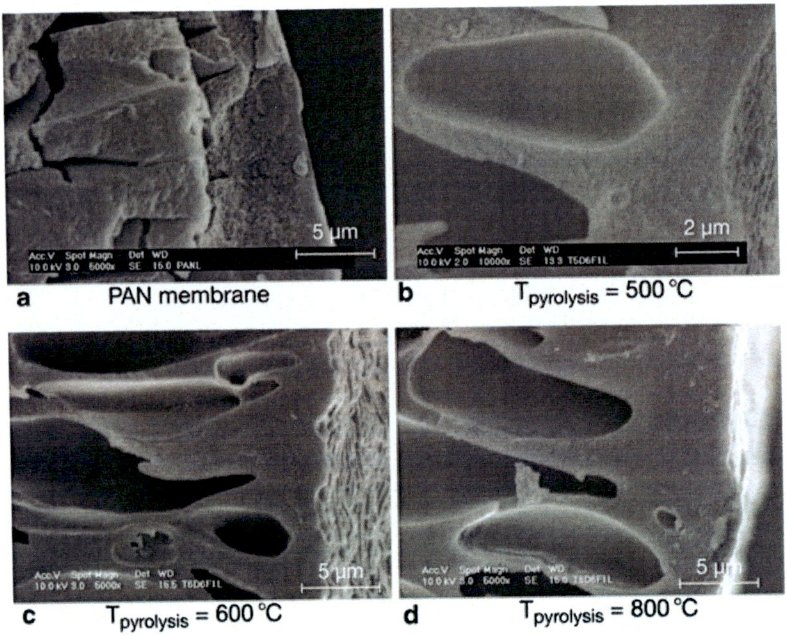

Fig. 4.32 Structural changes of the typical hollow fiber membrane skin layer at different pyrolysis temperatures. (From [84])

As mentioned above, during pyrolysis of a polymer, byproducts of different volatilities are produced and generally cause a large weight loss [87]. Typically volatile byproducts may include ammonia (NH_3), hydrogen cyanide (HCN), methane (CH_4), hydrogen (H_2), nitrogen (N_2), carbon monoxide (CO) and others, depending on the polymer [85]. The polymeric precursors are initially cross-linked or become cross-linked during pyrolysis. This prevents the formation of large graphite-like crystals during carbonization, and leads to the formation of disordered structure (non-graphitizing carbons) with a very narrow porosity [70].

The transport properties of the carbon membrane are dependent on its porous structure and pore size distribution [4]. During pyrolysis, amorphous carbon material is created and exhibits a distribution of micropore dimensions with only short-range order of specific pore sizes and also pores larger than the ultra-micropores. These larger pores that connect the ultra-micropores are required to exhibit molecular sieving properties and therefore allow high gas separation productivities [34]. The pore system of carbon membranes is generally non-homogeneous, as it is comprised of relatively wide openings with a few constrictions [86]. The pores may vary in size, shape and degree of connectivity, greatly depending on the nature of the polymeric precursor and the conditions of pyrolysis. Figure 4.33 depicts an idealized structure of a pore in a carbon material. The pore mouth "d" is often referred to as an ultra-micropore (<10 Å) that allows molecular sieving of the penetrating molecules, meanwhile the larger micropores, "D" of the material (6–20 Å) may allow the diffusion of gas molecules to occur through the carbon material [6, 13, 22, 84]. Therefore, it is believed that a properly prepared carbon material is able to simultaneously demonstrate the ability to perform molecular sieving and also allow a considerably high flux of the penetrating molecules through the material [87].

The pores of the starting material do not change uncontrollably when the pyrolysis process is being carried out. Instead, the pore structure is essentially retained and can be controlled selectively by adjusting the various process parameters [24]. A number of different variables can affect the pyrolysis process and small changes in pyrolysis parameters have been found to have a significant impact on the final properties of a carbon membrane [32, 88]. As mentioned above, among the factors that influence the pyrolysis process are the pyrolysis temperature, heating rate, ther-

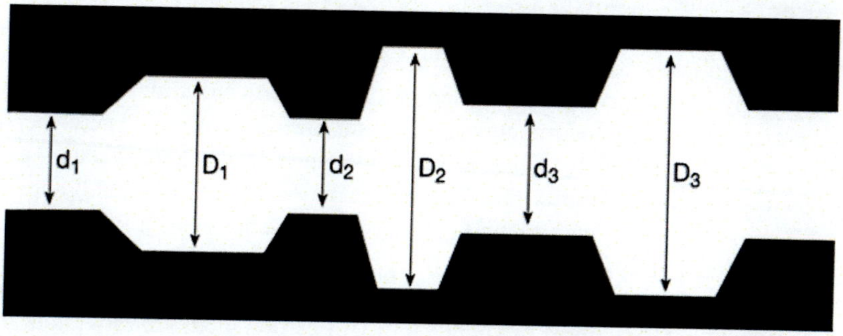

Fig. 4.33 Idealized structure of a pore in a carbon material. [87]

mal soak time, type of pyrolysis atmosphere and, most noticeably during inert gas pyrolysis, the gas flow rate, pressure and concentration.

Pyrolysis temperature has a remarkable influence on the properties of carbon membranes in terms of the membrane structure, separation performance (permeability and selectivity) and the transport mechanism for gas separation [11]. The pyrolysis process can be conducted within a wide range of temperatures, in between the decomposition temperature of the carbonaceous precursor and its graphitization temperature (generally above 3000°C). Usually pyrolysis is carried out in the range between 500 and above 1,000°C [85, 89]. Depending on the experimental conditions chosen, the pyrolysis process removes most of heteroatoms that are originally present in the polymeric precursors, while leaving a cross-linked and stiff carbon matrix behind [44]. Table 4.1 lists typial pyrolysis temperatures found in the literature for the production of carbon membranes.

Thus, pyrolysis temperature has a strong influence on the properties of carbon membranes in terms of membrane structure, separation performance (permeability and selectivity) and transport mechanism for gas separation. The optimum temperature depends very much on the type of the precursor used. The increase in pyrolysis temperature normally leads to a carbon membrane with higher compactness, a more turbostratic structure, higher crystallinity and density, and smaller average interplanar spacing between the graphite layers of the carbon [90]. Generally, an increase in the pyrolysis temperature will give rise to a decrease in gas permeability and an increase in selectivity [34, 85, 89].

The thermal soak time can be different depending on the final pyrolysis temperature [24]. This parameter may be used to fine-tune the transport properties of a carbon membrane using a particular final pyrolysis temperature [91]. Previous studies showed [41, 87, 89, 91, 92] that increments in thermal soak time would increase the selectivity of carbon membranes. It is believed that only microstructural rearrangement occurs during the thermal soak time, thus affecting the pore size distribution and average porosity of carbon membranes [34].

Heating rate will determine the evolution rate of volatile components from a polymeric membrane and consequently affect the formation of pores in the carbon membrane [2]. Widely different heating rates can be used, ranging from 1 to >10°C/min [89]. However, lower heating rates are preferable in order to produce carbon membranes of higher selectivity [41, 89]. Higher heating rates may lead to the formation of pinholes, microscopic cracks, blisters and distortions, which in extreme cases may render membranes useless (i.e. of low selectivity) for gas separation [2].

The pyrolysis must be controlled in order to prevent undesired burn off and chemical damage of the membrane precursor during pyrolysis. Therefore, the pyrolysis can be carried out either in vacuum or inert atmosphere. Vacuum pyrolysis was reported to yield more selective but less permeable carbon membranes (from a polyimide precursor) than an inert gas pyrolysis system [34, 85]. When dealing with the inert gas pyrolysis system, one must consider how the inert gas flow rate will affect the performance of the resulting carbon membranes. Generally, an increase in gas flow rate will improve the permeability of carbon membranes without interfering with their selectivity very much [85, 91].

In principle, one can determine which pyrolysis parameters are important and contribute most significantly to the structural changes of the material. In that case, it would be possible to predict the trends of transport properties for a given carbon material more effectively. Recently, the precursors were pyrolyzed in a step-by-step fashion (i.e. after reaching a certain temperature, holding the membrane at that temperature for a period of time and then heating it again) until the final pyrolysis temperature is reached [34, 44, 51, 92, 93].

Typical examples of pyrolysis studies are shown below in detail.

The effect of pyrolysis environment on CMSM's performance was studied in detail by Geiszler and Koros [85].

A polyimide (PI) synthesized from three monomers; 2,4,6-trimethyl-1,3-phenylenediamine, 3,4,3′4′-biphenyltetracarboxylic acid dianhydride, and 5,5′-[2,2,2-trifluoro-1-(trifluoromethyl)ethylidene]bis-1,3-isobenzofurandione whose structures are shown in Fig. 4.34 was used to spin hollow fibers. The hollow fibers were then pyrolyzed in vacuum or in an atmosphere of inert gas at different temperatures (550 and 800°C) and different gas flow rates (20 and 200 cm^3(STP)/min).

Table 4.13 compares the vacuum and inert gas pyrolysis at 550°C. Vacuum pyrolysis showed lower flux and higher selectivity than the inert gas pyrolysis. The change in inert gas did not change the gas separation performance significantly.

The effect of inert gas flow rate on the gas separation performance is presented in Table 4.14. Decrease in flow rate remarkably decreases the flux without significant change in the selectivity. When the non-volatile by-products are not removed quickly enough during pyrolysis, they can presumably degrade further and leave carbon deposits on the surface of the carbon, which can reduce the permeant gas flux.

The effect of pyrolysis temperature on the gas separation performance is shown in Table 4.15. For vacuum pyrolysis, the temperature increase resulted in decrease in flux and slight increase in selectivity. For the helium gas pyrolysis, the same was observed for the helium flow rate of 200 cm^3(STP)/min. When helium flow rate is 20 cm^3(STP)/min, by-product residue plugged the pores and the flux decreased remarkably. With an increase in pyrolysis temperature, the by-product residue was removed and the flux increased.

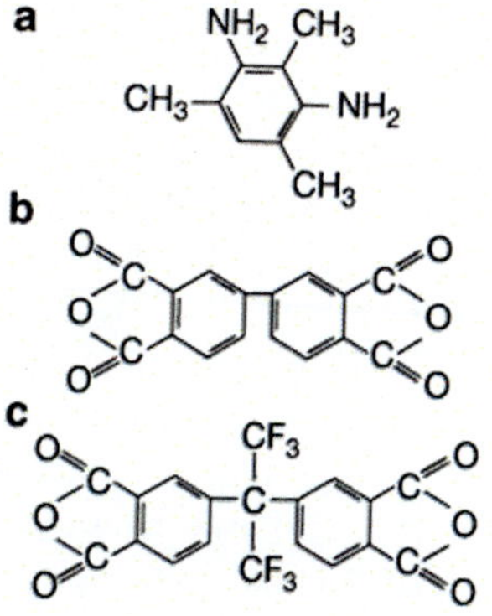

Fig. 4.34 Monomers to synthesize polyimide; (**a**) 2,4,6-trimethyl-1,3-phenylenediamine, (**b**) 3,4,3′,4′-biphenyltetracarboxylic acid dianhydride, (**c**) 5,5′-[2,2,2-trifluoro-1-(trifluoromethyl)ethylidene] bis-1,3-isobenzofurandione

Table 4.13 O_2/N_2 and H_2/N_2 separation performance for CMSMs produced at 550°C (and 200 cm^3(STP)/min for gas pyrolysis)

	O_2 flux (GPU)	O_2/N_2 selectivity	H_2 flux (GPU)	H_2/N_2 selectivity
Vacuum	25–50	7.4–9.0	372–473	64–110
Argon	71–284	2.8–6.1	451–713	6.8–31.2
Helium	73–140	4.7–6.1	428–676	15.2–35.7
Carbon dioxide	75–306	2.6–6.1	400–654	10.4–30.0

Table 4.14 Permselective properties of CMSMs pyrolyzed at 550°C in argon, helium and carbon dioxide at different gas flow rates

	Flow rate (cm^3(STP)/min)	O_2 flux (GPU)	O_2/N_2 selectivity
Argon	200	71–284	2.8–6.1
	20	0.050–0.54	2.4–7.0
Helium	200	73–140	4.7–6.1
	20	0.05–0.11	4.0–5.2
Carbon dioxide	200	75–306	2.6–6.1
	20	0.05–15	2.0–7.5

Table 4.15 Permselective properties of CMSMs produced by vacuum and helium pyrolysis at different temperatures

Pyrolysis temperature (°C)	He flow rate (cm^3(STP)/min)	O_2 flux (GPU)	O_2/N_2 selectivity
Vacuum pyrolysis			
550	–	25–50	7.4–9.0
800	–	1.6–4.2	6.8–10.3
Helium gas pyrolysis			
550	200	73–140	4.7–6.1
550	20	0.05–0.11	4.0–5.2
800	200	1.4–6.9	4.6–9.0
800	20	1.4–3.2	7.0–11.4

4.5 Post Treatment

Upon pyrolysis polymeric membranes are transformed into carbon membranes with varying degrees of porosity, structure and separation properties that depend to an extent on the pyrolysis conditions. In some cases, it is found to be advantageous that the pore dimension and its distribution in a carbon membrane are finely adjusted by a simple thermochemical treatment to meet different separation needs and objectives [80]. Therefore, various post treatment methods have been applied to meet the desired pore structure and separation properties of carbon membranes, and at the same time to repair the defects and cracks that exist in the carbon membrane.

Post-oxidation or activation is the favorite post-treatment used to alter the pore structure of carbon membranes. Typically, when a membrane is exposed to an oxidizing atmosphere after the pyrolysis step, the ensuing oxidation increases the average pore size [2, 24, 94, 95].

The oxidation of carbon membranes can be performed using pure oxygen, oxygen admixed with other gases, air, or other oxidizing agents such as steam, carbon dioxide, nitrogen oxides and chlorine oxides or solutions of oxidizing agents such as nitric acid, mixtures of nitric and sulfuric acids, chromic acid and peroxide solutions at elevated temperatures [95].

The selectivity of a carbon membrane may be increased through the introduction of organic species into the pore system of the carbon membrane and their pyrolytic decomposition (i.e. chemical vapor deposition, CVD) [39, 94–96]. Generally, to manufacture carbon molecular sieves, the inherent pore structure of the carbonaceous precursor is initially tailored into a suitable pore size range by controlling the thermal pretreatment, followed by a final adjustment of the pore structures by CVD [97]. Moreover, it is possible to produce asymmetric membranes by CVD, using homogeneous (symmetric) carbon membranes as the starting materials [94].

The organic species used for cracking and the conditions of their pyrolysis should be carefully chosen to produce sufficient and selective deposition of carbon at the pore apertures [98]. Among the organic molecules which may be used for this purpose are ethane, propane, ethylene, benzene and other hydrocarbons [95]. Aside from CVD, post pyrolysis is another treatment that can be used to decrease membrane pore size. Typically, post pyrolysis is applied after post oxidation in order to recover from an excessive pore enlargement. Sometimes post oxidation and post pyrolysis are repeated several times until the desired pore size distribution is achieved. However, this treatment is rarely used, because the first pyrolysis step at high temperature produces small pores efficiently due to the shrinkage of the carbon structure.

4.5.1 Post Oxidation

Kusakabe and co-workers [99] showed in their study that the post-oxidation of a PI carbon membrane in a mixture of O_2–N_2 increased its permeability to CO_2 and O_2 without damaging its permselectivity greatly. The outer surface of a porous alumina tube was dip-coated with BPDA-pp′ODA based polyacrylic acid (PAA) and then imidized in nitrogen at 300°C for 1 h. Coating was done two or three times. Then the precursor membrane was carbonized in a deoxygenated nitrogen stream at 700°C. The carbonized membrane was further oxidized at 300°C for 3 h. The permeance data for various single gases at either 65 or 100°C are shown in Fig. 4.35. The oxidation at 300°C for 3 h significantly increased the permeance without having adverse effect on the permselectivity. Figure 4.36 shows that the micropore volume increased without broadening the pore size distribution, when the carbonized membrane was oxidized.

For post-oxidation, different activation temperatures and dwell times have been applied to obtain desired pore structures in different materials, as illustrated in Table 4.16.

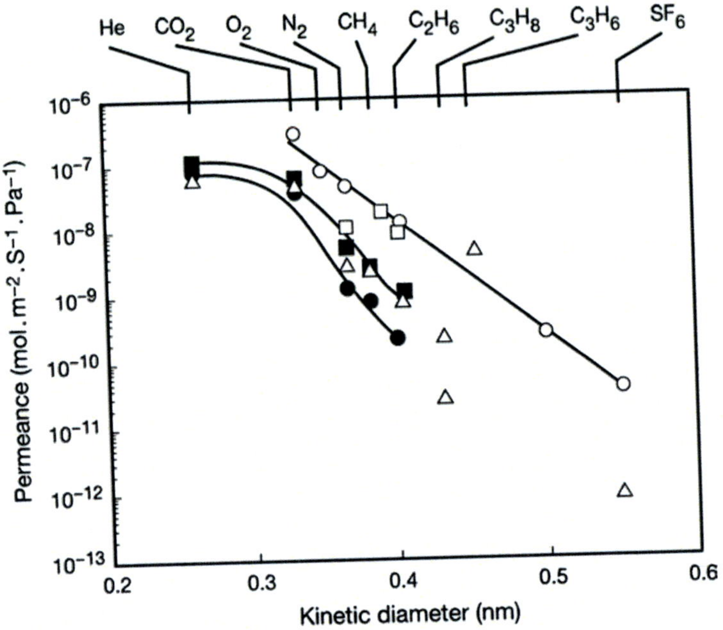

Fig. 4.35 Effect of oxidation at 300°C for 3 h on permeances of membranes carbonized at 700°C. Permeation temperature = 65°C; (●) as formed, (○) oxidized in O_2; Permeation temperature 100°C; (■) as formed, (Δ) oxidized in O_2–N_2 mixture (O_2 fraction = 0.1), (□) oxidized in O_2. (From [99])

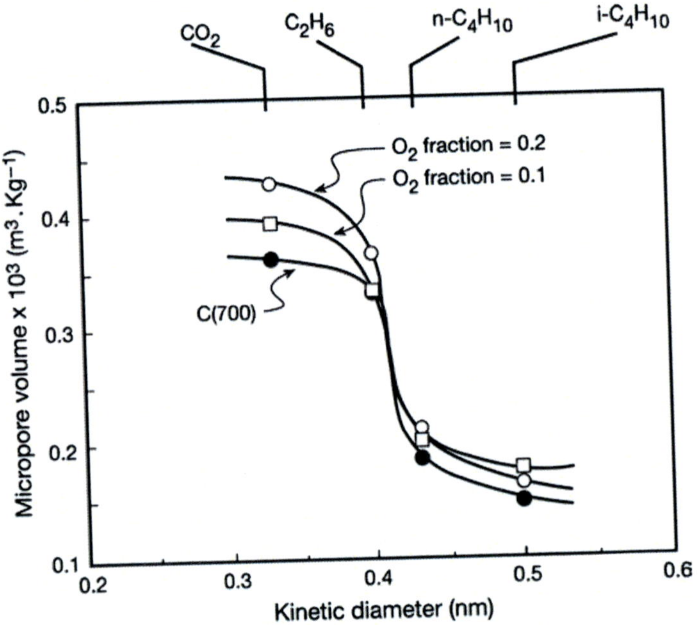

Fig. 4.36 Effect of oxidation at 300°C on the micropore volume of membrane carbonized at 700°C. (●) As formed, (□) oxidized in O_2–N_2 mixture (O_2 fraction = 0.1) at 300°C for 3 h, (○) oxidized in O_2–N_2 mixture (O_2 fraction = 0.2) at 300°C for 3 h. (From [99])

Table 4.16 Examples of post-oxidation treatments of selected carbon membranes

Precursor	Configuration	Post treatment conditions[a]	References
Cellulose	Hollow fiber or supported film	400; 15	[2]
Phenol formal-dehyde	Film supported on stainless steel plate	800; 60; CO_2	[110]
Phenol formal-dehyde	Flat	800–950; 30–60; 0.5–2.0% O_2–N_2; 80–90 mL/min	[108]
Phenol formal-dehyde	Film supported on outer surface of tubular membrane	800; –; CO_2	[109]
Phenolic resin	Film supported on inner surface of tubular membrane	100–475; 30 min–6 h	[47]
Phenolic resin	Film supported on outer surface of tubular membrane	300–400; 30	[48]
Phenolic resin	Film supported on inner surface of tubular membrane	75–350; 30	[14]
Phenolic resin	Film supported on inner surface of tubular membrane	800; –; CO_2	[113]
Polyimide	Film supported on outer surface of tubular membrane	300; 3 h; mixture O_2–N_2 or O_2	[99]
Polyimide	Hollow fiber	250–455; 0.2–50 h	[123]

[a] °C; min; air unless otherwise stated

As the oxidation temperature is raised, there is an increase in gas permeance for all gases (permanent and hydrocarbon) and a reduction in permselectivity of permanent gas pairs such as O_2/N_2 or CO_2/N_2 [47]. However, hydrocarbon gases (i.e. n-C_4H_{10}, C_2H_6, C_2H_4) exhibit higher permeances than permanent gases (i.e. He, N_2), which is a consequence of enhanced transport via hydrocarbon adsorption in the micropores [47]. It is believed that the activation process increases the permeability by expelling carbon atoms from the pore wall, thus enlarging them [94]. The variation observed in perm-selectivities suggests that for permanent gases (i.e. O_2, N_2, etc.), the transport mechanism through the membrane changes as oxidation progresses, from a molecular sieving mechanism to surface diffusion for highly oxidized samples.

4.5.2 Chemical Vapor Deposition

Verma and Walker [97] selected propylene as the preferred organic source of pyrolytic carbon since its heat of cracking (ΔH) is low, and the gas is easy to handle under ambient temperature. A few additional organic sources for chemical vapor deposition (CVD) and their application conditions are summarized in Table 4.17.

The CVD of carbon onto a carbon membrane may bring about three distinct results, namely homogeneous deposition, adlayer deposition and in-layer deposition as depicted in Fig. 4.37a–c. The preferred mode is homogeneous decomposition but, in practice, the three modes need not necessarily be completely distinct. For

Table 4.17 Chemical vapor deposition (CVD) of carbon membrane

Precursor	CVD conditions[a]	References
Cellulose	1-1-1-trichloroethane saturated in Ar; 600; 5	[94]
Phenolic resin	N_2 saturated with trichloroethane (TCE); 500; 1–10	[14]
Polyfurfuryl alcohol	20% propylene in He; 600; 2 h	[54]
Polyimide	0.05 mol fraction propylene in carrier gas; 650; 1 h	[68]

[a] Atmosphere; °C; min unless otherwise stated

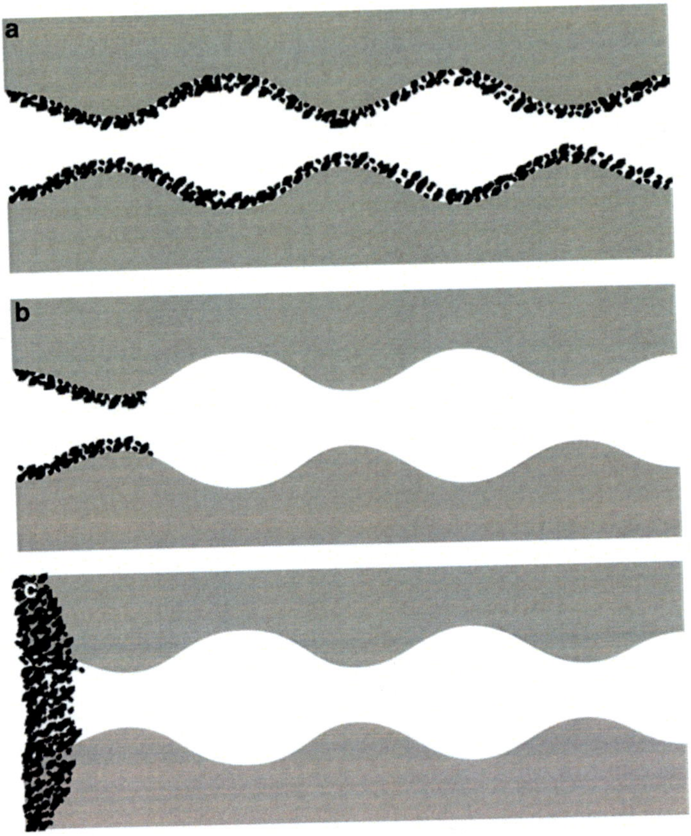

Fig. 4.37 Mechanism of carbon deposition on the pore system of carbon membrane: **(a)** homogeneous carbon deposition on membrane pore walls; **(b)** in-layer carbon deposition on membrane pore wall entrances; **(c)** adlayer carbon deposition outside membrane pores. (From [94])

instance, there may be an adlayer on top of the in-layer or some deposition in depth in addition to the in-layer.

An example of CVD treatment is shown in the following. BPDA-pp′ODA PI was coated by Hayashi et al. [68] on the shell side of a porous α-alumina tube and carbonized in deoxygenated argon atmosphere at a heating rate of 5°C/min to 700°C and then cooled down to 650°C in a reactor system shown in Fig. 4.38. Then,

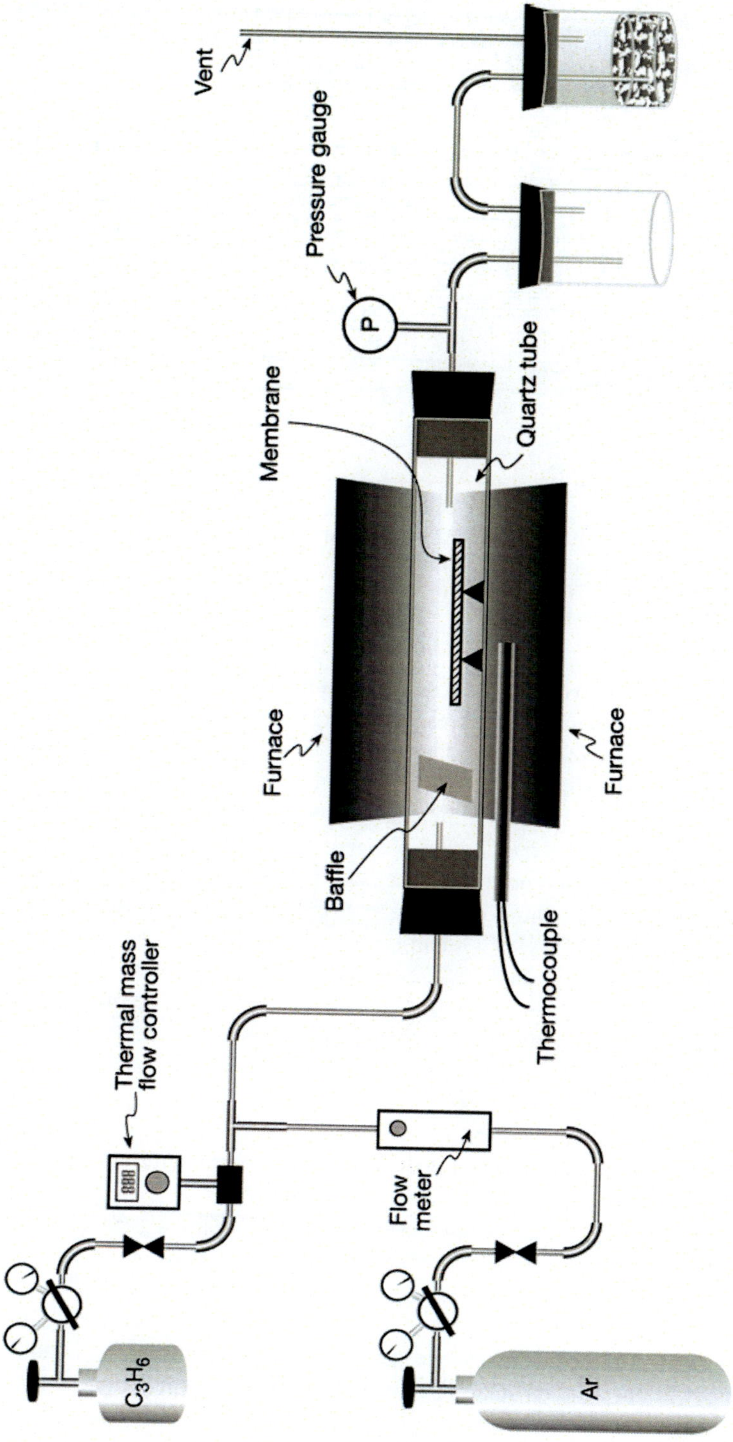

Fig. 4.38 Schematic diagram of the CVD system. (From [68])

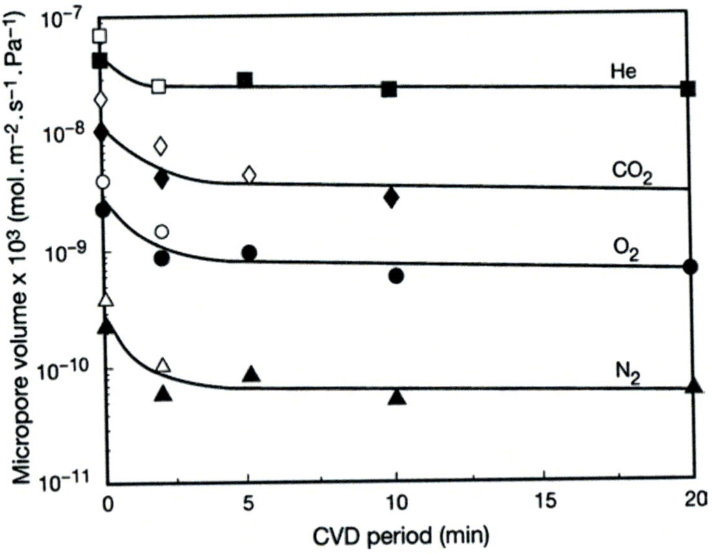

Fig. 4.39 Effect of CVD period on permeance for various permeant gases (data at 35°C). (From [68])

propylene was mixed to the carrier gas at a mole fraction of 0.05. The mixture was introduced into reactor for a prescribed period to pyrolyze the propylene. Carbonaceous solids were thus formed in pores of the membrane by CVD. The membrane was kept at 650°C for 1 h after stopping the propylene feed. Figure 4.39 shows the effect of the CVD period on the permeance of the gaseous molecules. The permeances of all the tested membranes decrease with an increase in the CVD period. The effect of CVD period on the permselectivity is shown in Fig. 4.40. It is interesting to note that a maximum is observed for O_2/N_2 and CO_2/N_2 selectivity while there is no such maximum in He/CO_2 and He/N_2 selectivity. O_2/N_2 selectivity reached 14.6 at the maximum. Similarly, CO_2/N_2 selectivity reached 73, which was much higher than the selectivity shown by polyimide membrane. The permeance of O_2 and N_2 decreased considerably in the presence of water vapor in the feed. Oxidized pores provide adsorption sites for water molecules whose permeance is enhanced by the surface diffusion mechanism.

Hayashi and co-workers also proposed the following mechanism for the narrowing of pore size distribution. When micro-pores with a size distribution are narrowed with carbon deposition, the narrowing rate of each pore depends on the size of the pore. The pores are classified into the following categories, A, B, C, D, depending on the pore size. The pore size is assumed to increase progressibly from A to D. The D pores are much larger than N_2 molecule as shown in Fig. 4.41a. The B and C pores have pore sizes similar to CO_2, O_2 and N_2. Carbon clusters are formed in the gas phase and deposited on the mouth of the D pores. The pore size thus becomes as small as that of C. Further accumulation of carbon clusters decreases the permeance but does not change the selectivity. To decrease the size of

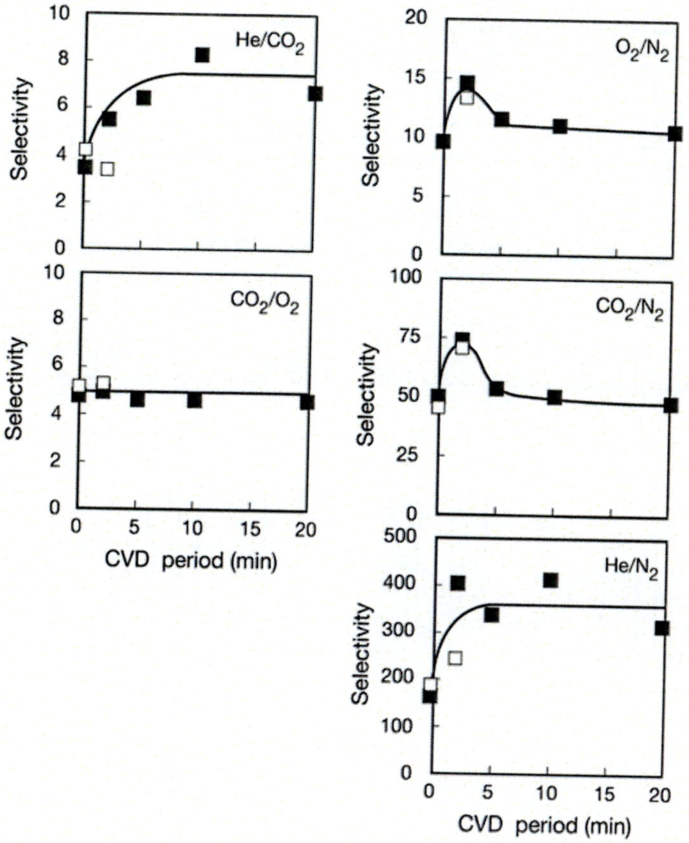

Fig. 4.40 Effect of CVD periods on permselectivity for various gas pairs (data at 35°C). (From [68])

C pores to B pores, carbon deposits should be densified or adsorbed carbon should be coked. The densification may occur at a relatively slow rate because it requires the change in the structure of the carbon deposit. The size of the pore B is assumed to be close to those of CO_2 and O_2 molecules. Thus, the B pores are reduced to the A pores faster than that of the C pores. Suppose the populations of pores A, B, C and D are 1, 2, 3 and 2, respectively. And suppose the relative speeds of the pore size reduction is 1.0, 0.5 and 1.5 for B to A, C to B, and D to C, the change in pore size distribution in time becomes as shown in Fig. 4.41b. The average pore size distribution is given by

$$\frac{\text{Sum of (pore size) (number of pores)}}{\text{Sum of (number of pores)}}$$

The average pore size then decreases and then starts to increase as the function of CVD period. This explains the trend observed in Fig. 4.40 for some of gas pairs.

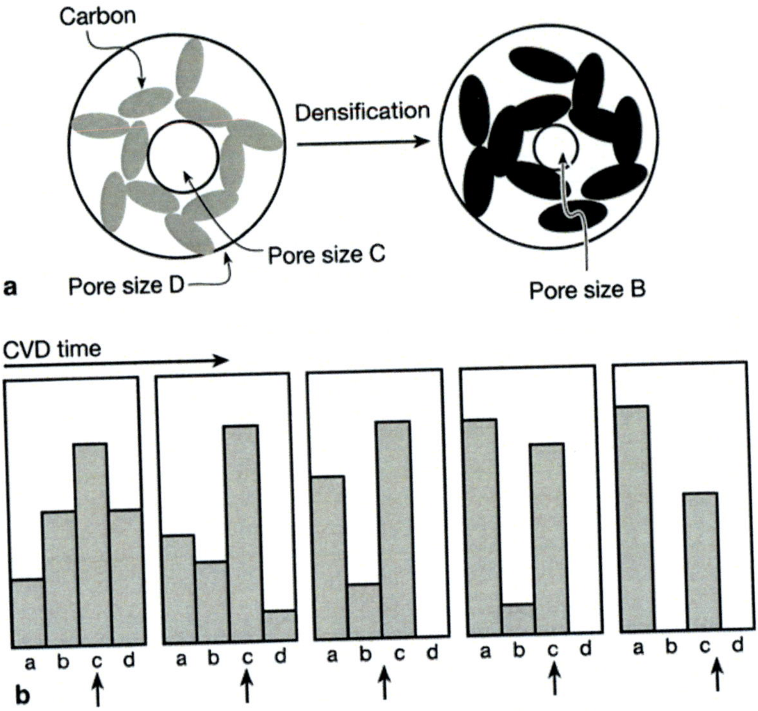

Fig. 4.41 Mechanism of carbon deposition: **(a)** Schematic view of carbon deposition, **(b)** the change in pore size distribution as the result of carbon deposition. (From [68])

4.5.3 Post Pyrolysis

The process of sintering is one way to decrease the pore volume. Sintering involves pore collapse, which proceeds more readily on the smaller pores. Heating the carbon membranes in an inert or reducing atmosphere to above 400°C (typically, 800°C) is one way to induce pore sintering. The more facile collapse of smaller pores is due to the fact that the surface energy of high curvature (small pore size) surfaces is higher. Thus, the contractive surface forces acting to diminish the pore wall surface areas and to close the pores are greater for the smaller pores. Therefore the smaller pores are the first to become closed by sintering [94].

Menendez and Fuertes [46] applied post-pyrolysis treatments to recover an aged carbon membrane used in their research. Carbon membranes were fabricated by the carbonization of a thin phenolic resin (PR) film deposited on the inner surface of porous alumina tubes. The membranes were stored in different storage conditions; i.e., storage in the air under the laboratory condition, in the air of hundred per cent humidity, in the dried air, in closed glass bottles filled with nitrogen and in closed glass bottles filled with propylene. Aging effect of the membranes under different storage conditions were monitored by the change in permeance of single gases

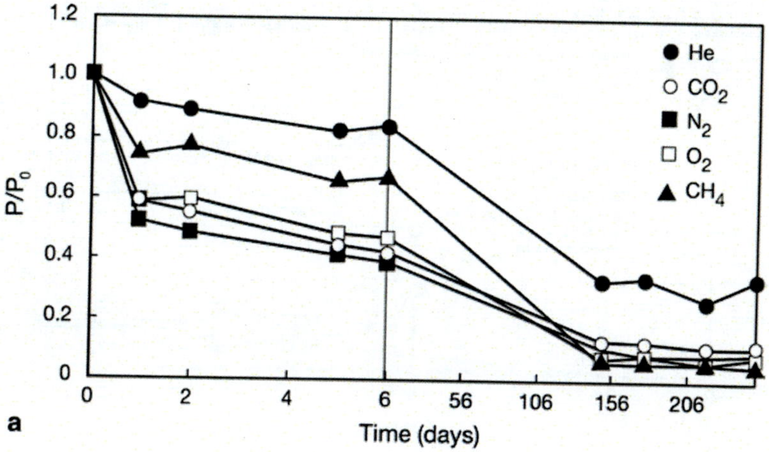

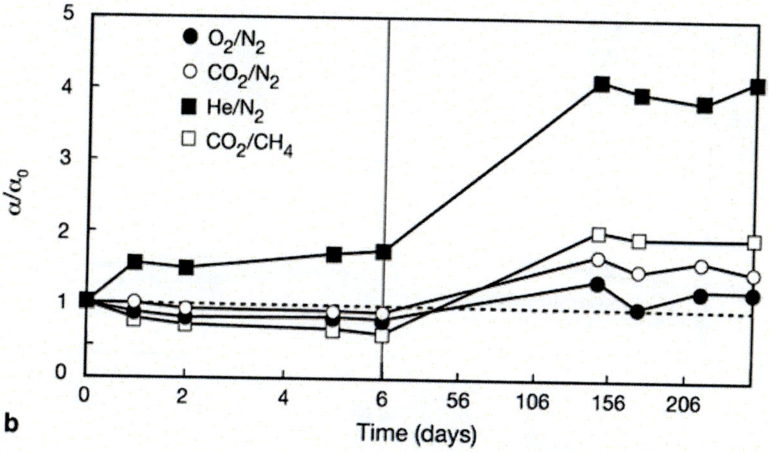

Fig. 4.42 Modification with time of (**a**) normalized permeance (P/P_0) and (**b**) the normalized permselctivity (α/α_0) of a carbon membrane stored under air laboratory environment: Permeances of fresh membrane in 10^{-10} (mol/m^2 s Pa at 20°C) are: He, 776; CO_2, 409; O_2, 136; N_2, 15; CH_4, 6; permselectivities of fresh membranes are: He/N_2, 51; O_2/N_2, 9; CO_2/N_2, 27; and CO_2/CH_4, 67. (From [46])

and the permselectivity with time. Drastic decrease in permeance and increase in permselectivity were observed when the membranes were stored, either in the laboratory condition or in the dried condition (Figs. 4.42 and 4.43). The air with 100% humidity did not enhance the aging effect. It is speculated that the exposure to the air during the storage period resulted in the formation of C–O bond and the reduction in the effective size of the micro-pores. Storage in nitrogen did not change the permeance as well as permselctivity much. On the other hand, the storage in propylene even increased the permeance and decreased the selectivity. To regenerate the membrane, they have heated the membrane after 418 days of aging. Table 4.18

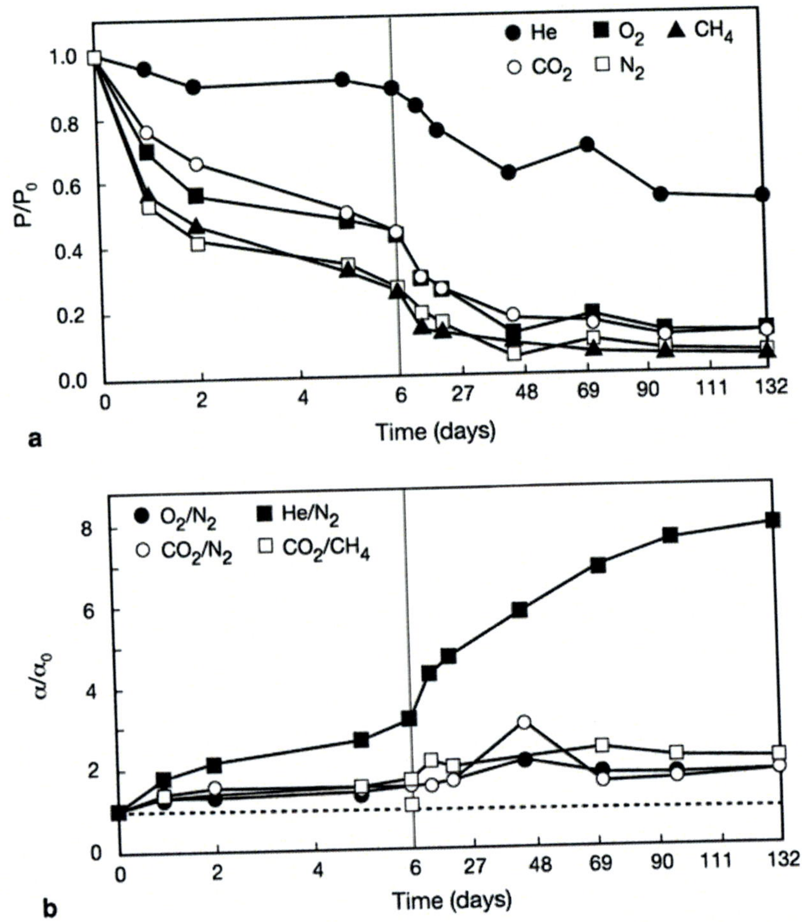

Fig. 4.43 Modification with time of (**a**) normalized permeance (P/P_0) and (**b**) the normalized permselctivity (α/α_0) of a carbon membrane stored under dry air: Permeances of fresh membrane in 10^{-10} (mol/m^2 s Pa at 20°C) are: He, 231; CO_2, 212; O_2, 60; N_2, 9; CH_4, 4; permselectivities of fresh membranes are: He/N_2, 26; O_2/N_2, 7; CO_2/N_2, 24 and CO_2/CH_4, 49. (From [46])

Table 4.18 Permeances and permselectivities of a carbon membrane, aged in the air in the laboratory air condition for 418 days, before and after treatment in vacuum at 120°C

	Permeance (mol/m^2 s Pa × 10^{10})				Permselectivity		
	He	CO_2	O_2	N_2	O_2/N_2	CO_2/N_2	He/N_2
Before regeneration	98	8.7	2.5	0.31	8.1	28	316
After regeneration	195	19.4	5.3	0.52	10.2	37	375

shows that the permeances were restored partially for each gas but the permeances after the heat treatment are still very small compared with the permeances of the fresh membrane given in Fig. 4.42a. When the aged carbon membrane (stored 245 days) was regenerated at 600°C for 1 h, better regeneration results were obtained

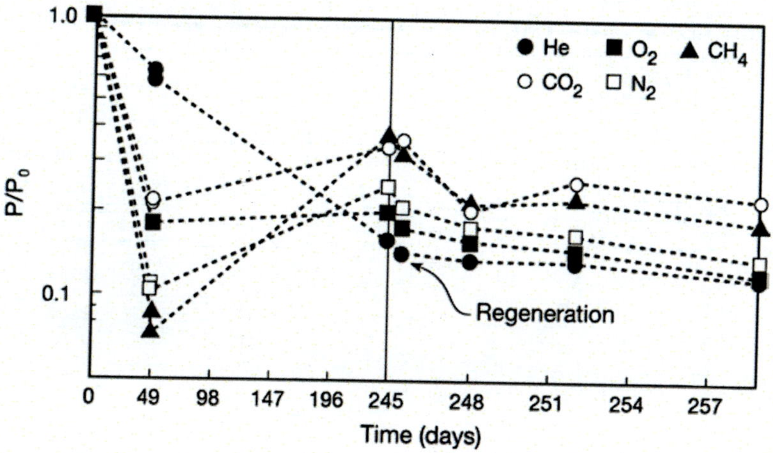

Fig. 4.44 Modification with time of normalized permeance (P/P_0) of a carbon membrane stored air for 245 days and after regeneration at 600°C. (From [46])

as shown in Fig. 4.44. From day 49 to day 245 a large increase in permeance was observed for each tested gas.

Suda and Haraya made post treatment of a CMS membrane, called original membrane, fabricated by the pyrolysis of Kapton PI film at 1,273 K for 2 h [100]. The original membrane was heated to 673 K at a heating rate of 10 K/min under argon stream (post pyrolysis) and then in helium stream containing water vapor. After 10 min helium was replaced by argon and cooled down to room temperature. The membrane is called the modified membrane. The gas permeation characteristics of the original and modified membranes are given in Fig. 4.45. The original membrane showed a sieve curve except for the reversed order between He and H_2. But the modified membrane showed much larger scatter due to the formation of larger pores during the post heat treatment process, which is also evidenced by the much greater permeabilities (10–700 times higher) of the modified membrane. It is particularly interesting to note that the permselectivity of propene and propane is more than 100 at 308 K. Thus this modified membrane seems suitable for the alkene/alkane separation.

4.5.4 *Fouling Reduction*

Carbon membranes generally have nonpolar surface and therefore organophilic. Strong adsorption of organic compounds results in significant problems in membrane application. Jones and Koros presented a new technique of regeneration of carbon membranes by exposing them to propyrene vapor [101]. CMS hollow fibers were fabricated by the vacuum pyrolysis of aromatic polyimide (PI) hollow fibers at temperatures of 500 and 550°C. Membrane performance changes resulting from

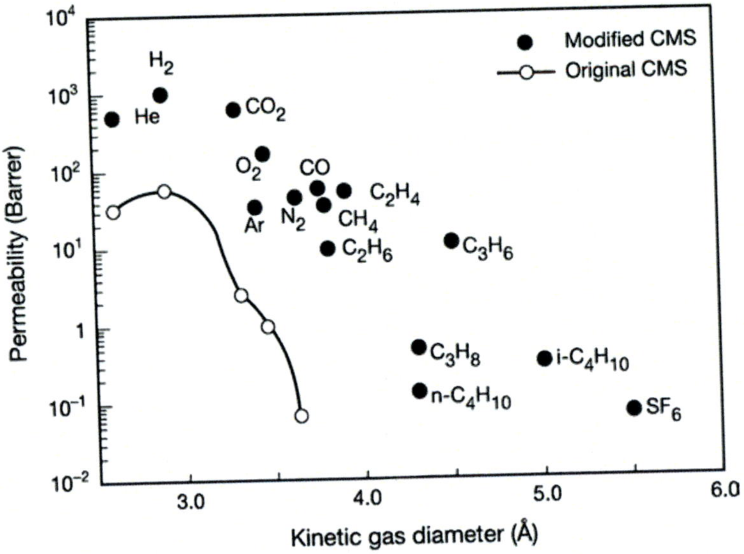

Fig. 4.45 Permeabilities of selected gases for the original and modified CMSMs (data at 373 K). (From [100])

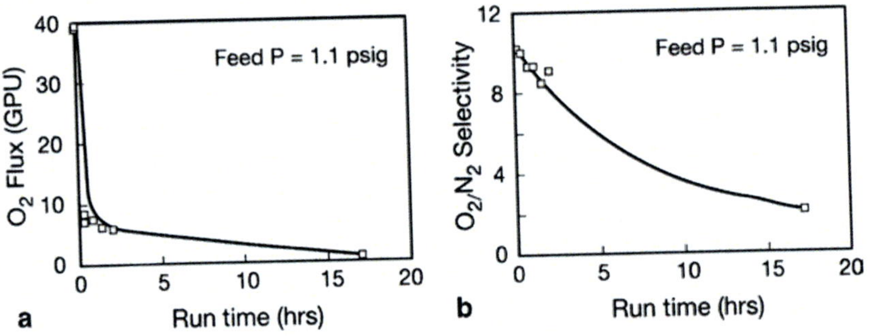

Fig. 4.46 Change in (**a**) O_2 flux and (**b**) O_2/N_2 selectivity with time when the membrane was exposed to hexane-saturated air feed. (From [101])

exposure to organics as can be seen in O_2 flux and O_2/N_2 selectivity. Figure 4.46a and b show the change in O_2 flux and O_2/N_2 selectivity in the presence of hexane in the feed stream of air. The O_2 flux went down from 39.0 to 7.6 GPU within 9 min. The selectivity also dropped significantly. The hollow fiber module was shut down after the experiment with hexane. The module was then exposed to dry air feed. The module was further exposed to pure propylene at feed pressures between 10 and 150 psig for time periods ranging 2–3 h. After each exposure, dry air feed was resumed and change in the hollow fiber performance was checked. Figure 4.47 shows the results from the hollow fiber whose flux went down after the exposure to hexane. It shows a dramatic increase in O_2 flux after the first exposure to pro-

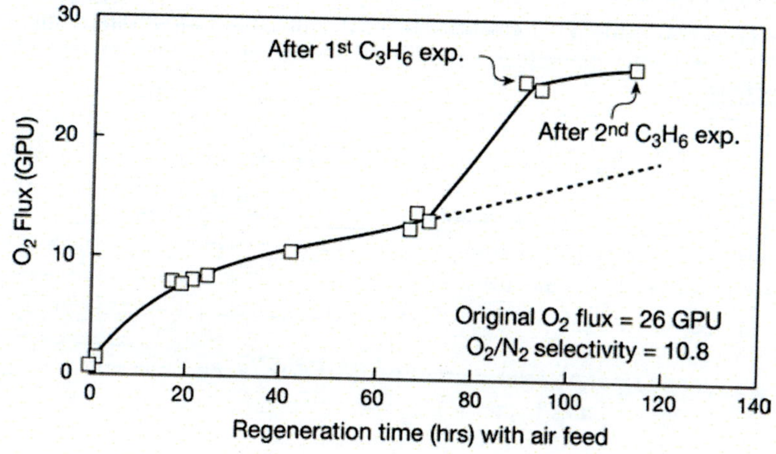

Fig. 4.47 Membrane regeneration after hexane exposure. (From [101])

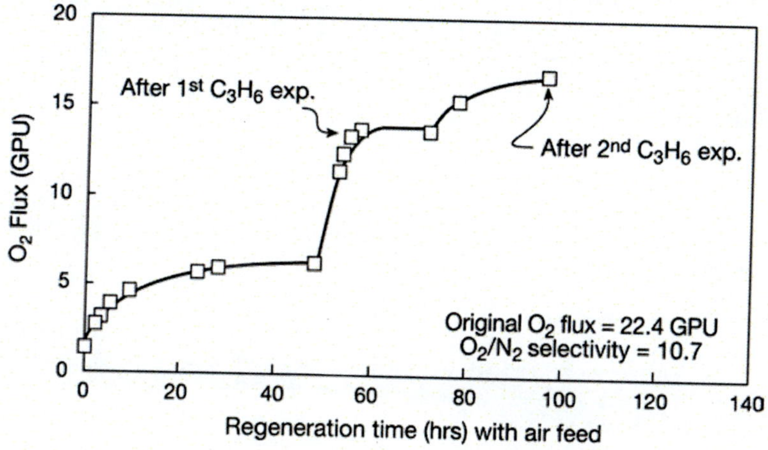

Fig. 4.48 Membrane regeneration after hexadecane exposure. (From [101])

pylene and after the second exposure, the initial O_2 flux (26 GPU) was completely recovered. Similar results were obtained for the hollow fiber module contaminated by hexadecane (Fig. 4.48).

4.5.5 *Coating*

The coating technique has been applied to repair defects in their carbon membrane by coating a thin layer of polydimethylsiloxane (PDMS) to prevent gas flow through defect structures. The membrane was immersed in a 2% solution of PDMS (WACKER E41) in heptane, and then withdrawn to harden the coated layer.

Alternatively, a method entailing surface coating with an alcohol solution of 60% B rank phenol formaldehyde resin (PFR) and dispersant, followed by recarbonization (700°C, 60 min), was used to prepare carbon–carbon composite membranes with improved separation performance by Liang and co-workers [80]. On the carbon–carbon composite membrane coated with PFR alcohol solution and added dispersant, the gas permeation rates decreased by a factor of 10, and the ideal separation factor increased, probably due to the PFR's dispersing effect.

Jones and Koros [102, 103] coated their carbon membrane to counteract the decrease in carbon membrane performance when exposed to water vapor. They used poly(4-methyl-1-pentene) PMP from Scientific Polymer Products, and DuPont's Teflon AF 1600 and AF 200, as coating agents. They managed to provide a successful protection barrier, which significantly limited the permeation of water vapor or other impurities such as hydrocarbons without significantly inhibiting the permeation of the faster fluid component or lowering its selectivity. Accordingly, they have patented [102] composite membranes that retain good fluid separation properties and are resistant to the adverse effects on membrane performance commonly observed in environments having high humidity.

References

1. Morthon-Jones, DH (1989) Polymer Processing. Chapman and Hall, London, UK
2. Soffer A, Rosen D, Saguee S, Koresh J (1989) Carbon membrane. GB Patent 2,207,666.
3. László K, Bóta A, Nagy LG (2000) Comparative adsorption study on carbons from polymer precursors. Carbon 38 (14): 1965-1976
4. Lua AC, Su J (2006) Effects of carbonisation on pore evolution and gas permeation properties of carbon membranes from Kapton® polyimide. Carbon 44 (14): 2964-2972
5. Kim S, Pechar TW, Marand E (2006) Poly(imide siloxane) and carbon nanotube mixed matrix membranes for gas separation. Desalination 192 (1-3): 330-339
6. Steel KM, Koros WJ (2005) An investigation of the effects of pyrolysis parameters on gas separation properties of carbon materials. Carbon 43 (9): 1843-1856
7. Park HB, Lee SY, Lee YM (2005) Pyrolytic carbon membranes containing silica: Morphological approach on gas transport behavior. J Mol Struct 739 (1-3): 179-190
8. Shao L, Chung T-S, Wensley G, Goh SH, Pramoda KP (2004) Casting solvent effects on morphologies, gas transport properties of a novel 6FDA/PMDA–TMMDA copolyimide membrane and its derived carbon membranes. J Membr Sci 244 (1-2): 77-87
9. Park HB, Jung CH, Kim YK, Nam SY, Lee SY, Lee YM (2004) Pyrolytic carbon membranes containing silica derived from poly(imide siloxane): The effect of siloxane chain length on gas transport behavior and a study on the separation of mixed gases. J Membr Sci 235 (1-2): 87-98
10. Kim YK, Park HB, Lee YM (2005) Gas separation properties of carbon molecular sieve membranes derived from polyimide/polyvinylpyrrolidone blends: Effect of the molecular weight of polyvinylpyrrolidone. J Membr Sci 251 (1-2): 159-167
11. Kim YK, Park HB, Lee YM (2005) Preparation and characterization of carbon molecular sieve membranes derived from BTDA–ODA polyimide and their gas separation properties. J Membr Sci 255 (1-2): 265-273
12. Centeno TA, Fuertes AB (1999) Supported carbon molecular sieve membranes based on phenolic resin. J Membr Sci 160 (2): 201-211

13. Centeno TA, Fuertes AB (2001) Carbon molecular sieve membranes derived from a phenolic resin supported on porous ceramic tubes. Sep Purif Technol 25 (1-3): 379-384
14. Fuertes AB, Menendez I (2002) Separation of hydrocarbon gas mixtures using phenolic resin-based carbon membranes. Sep Purif Technol 28 (1): 29-41
15. Wu J, Chung DDL (2002) Increasing the electromagnetic interference shielding effectiveness of carbon fiber polymer–matrix composite by using activated carbon fibers. Carbon 40 (3): 445-467
16. Song C, Wang T, Wang X, Qiu J, Cao Y (2008) Preparation and gas separation properties of poly(furfuryl alcohol)-based C/CMS composite membranes. Sep Purif Technol 58 (3): 412-418
17. Zhang X, Hu H, Zhu Y, Zhu S (2007) Carbon molecular sieve membranes derived from phenol formaldehyde novolac resin blended with poly(ethylene glycol). J Membr Sci 289 (1-2): 86-91
18. Wei W, Qin G, Hu H, You L, Chen G (2007) Preparation of supported carbon molecular sieve membrane from novolac phenol–formaldehyde resin. J Membr Sci 303 (1-2): 80-85
19. Lie JA, Hägg MB (2005) Carbon membranes from cellulose and metal loaded cellulose. Carbon 43 (12): 2600-2607
20. Chen JC (1998) Modification of polyacrylonitrile (PAN) precursor fiber via post-spinning plasticization and stretching. PhD Thesis, The Pennsylvania State University, University Park, PA, USA
21. Gupta A, Harrison IR (1996) New aspects in the oxidative stabilization of PAN-based carbon fibers. Carbon 34 (11): 1427-1445
22. Donnet JB, Bansal RC (1984) Carbon Fiber. Marcel Dekker, New York, NY, USA
23. Saufi SM, Ismail AF (2002) Development and characterization of polyacrylonitrile (PAN) based carbon hollow fiber membrane. Songklanakarin J Sci Technol 24 (Suppl): 843-854
24. Schindler E, Maier F (1990) Manufacture of porous carbon membranes. US Patent 4,919,860
25. Yoneyama H, Nishihara Y (1990) Porous hollow carbon fiber film and method of manufacturing the same. EP Patent 0,394,449
26. Linkov VM, Sanderson RD, Jacobs EP (1994) Highly asymmetrical carbon membranes. J Membr Sci 95 (1): 93-99
27. Linkov VM, Sanderson RD, Jacobs EP (1994) Carbon membranes from precursors containing low-carbon residual polymers. Polym Inter 35 (3): 239-242
28. Smith SPJ, Linkov VM, Sanderson RD, Petrik LF, O'Connor CT, Keiser K (1995) Preparation of hollow-fiber composite carbon-zeolite membranes. Microporous Mater 4 (5): 385-390
29. Linkov VM, Sanderson RD, Rychkov BA (1994) Composite carbon-polyimide membranes. Mater Lett 20 (1-2): 43-46
30. Linkov VM, Bobrova LP, Timofeev SV, Sanderson RD (1995) Composites membranes from perfluoro-sulfonated ionomer on carbon and ceramic supports. Mater Lett 24 (1-3): 147-151
31. David LIB, Ismail AF (2003) Influence of the thermostabilization process and soak time during pyrolysis process on the polyacrylonitrile carbon membranes for O_2/N_2 separation. J Membr Sci 213 (1-2): 285-291; Geiszler VC (1997) Polyimide precursors for carbon membranes. PhD Thesis, University of Texas at Austin, TX, USA
32. Jones CW, Koros WJ (1994) Carbon molecular sieve gas separation membranes-I. Preparation and characterization based on polyimide precursors. Carbon 32 (8): 1419-1425
33. Hatori H, Kobayashi T, Hanzawa Y, Yamada Y, Iimura Y, Kimura T, Shiraishi M (2001) Mesoporous carbon membranes from polyimide blended with poly(ethylene glycol). J Appl Polym Sci 79 (5): 836-841
34. Vu DQ, Koros WJ, Miller SJ (2002) High pressure CO_2/CH_4 separation using carbon molecular sieve hollow fiber membranes. Ind Eng Chem Res 41 (3): 367-380
35. Yamamoto M, Kusakabe K, Hayashi J, Morooka S (1997) Carbon molecular sieve membrane formed by oxidative carbonization of a copolyimide film coated on a porous support tube. J Membr Sci 133 (2): 195-205

36. Shiflett MB (2002) Synthesis, characterization and application of nanoporous carbon membranes. PhD Thesis, University of Delaware, Newark, DE, USA
37. Fuertes AB, Centeno TA (1998) Preparation of supported asymmetric carbon molecular sieve membranes. J Membr Sci 144 (1-2): 105-111
38. Ogawa M, Nakano Y (1999) Gas permeation through carbonized hollow fiber membranes prepared by gel modification of polyamic acid. J Membr Sci 162 (1-2): 189-198
39. Hayashi J, Yamamoto M, Kusakabe K, Morooka S (1995) Simultaneous improvement of permeance and permselectivity of 3,3′,4,4′-biphenyltetracarboxylic dianhydride-4,4′-oxydianiline polyimide membrane by carbonization. Ind Eng Chem Res 34 (12): 4364-4370
40. Hattori H, Yamada Y, Shiraishi M, Nakata H, Yoshitomi S (1992) Carbon molecular sieve films from polyimide. Carbon 30 (4): 719-720
41. Petersen J, Matsuda M, Haraya K (1997) Capillary carbon molecular sieve membranes derived from kapton for high temperature gas separation. J Membr Sci 131 (1-2): 85-94
42. Fuertes AB, Nevskaia DM, Centeno TA (1999) Carbon composite membranes from matrimid and kapton polyimide. Microporous Mesoporous Mater 33 (1-3): 115-125
43. Fuertes AB, Centeno TA (1998) Carbon molecular sieve membranes from polyetherimide. Microporous Mesoporous Mater 26 (1-3): 23-26
44. Sedigh MG, Xu L, Tsotsis TT, Sahimi M (1999) Transport and morphological characteristics of polyetherimide-based carbon molecular sieve membranes. Ind Eng Chem Res 38 (9): 3367-3380
45. Barbosa-Coutinho E, Salim VMM, Borges CP (2003) Preparation of carbon hollow fiber membranes by pyrolysis of polyetherimide. Carbon 41 (9): 1707-1714
46. Menendez I, Fuertes AB (2001) Aging of carbon membranes under different environments. Carbon 39 (5):733-740
47. Fuertes AB (2001) Effect of air oxidation on gas separation properties of adsorption-selective carbon membranes. Carbon 39 (5): 697-706
48. Fuertes AB (2000) Adsorption-selective carbon membrane for gas separation. J Membr Sci 177 (1-2): 9-16
49. Zhou W, Yoshino M, Kita H, Okamoto K (2001) Carbon molecular sieve membranes derived from phenolic resin with a pendant sulfonic acid group. Ind Eng Chem Res 40 (22): 4801-4807
50. Sedigh MG, Onstot WJ, Xu L, Peng WL, Tsotsis TT, Sahimi M (1998) Experiments and simulation of transport and separation of gas mixtures in carbon molecular sieves membranes. J Phys Chem A 102 (44): 8580-8589
51. Acharya M (1999) Engineering design and theoretical analysis of nanoporous carbon membranes for gas separation. PhD Thesis, University of Delaware, Newark, DE, USA
52. Shiflett MB, Foley HC (2000) On the preparation of supported nanoporous carbon membranes. J Membr Sci 179 (1-2): 275-282
53. Shiflett MB, Foley HC (1999) Ultrasonic deposition of high-selectivity nanoporous carbon membranes. Science 285 (5435): 1902-1905
54. Acharya M, Foley HC (1999) Spray-coating of nanoporous carbon membranes for air separation. J Membr Sci 161 (1-2): 1-5
55. Corbin DR, Foley HC, Shiflett MB (2004) Mixed matrix nanoporous carbon membranes. US Patent 6,740,143
56. Shiflett MB, Foley HC (2001) Reproducible production on nanoporous carbon membranes. Carbon 39 (9): 1421-1446
57. Strano MS, Foley HC (2001) Synthesis and characterization of catalytic nanoporous carbon membranes. AIChE J 47 (1): 66-78
58. Strano MS, Foley HC (2002) Temperature- and pressure-dependent transient analysis of single component permeation through nanoporous carbon membranes. Carbon 40 (7): 1029-1041
59. Chen YD, Yang RT (1994) Preparation of carbon molecular sieve membrane and diffusion of binary mixtures in the membrane. Ind Eng Chem Res 33 (12): 3146-3153

60. Wang H, Zhang L, Gavalas GR (2000) Preparation of supported carbon membranes from furfuryl alcohol by vapour deposition polymerization. J Membr Sci 177 (1-2): 25-31
61. Xiao Y, Chung T-S, Guan HM, Guiver MD (2000) Synthesis, cross-linking and carbonization of co-polyamides containing internal acetylene units for gas separation. J Membr Sci 302 (1-2): 254-264
62. Kim YK, Park HB, Lee YM (2004) Carbon molecular sieve membranes derived from thermally labile polymer containing blend polymers and their gas separation properties. J Membr Sci 243 (1-2): 9-17
63. Centeno TA, Fuertes AB (2000) Carbon molecular sieve gas separation membranes based on poly(vinylidene chloride-co-vinyl chloride). Carbon 38 (7): 1067-1073
64. Kusakabe K, Gohgi S, Morooka S (1998) Carbon molecular sieving membranes derived from condensed polynuclear aromatic (COPNA) resins for gas separations. Ind Eng Chem Res 37 (11): 4262-4266
65. Yoshimune M, Fujiwara I, Haraya K (2007) Carbon molecular sieve membranes derived from trimethylsilyl substituted poly(phenylene oxide) for gas separation. Carbon 45 (3): 553-560
66. Lie JA, Hägg M-B (2006) Cellulose membrane from cellulose: Synthesis, performance and regeneration. J Membr Sci 284 (1-2): 79-86
67. Ismail AF, David LIB (2001) A review on the latest development of carbon membranes for gas separation. J Membr Sci 193 (1):1-18
68. Hayashi J, Mizuta H, Yamamoto M, Kusakabe K, Morooka S (1997) Pore size control of carbonized BPTA-pp'ODA polyimide membrane by chemical vapor deposition of carbon. J Membr Sci 124 (2): 243-251
69. Linkov VM, Sanderson RD, Lapidus AL, Krylova AJ (1994) Carbon membrane-based catalysts for hydrogenation of CO. Catal Lett 27 (1-2): 97-101
70. Fuertes AB, Centeno TA (1999) Preparation of supported carbon molecular sieve membrane. Carbon 37 (4): 679-684
71. Clausi DT, Koros WJ (2000) Formation of defect-free polyimide hollow fiber membranes for gas separation. J Membr Sci 167 (1): 79-89
72. Puri PS (1990) Fabrication of hollow fiber gas separation membranes. Gas Sep Purif 4 (1): 29-36
73. Baker RW (2002) Future directions of membrane gas separation technology. Ind Eng Chem Res 41 (6): 1393-1411
74. Kusuki Y, Shimazaki H, Tanihara N, Nakanishi S, Yoshinaga T (1997) Gas permeation properties and characterization of asymmetric carbon membranes prepared by pyrolyzing asymmetric polyimide hollow fiber membrane. J Membr Sci 134 (2): 245-253
75. Tanihara N, Kusuki Y (2000) Partially carbonized polyimide membrane, preparation thereof, and separation of halogen compound gases therewith. EP Patent Appl 1,034,836
76. Okamoto K, Kawahara S, Yoshino M, Kita H, Hirayama Y, Tanihara N, Kusuki Y (1999) Olefin/paraffin separation through carbonized membranes derived from an asymmetric polyimide hollow fiber membrane. Ind Eng Chem Res 38 (11): 4424-4432
77. Itoh N, Haraya K (2000) A carbon membrane reactor. Catal Today 56 (1-3): 103-111
78. Centeno TA, Fuertes AB (2001) Carbon molecular sieve membranes from a phenolic resin supported on porous ceramic tubes. Sep Purif Technol 25 (1-3): 379-384
79. Centeno TA, Fuertes AB (2000) Carbon molecular sieve gas separation membranes based on poly(vinylidene chloride-co-vinyl chloride). Carbon 38 (7): 1067-1073
80. Liang C, Sha G, Guo S (1999) Carbon membrane for gas separation derived from coal tar pitch. Carbon 37 (9): 1391-1397
81. Soffer A, Gilron J, Saguee S, Hed-Ofek R, Cohen H (1995) Process for the production of hollow carbon fiber membranes. EP Patent 0,671,202
82. Tin PS, Chung T-S, Hill AJ (2004) Advanced fabrication of carbon molecular sieve membranes by nonsolvent pretreatment of precursor polymers. Ind Eng Chem Res 43 (20): 6476-6483

83. Lagorsse S Magalhaes FD, Mendes A (2007) Xenon recycling in an anaesthetic closed-system using carbon molecular sieve membranes. J Membr Sci 301 (1-2): 29-38
84. Ismail AF, Li K (2008) From polymeric precursors to hollow fiber carbon and ceramic membranes. In: Inorganic Membranes: Synthesis, Characterization and Applications, Mallada R, Menéndez M (Eds.) Membrane Science Technology Ser, Elsevier, Amsterdam, The Netherlands, Vol 13, Ch 3, 81-119
85. Geiszler VC, Koros WJ (1996) Effect of polyimide pyrolysis conditions on carbon molecular sieve membrane properties. Ind Eng Chem Res 35 (9): 2999-3003
86. Koresh JE, Soffer A (1980) Study of molecular sieve carbons. Part 1. Pore structure, gradual pore opening, and mechanism of molecular sieving. J Chem Soc Faraday Trans I 76 (12): 2457-2471
87. Steel KM, Koros WJ (2003) Investigation of porosity of carbon materials and related effects on gas separation properties. Carbon 41 (2): 235-266
88. Jones CW, Koros WJ (1995) Characterization of ultramicroporous carbon membranes with humidified feeds. Ind Eng Chem Res 34 (1): 158-163
89. Suda H, Haraya K (1997) Gas permeation through micropores of carbon molecular sieve membranes derived from kapton polyimide. J Phys Chem B 101 (20): 3988-3994
90. Tanihara N, Shimazaki H, Hirayama Y, Nakanishi S, Yoshinaga T, Kusuki Y (1999) Gas permeation properties of asymmetric carbon hollow fiber membranes prepared from asymmetric hollow fiber. J Membr Sci 160 (2): 179-186
91. David LIB (2001) Development of asymmetric polyacrylonitrile carbon hollow fiber membrane for oxygen/nitrogen gas separation. MSc Thesis, Universiti Teknologi Malaysia, Skudai, Johor, Malaysia
92. Singh-Ghosal A, Koros WJ (2000) Air separation properties of flat sheet homogeneous pyrolytic carbon membranes. J Membr Sci 174 (2): 177-188
93. Barsema JN, Van Der Vegt NFA, Koops GH, Wessling M (2002) Carbon molecular sieve membranes prepared from porous fiber precursor. J Membr Sci 205 (1-2): 239-246
94. Soffer A, Azariah A, Amar A, Cohen H, Golub D, Saguee S, Tobias H (1997) Method of improving the selectivity of carbon membranes by chemical vapor deposition. US Patent 5,695,818
95. Soffer A, Koresh J, Saggy S (1987) Separation device. US Patent 4,685,940
96. Cabrera AL, Zehner JE, Coe CG, Gaffney TR, Farris TS (1993) Preparation of carbon molecular sieves, I. Two-step hydrocarbon deposition with a single hydrocarbon. Carbon 31 (6): 969-976
97. Verma SK, Walker Jr PL (1992) Preparation of carbon molecular sieves by propylene pyrolysis over microporous carbons. Carbon 30 (6): 829-836
98. Chihara K, Suzuki M (1979) Control of micropore diffusivities of molecular sieving carbon by deposition of hydrocarbons. Carbon 17 (4): 339-343
99. Kusakabe K, Yamamoto M, Morooka S (1998) Gas permeation and micropore structure of carbon molecular sieving membranes modified by oxidation. J Membr Sci 149 (1): 59-67
100. Suda H, Haraya K (1997) Alkene/alkane permselectivityies of a carbon molecular sieve membrane. J Chem Soc Chem Commun: 93-94
101. Jones CW, Koros WJ (1994) Carbon molecular sieve gas separation membranes.II. Regeneration following organic exposure. Carbon 32 (8): 1427-1432
102. Koros WJ, Jones CW (1994) Composite carbon fluid separation membranes. US Patent 5,288,304
103. Jones CW, Koros WJ (1995) Carbon composite membranes: A solution to adverse humidity effects. Ind Eng Chem Res 34 (1): 164-167
104. Yoneyama H, Nishihara Y (1992) Carbon based porous hollow fiber membrane and method for producing same. US Patent 5,089,135
105. Gilron J, Soffer A (2002) Knudsen diffusion in microporous carbon membranes with molecular sieving character. J Membr Sci 209 (2): 339-352
106. Hatori H, Yamada Y, Shiraishi M (1992) Preparation of macroporous carbon films from polyimide by phase inversion method. Carbon 30 (2): 303-304

107. Suda H, Haraya K (1995) Molecular sieving effect of carbonized Kapton polyimide membrane. J Chem Soc Chem Commun: 1179-1180
108. Wang S, Zeng M, Wang Z (1996) Asymmetric molecular sieve carbon membranes. J Membr Sci 109 (2): 267-270
109. Steriotis TA, Beltsios K, Mitropoulos ACh, Kanellopoulos N, Tennison S, Wiedenman A, Keiderling U (1997) On the structure of an asymmetric carbon membrane with a novolac resin precursor. J Appl Polym Sci 64 (12): 2323-2345
110. Katsaros FK, Steriotis TA, Stefanopoulos KL, Kanellopoulos NK, Mitropoulos AC, Meissner M, Hoser A (2000) Neutron diffraction study of adsorbed CO_2 on a carbon membrane. Physica B 267-278: 901-902
111. Sakata Y, Muto A, Uddin MA, Suga H (1999) Preparation of porous carbon membrane plates for pervaporation separation applications. Sep Purif Technol 17 (2): 97-100
112. Clint JH, Lear AM, Oliver LF, Tennison SR (1992) Membranes. EP Patent 0,474,424
113. Katsaros FK, Steriotis TA, Stubos AK, Mitropoulos A, Kanellopoulos NK, Tennison S (1997) High pressure gas permeability of microporous carbon membranes. Microporous Mater 8 (3): 171-176
114. Lee L-L, Tsai D-S (2001) Synthesis and permeation properties of silicon-carbon-based inorganic membrane for gas separation. Ind Eng Chem Res 40 (2): 612-616
115. Rao MB, Sircar S (1993) Nanoporous carbon membranes for separation of gas mixtures by selective surface flow. J Membr Sci 85 (3): 253-264
116. Rao MB, Sircar S, Golden TC (1992) Gas separation by adsorbent membranes. US Patent 5,104,425
117. Rao MB, Sircar S (1996) Performance and pore characterization of nanoporous carbon membrane for gas separation. J Membr Sci 110 (1): 109-118
118. Thaeron C, Parrillo DJ, Sircar S, Clarke PF, Paranjape M, Pruden BB (1999) Separation of hydrogen sulfide-methane mixtures by selective surface flow membrane. Sep Purif Technol 15 (2): 121-129
119. Sedigh MG, Jahangiri M, Liu PKT, Sahimi M, Tsotsis TT (2000) Structural characterization of polyetherimide-based carbon molecular sieve membranes. AIChE J 46 (11): 2245-2255
120. Hayashi J, Yamamoto M, Kusakabe K, Morooka S (1997) Effect of oxidation on gas permeation of carbon molecular sieve membranes based on BPDA-pp'/ODA polyimide. Ind Eng Chem Res 36 (6): 2134-2140
121. Hayashi J, Mizuta H, Yamamoto M, Kusakabe K, Morooka S (1996) Separation of ethane/ethylene and propane/propylene system with a carbonized BPDA-pp'ODA polyimide. Ind Eng Chem Res 35 (11): 4176-4181
122. Koros WJ, Vu DQ (2003) High carbon content filamentary membrane and method of making the same. US Patent 6,565,631
123. Yoshinaga T, Shimazaki H, Kusuki Y, Sumiyama Y (1991) Asymmetric hollow filamentary carbon membrane and process for producing the same. EP Patent 0,459,623
124. Ogawa M, Nakano Y (2000) Separation of CO_2/CH_4 mixture through carbonized membrane prepared by gel modification. J Membr Sci 173 (1): 123-132
125. Kita H, Yoshino M, Tanaka K, Okamoto K (1997) Gas permselectivity of carbonized polypyrrolone membrane. J Chem Soc Chem Commun: 1051-1052

Chapter 5
Examples of CMSM Preparation, Characterization and Testing

5.1 Hollow Fiber CMSM Membrane from Polyacrylonitrile (PAN)

5.1.1 *Polymer Solution Preparation*

PAN solution is prepared using the round bottom vessel shown in Fig. 5.1. PAN in powder form is heated in an oven for one day to dry completely before use, followed by heating in a vacuum oven before being used for solution preparation. The vessel is first loaded with dimethyllformamide (DMF), the solvent. Then, PAN powder is added to the solvent slowly under vigorous stirring. The temperature of the vessel should be below the boiling point of DMF (153°C) but sufficiently high to facilitate the dissolution of PAN powder. Normally, it is kept between 80 and 90°C. It takes 7–9 h to obtain homogeneous solution.

After the polymer is fully dissolved, the solution is degassed for 48 h by the ultrasonic method (Branson Ultrasonics) to remove tiny air bubbles. Finally, the homogeneous solution is stored in a bottle and is ready for hollow fiber spinning.

5.1.2 *Hollow Fiber Spinning Process*

PAN hollow fibers are spun by the dry/wet spinning process using the equipment shown in Fig. 5.2.

The PAN polymer solution and the bore fluid are extruded simultaneously from a spinneret (dimension OD:ID = 600:300 μm) to form a nascent hollow fiber membrane at ambient temperature. The ratio of polymer solution and bore fluid flow rate is 3 to1. A gear pump is used to deliver the polymer solution to the spinneret smoothly at a certain dope extrusion rate (DER) ranging from 1.5 to 3.0 cm^3/min. The fiber is then directed to a chamber with a forced flow of nitrogen for dry phase

A. F. Ismail et al., *Carbon-based Membranes for Separation Processes*,
DOI 10.1007/978-0-387-78991-0_5,

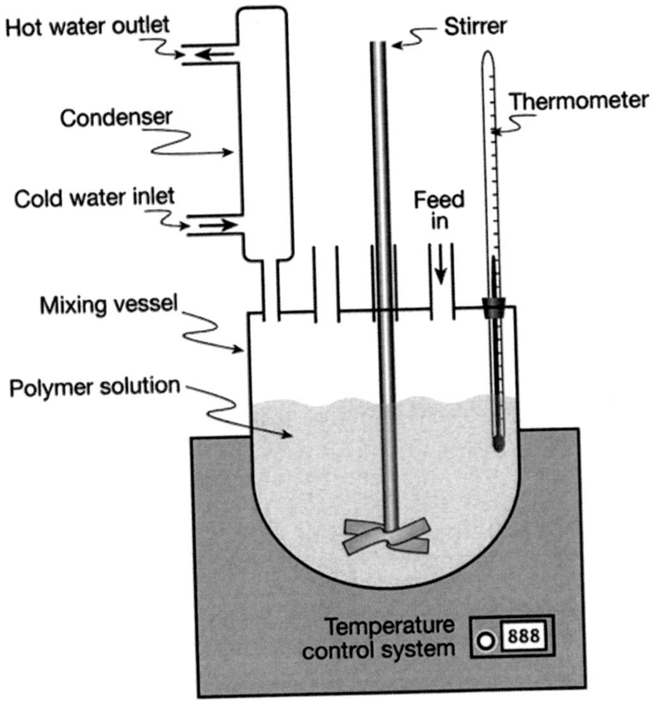

Fig. 5.1 Solution preparation vessel. [1]

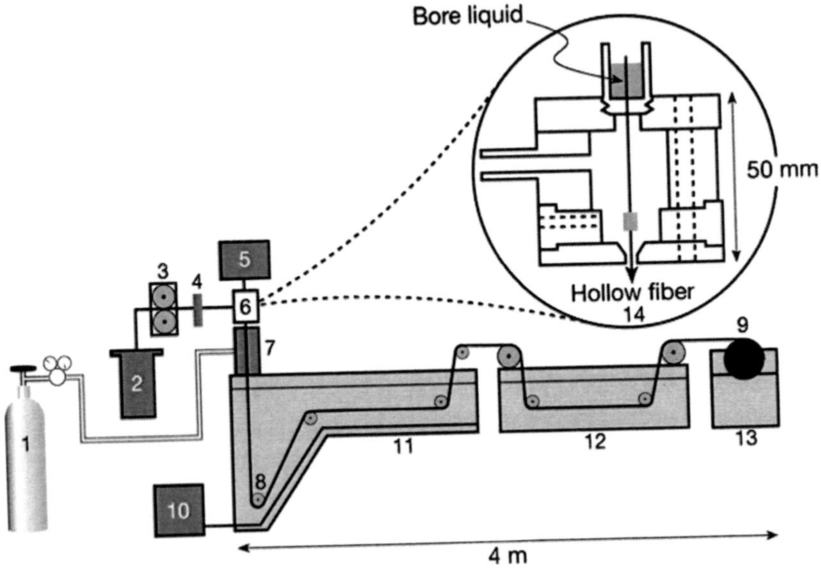

Fig. 5.2 Schematic diagram of hollow fiber spinning system: (*1*) nitrogen cylinder; (*2*) dope reservoir; (*3*) gear pump; (*4*) on-line filter; (*5*) syringe pump; (*6*) spinneret; (*7*) forced convective tube; (*8*) roller; (*9*) take-up drum; (*10*) refrigeration/heating unit; (*11*) coagulation bath; (*12*) washing/treatment bath; (*13*) take-up bath; (*14*) schematic spinneret. [1]

Table 5.1 Dry/wet spinning condition

Dry/wet spinning condition	Value/type
Polymer solution composition	15 wt% PAN 85 wt% DMF
Polymer solution temperature	Ambient temperature (27°C)
Coagulation bath	
Type	Water
Temperature	14°C
Bore fluid	
Type	20.0 wt% potassium acetate in water
DER:bore fluid flow rate	3:1
Spinneret dimension	
OD:ID	600:300 μm
Air gap distance	0–9 cm
Dope extrusion rate (DER)	1.5–3.0 cm^3/min
Jet stretch	1
Take-up drum diameter	17 cm

separation. The distance between the spinneret to the coagulation bath, called nitrogen gap (normally called air gap) is between 0 and 9 cm.

The hollow fiber is then introduced into the nonsolvent (water) coagulation bath for further wet phase separation and then guided to the washing/treatment bath. The coagulation bath temperature is controlled at 14°C by a refrigeration/heating unit to ensure rapid solidification, while the washing bath is maintained at ambient temperature. Then, the hollow fiber is continuously collected onto a take-up drum of 17 cm diameter. After removing from the take-up drum, the hollow fiber is kept in water at room temperature for two days to remove the residual solvent before being subjected to the solvent exchange process.

The ratio of initial fiber velocity to the take-up drum velocity, called the draw ratio or jet stretch (JS), is equal to unity. First part of this study is focused on the preparation of PAN hollow fiber membrane during dry/wet spinning process. Table 5.1 shows the condition of dry/wet spinning process used during this study.

5.1.3 *Solvent Exchange Drying Process*

For the solvent exchange, methanol and hexane are used as the first and the second solvent, respectively. The hollow fiber is removed from the water bath, and subjected to successive immersion into a methanol bath and a hexane bath for two days in each solvent. Then, the hollow fiber membrane is allowed to dry at room temperature while both ends of the fiber are fixed to prevent serious shrinkage.

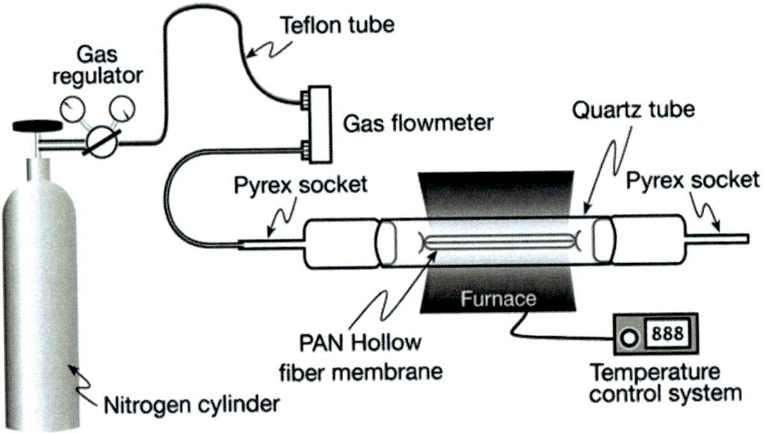

Fig. 5.3 Pyrolysis system using nitrogen as purge gas. [1]

5.1.4 *Pyrolysis System*

Nitrogen gas pyrolysis system was designed and set up to transform the PAN hollow fiber membrane into PAN-based carbon hollow fiber membrane. The instrument involved in this system is illustrated in Fig. 5.3. Carbolite (Model CTF 12/65/550) wire wound tube furnace with the Eurotherm 2416CC temperature control system is used for the pyrolysis of the PAN hollow fiber membrane. The furnace can be operated at a maximum temperature of 1200°C and has the ability to control the heating rate and the thermal soak time during the pyrolysis process.

A PAN hollow fiber bundle is inserted into a quartz tube, where both ends of the tube are fitted with Pyrex socket. The front Pyrex socket is used to channel nitrogen gas into the tube and the back socket is used to purge the volatile compounds evolved during pyrolysis. Then the quartz tube is inserted into the Carbolite furnace. All the connections must be properly tightened up in order to prevent the air from entering the quartz tube, which can interrupt the inert gas pyrolysis process.

However, before pyrolysis, the PAN hollow fber must be subjected to the thermo-stabilization or preoxidation process in air or oxygen. This is necessary in order to cross-link the PAN chains and to prepare a PAN aromatic ladder structure that can withstand high temperatures during the pyrolysis process. For this purpose air is introduced to the quartz tube and the hollow fiber is heated at 250°C for 30 min in air.

After thermo-stabilization, air in the tube is purged by nitrogen to prevent the oxidation from occurring during the high temperature pyrolysis process. The hollow fiber is then heated to a required pyrolysis temperature and under the required conditions. The resulting carbon membrane is cooled down to ambient temperature in nitrogen atmosphere. Table 5.2 shows the pyrolysis conditions required for the preparation of the PAN carbon hollow fiber membrane.

Table 5.2 Nitrogen pyrolysis system for producing PAN carbon hollow fiber membrane

Pyrolysis condition	Value/type
Stabilization condition	
Gas type	Compresses air
Flow rate	200 cm^3/min
Temperature	250°C
Heating rate	5°C/min
Soak time	30 min
Pyrolysis process	
Gas type	Inert gas (nitrogen)
Flow rate	200 cm^3/min
Temperature	500, 600, 700, and 800°C
Heating rate	3°C/min
Soak time	30 min
Cooling process	
Gas type	Nitrogen
Flow rate	200 cm^3/min
Temperature	30°C
Cooling rate	10°C/min

5.1.5 Membrane Characterization

5.1.5.1 Scanning Electron Microscopy (SEM)

Scanning Electron Microscopy (SEM) is used to investigate the morphology of the membrane. Images of the fiber surface, the skin layer structure and the cross-section of the hollow fibers prepared under different spinning conditions and pyrolysis temperatures can be observed clearly by SEM. Before being subjected to SEM observation, the membrane samples had to undergo the gold coating process. Then, the coated samples are imaged and photographed by employing Philips-XL40 SEM with potentials of 10 kV in achieving magnification up to 10,000X.

5.1.5.2 Fourier Transform Infrared Spectroscopy (FTIR)

Fourier Transform Infrared Spectroscopy (FTIR) is a useful tool to detect functional groups in a carbon membrane. Samples of the carbon membrane are ground into fine powder and then mixed with a small amount of KBr. A hydraulic die is loaded with the powder mixture to form a transparent disc. The IR spectrum of the sample disc is directly recorded using Perkin Elmer 2,000 FTIR spectrometer.

5.1.5.3 Attenuated Total Reflection-Fourier Transform Infrared Spectroscopy (ATR-FTIR)

Molecular orientation in the active layer of the membrane is directly detected using plane polarized reflectance infrared spectroscopy. This technique can reveal anisotropy on the molecular level within a sample. Pronounced infrared dichroism (the difference in absorption between parallel and perpendicularly polarized light) shows alignment of molecules, whereas the absence of infrared dichroism shows randomly oriented molecules [2]. The IR spectra are recorded by a Magna IR-560 (Nicolet) FTIR spectroscope fitted with a horizontal ATR accessory called Thunderdome (Termo Nicolet Corporation and Spectra Tech, USA). Samples of PAN hollow fiber membranes are mounted on the Thunderdome accessory jig. For each PAN hollow fiber sample, IR spectra are obtained with the plane of polarization parallel and perpendicular to the shear direction.

5.1.5.4 Elemental Analysis

Elemental analysis is used to determine the composition of the carbon membrane prepared at different pyrolysis temperatures. The percentage composition of carbon, hydrogen, nitrogen and oxygen in the PAN carbon membrane is measured by Elementar Vario EL 111 elemental analytical equipment.

5.1.6 Gas Permeation Test

Before the gas permeation test is carried out, the PAN hollow fiber membranes are potted at both ends to form a bundle consisting of 5–10 fibers. One end of the fiber bundle is sealed into a stainless steel tube of 5/8 in. outer diameter, while the other end is potted in an aluminium cap. Loctite E-30CL epoxy adhesive is used as a potting resin. This bundle is then inserted in a suitable module as shown in Fig. 5.4. All

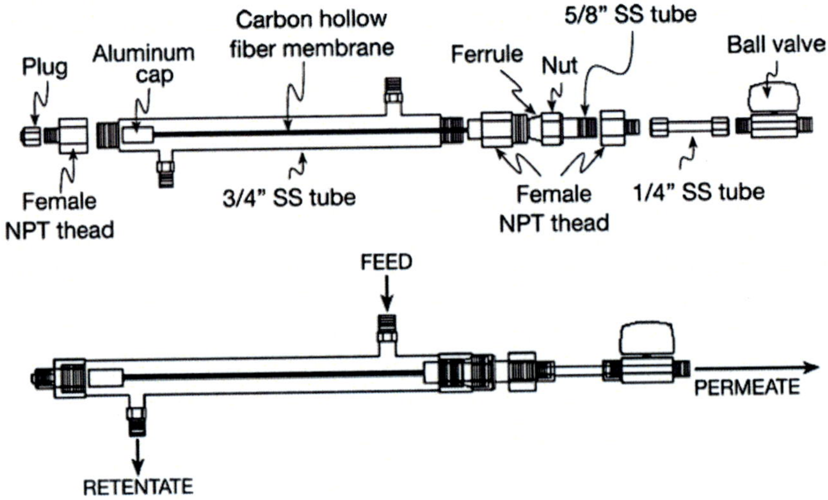

Fig. 5.4 Hollow fiber for the laboratory scale permeation experiments. [1]

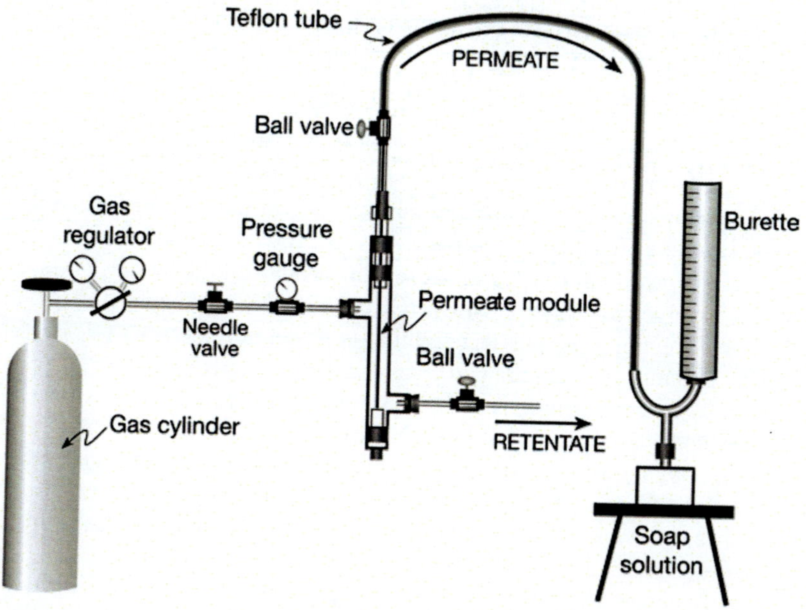

Fig. 5.5 Schematic diagram of pure gas permeation testing system. [1]

fittings and nuts used are of Swagelok type for better prohibition of leakage during the gas permeation test.

The single gas permeation system is illustrated in Fig. 5.5. Pure O_2 and pure N_2 are introduced into the system at feed pressures of 1–5 bar at ambient temperature. The permeate side of membrane is maintained at atmospheric pressure. A soap film flow meter is used to measure the gas permeation rate.

The permeance and ideal separation factor (or selectivity) of the carbon membrane are calculated from the permeation rate by using Eqs. (5.1) and (5.2) respectively.

$$\left(\frac{P}{l}\right)_i = \frac{Q_i}{\Delta p.A} \tag{5.1}$$

$$\alpha_{A/B} = \frac{P_A}{P_B} = \frac{\left(\frac{P}{l}\right)_A}{\left(\frac{P}{l}\right)_B} \tag{5.2}$$

Where; $(P/l)_i$ is permeance, Q_i is permeation rate of gas at standard temperature and pressure (STP), Δp is transmembrane pressure drop, A is area of membrane and $\alpha_{A/B}$ is ideal separation factor or selectivity.

Gas permeation unit (GPU) is often used as the unit for permeance (1 GPU = 1×10^{-6} cm^3(STP)/cm^2 s cmHg). When the membrane thickness l is known, the permeability P can also be obtained. Barrer is often used as the unit for permeability (1 Barrer = 1×10^{-10} cm^3(STP) cm/cm^2 s cmHg).

5.2 Flat Sheet CMSM

5.2.1 Precursor Membrane Formation

Polymeric membranes are prepared by the solution casting technique. Polymers are polyimides (PIs, 6FDA:BPDA (m:n):DAM, 6FDA:6FpDA, 6FDA:6FmDA), whose formula are given in Fig. 5.6.

Polymer powder is dried at 115°C to remove any adsorbed water. Then, the polymer is dissolved in dichloromethane (CH_2Cl_2, 99.9% from Sigma Aldrich) to make 2–5 wt% solution. The polymer solution is stored in a 20 or 40 mL ICHEM vial (Fisher Scientific) which are placed in a Burrel Model 75 Wrist Action Shaker or a roller system for mixing for 1 h. The solution is then loaded in a glass syringe, filtered under a pressure through a Millex®-RH 0.45 μm PTFE filter (Millipore Corporation) and slowly dripped into a stainless steel ring that is mounted onto the top surface of a glass mirror. An inverted glass funnel is placed over the solution to minimize the evaporation rate of the solvent and to prevent any dust from landing on the polymeric film (see Fig. 5.7).

Rapid solvent evaporation can cause temperature fluctuations leading to convective motion in the solution (Marangoni/Benard flows). This motion can result in uneven evaporation and thickness variations in the membrane. The solvent is allowed to evaporate until the membrane is fully vitrified, which is determined by a change in the appearance of the membrane. The resulting polymeric film is removed by lifting the edge of the film with a razor blade to ease the film off of the surface. If

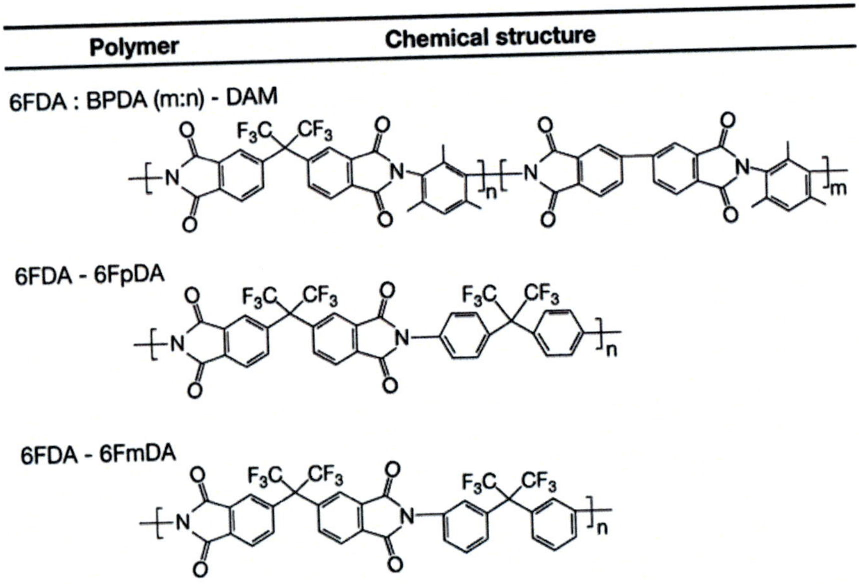

Fig. 5.6 Chemical structure of polyimides used for CMSMs formation. [3]

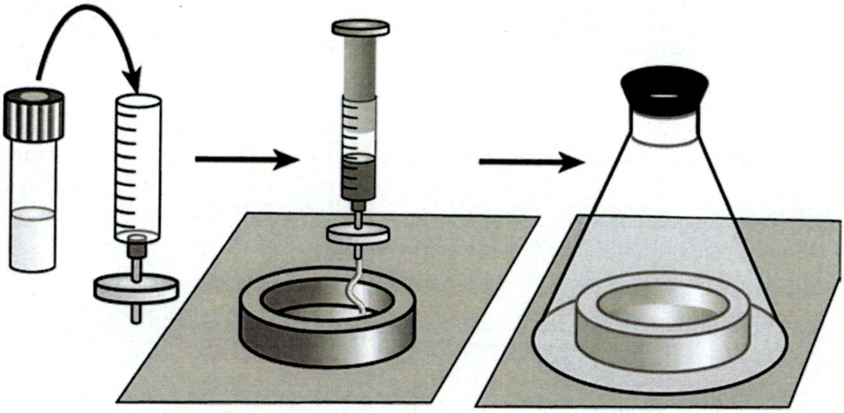

Fig. 5.7 Membrane casting procedure. [4]

the film is difficult to come off, a few drops of water is squirted under the exposed edge, which causes the film to lift off of the glass plate. The polymeric film is dried at room temperature for several hours and then at 115°C under vacuum to remove any residual solvent and water. Any areas that have visual stress points are not used for permeation experiments.

5.2.2 *Pyrolysis of Flat Sheet CMSMs*

From the dry polymeric film so obtained, circular films of 1″ diameter are punched out by a stainless steel die. When the die can not punch out the coupons, scissors are used to separate the coupons from the film sheet. The coupons so produced are weighed and placed on a ribbed quartz plate as shown in Fig. 5.8. The quartz plate has 1/4″ wide and 1/8″ deep grooves which allow decomposition products to evolve from both sides of the film equally.

The coupon is then placed in a pyrolysis furnace, a quartz tube that is shown in Fig. 5.9.

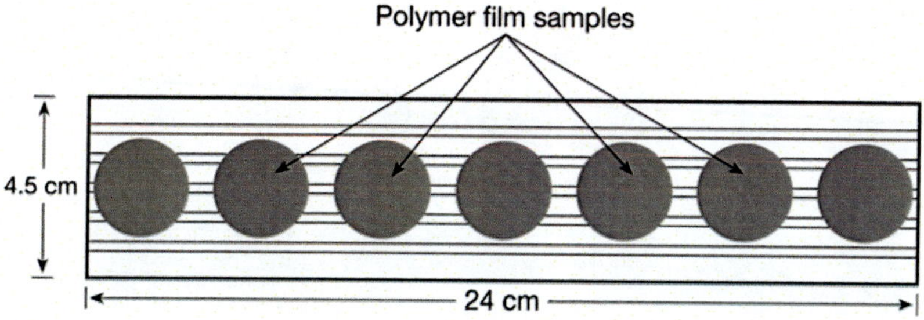

Fig. 5.8 Schematic of quartz plate with precursor films used during pyrolysis. [3]

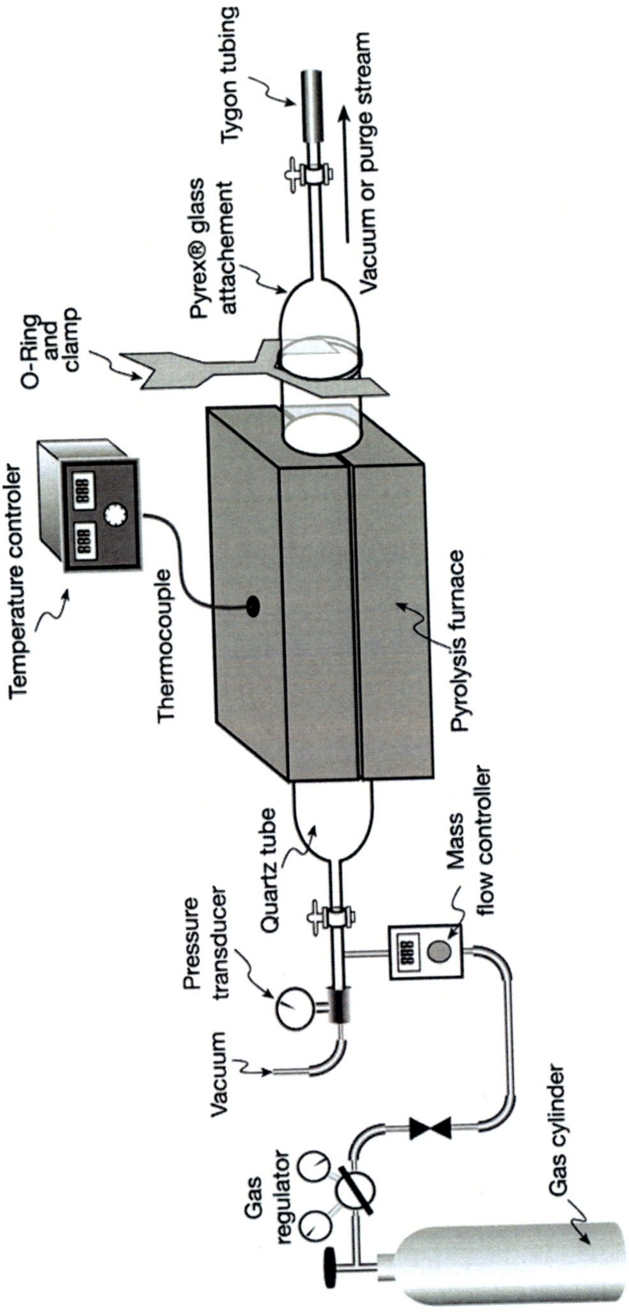

Fig. 5.9 Pyrolysis furnace used to produce CMSMs. [3]

The quartz tube has a dimension of 50 mm ID × 54 mm OD × 115 cm long. One end of the quartz tube is open to allow insertion of the quartz plate while the other end is sealed with glass. After inserting the quartz plate, the open end is covered by a Pyrex® glass cap that is sealed against an O-ring. The polymeric film is pyrolysed either under vacuum or under nitrogen atmosphere. For pyrolysis under vacuum, an Edwards Model 3 vacuum pump is connected to a stop-open valve of the Pyrex® glass cap (right side of the quartz tube) via a cold trap cooled by liquid nitrogen, which prevents any pump oil from entering into the quartz tube and catches any heavy by-products of pyrolysis. The pressure of the system is monitored by a MKS 628B Capacitance Manometer attached to a MKSPRD 2000 readout. The pressure during the pyrolysis is adjusted to 0.005–0.05 Torr. For pyrolysis in an inert gas, a compressed air cylinder is attached to the left side of the quartz tube instead of the pressure transducer. Instead of connecting to the vacuum pump the stop-open valve (right side of the quartz tube) is opened to allow the purge gas to flow through the quartz tube. The flow of the inert gas is regulated with a needle valve and/or a MKS mass flow controller and the flow rate is measured at the end of the quartz tube using a Hewlett Packard 0103-0113 Soap Film flow meter. Before beginning the heating cycle, inert gas equivalent to 20 quartz tube volumes is allowed to flow through the system. The quartz tube is heated by a Thermacraft Model 23-24-1ZH three zone tubular furnace. The heating cycle is controlled by an Omega Engineering Model CN-2010 temperature controller. As shown in Fig. 5.10, the pyrolysis protocol consisted of slow ramps to 550–800°C followed by a constant temperature soak for 2 h at the ultimate temperature, as used by several researchers. After pyrolisis the film is allowed to cool to below 50°C before being removed from the quartz tube.

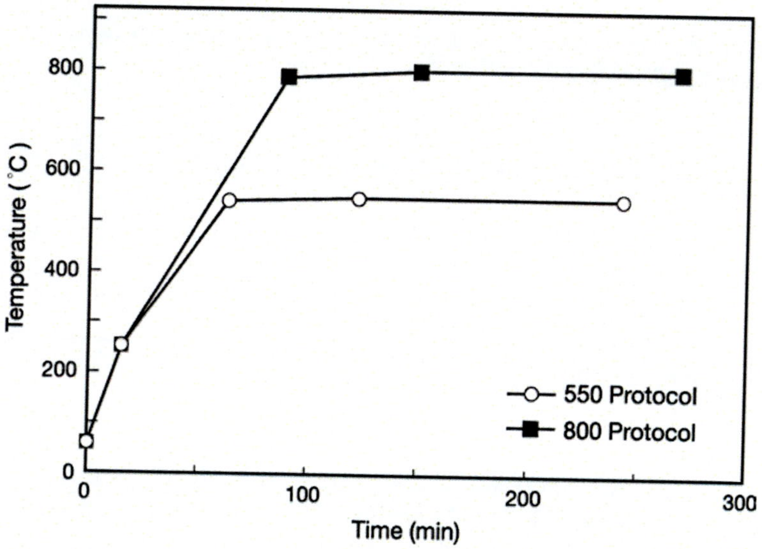

Fig. 5.10 Pyrolysis protocols to be used to form CMSMs. [3]

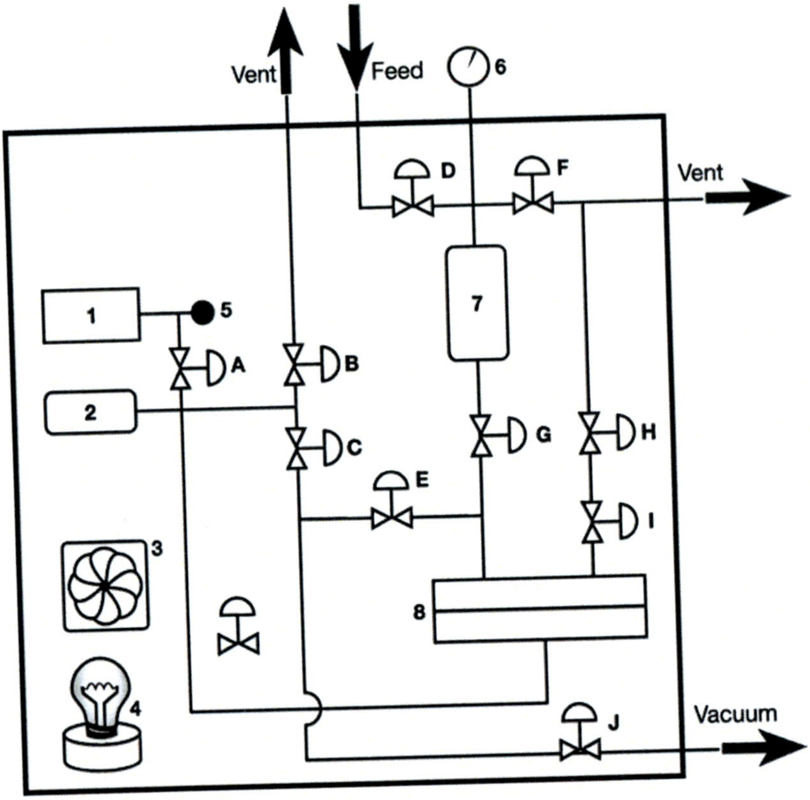

Fig. 5.11 Schematic of the permeation system [3]. (*1*) Down stream pressure transducer, (*2*) permeate reservoir, (*3*) fan, (*4*) heater, (*5*) rupture device, (*6*) upstream pressure transducer, (*7*) feed reservoir, (*8*) permeation cell. (*A*) Down stream pressure transducer isolation valve, (*B*) GC valve, (*C*) down stream vacuum valve, (*D*) feed valve, (*E*) "middle" valve, (*F*) vent valve, (*G*) cell isolation valve, (*H*) retentate (metering) valve, (*I*) retentate shutoff valve, (*J*) vacuum shutoff valve

5.2.3 Gas Permeation Experiment Preparation

Permeability and selectivity of membranes were determined by using a constant volume system shown in Fig. 5.11.

According to the method the permeability of a membrane is proportional to the rate of pressure rise and can be calculated by

$$P = \frac{2.94 \times 10^4 V l \frac{dp}{dt}}{T A \Delta p} \tag{5.3}$$

where P is permeability, (Barrer (= 10^{-10} cm^3(STP) cm)/(cm^2 s cmHg)), V is downstream volume (cm^3), l is film thickness (mil = 10^{-3} in.), (dp/dt) is the rate of pressure rise (torr/min), T is absolute temperature (K), A is film area (cm^2), and Δp is the cross membrane pressure drop (psia).

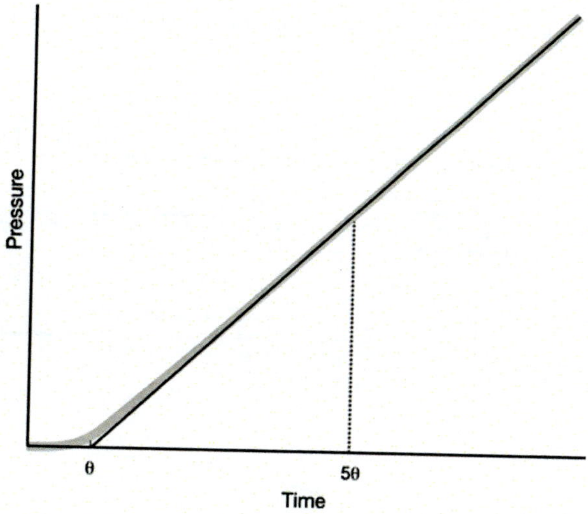

Fig. 5.12 Method to obtain θ_A and *dp/dt*. [4]

The time lag, θ_A, is obtained by extrapolating the p versus t straight line to the x axis. The time lag is related to the diffusion coefficient, D_A (cm^2/s) of gas through the membrane by the following equation.

$$\theta_A = \frac{l^2}{6D_A} \tag{5.4}$$

Steady state permeation rate *dp/dt* is obtained from the slope of p versus t plot for $t > 10\ \theta_A$. Since the time lag is very short ($\theta_A < 1$ min), it is difficult to determine precise diffusion coefficient.

The cross-section of the test permeation cell used to measure the gas permeation rate is schematically shown in Fig. 5.12. The permeation cell consists of two stainless steel blocks (upstream side and down stream side) to sandwich a membrane in between. The two O-rings seal the gas leak through any space between these two blocks.

Flat sheet CMSMs are masked with adhesive backed aluminum tape to prepare them for testing. Figure 5.13 shows the cross-section of the masked film in a permeation cell with a detailed view of the masked film. The membrane sample is sandwiched between two pieces of 1–5/8 in. aluminum tape. The piece of the tape used for the upstream side of the membrane had a 1/2 in. diameter hole cut out of the center while the center used for the down stream side of the membrane had a 5/8 in. diameter hole in the center. The larger hole on the downstream side of the membrane is used to provide a more accurate measurement of the area that is based on the smaller hole on the upstream side only. This precaution is necessary to prevent the holes of the two sides of the mask from being aligned visually. Once the membrane is sandwiched in the aluminum tape, the sample is ready to be masked onto the permeation cell.

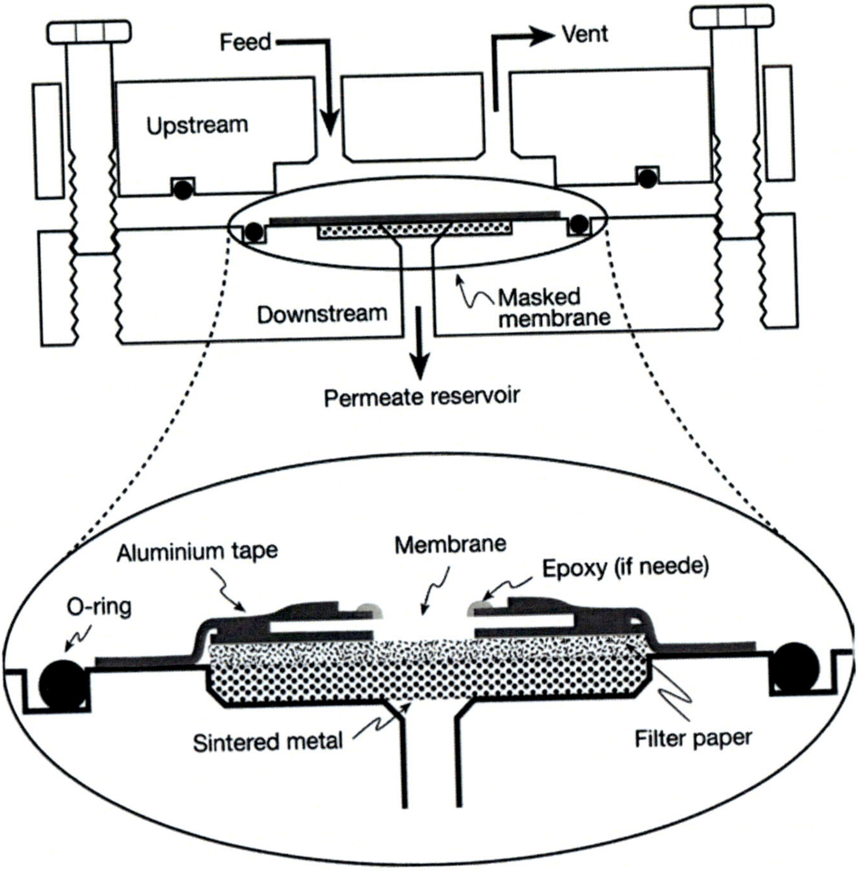

Fig. 5.13 Permeation cell and a masked CMSM loaded in the cell. [4]

With the smaller circle facing the upstream side of the permeation cell, the sandwiched membrane is placed on the permeation cell on top of two pieces of 1–5/8 in. diameter paper that sit directly over the sintered metal support disk in the permeation cell. These tapes are then masked using a 2–1/4 in. piece of aluminum tape with a 3/4 in. diameter cut out of the center. The entire tape is then pressed into place as well as possible to reduce the potential for leaks. If visible defects or small sample sizes prevent a good seal to be formed in the masked membrane, Devcon 5 min epoxy is used to cover the problem areas. Because the permeability of epoxy is very slow, it is considered as an impermeable barrier compared to the membranes tested. With the epoxy cured and the membrane fully masked to the permeation cell, the upstream side of the cell is attached to the down stream side of the cell using six bolts. The cell is then connected to the permeation system using 1/4 in. Swagelok® VCR connections.

5.2.4 Gas Permeation Experiment

After placing the permeation cell in the permeation system, the down stream of permeation cell (valve C) is opened to vacuum for ca 15 min. The downstream is then closed off and the upstream of the apparatus is opened to vacuum (valves E and G). Once the upstream and downstream pressures become almost similar, both the downstream and upstream are opened to vacuum for ca 24 h to completely degas the membrane and the mask. It is important to maintain the downstream pressure lower than or equal to the upstream or the mask can be pushed away from the cell, leading to the leak. After the membrane is degassed, the leak rate of the atmospheric gases into the permeate reservoir is measured by closing valves B, C, E and G and I to isolate the cell. The leak rate is quantified for further analyses. To begin an experiment, gas is first introduced into the feed side with the cell isolation valve closed. After that the system is allowed to reach equilibrium. After the temperature is allowed to equilibrate, the middle valve (E) and the downstream valves (B and C) are closed and the cell isolation valve (G) is opened to allow the feed gas to contact the membrane. The pressure rise of the permeate reservoir due to the permeation through the membrane is initiated. After the first gas is tested, the system is again equilibrated for at least ten time lags before testing the second gas. To make sure the permeation results are repeatable, one of the gases is retested after all of the other gases.

After all the gases are tested, the permeation cell is opened and the mask removed from the permeation cell. The digital image of the masked CMS is shown in Fig. 5.14.

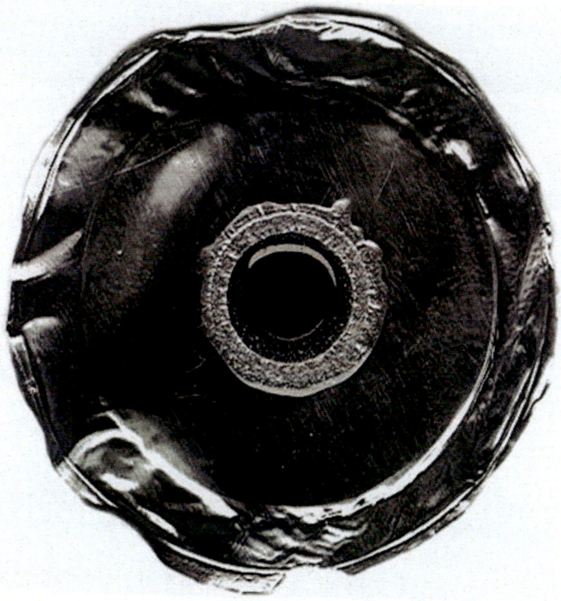

Fig. 5.14 CMSM image. [3]

References

1. Syed Mohd STC (2004) Foramtion and characterization of polyacrylonitrile (PAN) carbon hollow fiber membranes for gas separation. Master of Engineering Thesis, University Technology Malaysia, Skudai, Johor, Malaysia
2. Ismail AF, Dunkin IR, Gallivan SL, Shilton SJ (1999) Production of super selective polysulfone hollow fiber membranes for gas separation. Carbon 40 (23): 6499-6506
3. Williams PJ (2006) Analysis of factors influenceingthe performance of CMS membranes for gas separation. PhD thesis, Georgia Institute of Technology, Atlanta, GA, USA
4. Perry JD (2007) Formation and characterization of hybrid membranes utilizing high-performance polyimides and carbon molecular sieves. PhD thesis, Georgia Institute of Technology, Atlanta, GA, USA

Chapter 6
Membrane Characterization

6.1 Permeability Measurement

6.1.1 Liquid Permeability

Gas permeation tests have been carried out for many carbon membranes. However, there is very limited data for the liquid permeation tests, especially for the application of carbon membranes for ultrafiltration (UF) or microfiltration (MF). Some studies [1, 2] demonstrated that carbon UF membranes could be made by adding non-carbonizing polymers such as poly(ethylene glycol) (PEG) to the precursor solution. Encouraged by the findings of [3] and [4], Shah and co-workers successfully made nanoporous carbon UF membranes suitable for bioprocessing applications [5].

Carbon membranes were fabricated by pyrolysis of a mixture of poly(furfuryl alcohol) (PFA) and poly(ethylene glycol) (PEG) that was spin-coated on a macroporous stainless steel support. The stainless steel support was first modified by physical deposition of sub-micron-sized silica particles within the micropores and then PFA/PEG/acetone solution was spin-coated onto the silica-modified support. Coated support was heated in a stream of Ar to 600°C for 4 h for pyrolysis. The process was repeated. Hydraulic permeability was obtained under 40–50 psig. UF experiments with blue dextran as the solute was also conducted. Table 6.1 shows the hydraulic permeability and blue dextran sieving data.

Dextran sieving curves were also evaluated by filtration of dextrans of different average molecular weights in kDa, i.e. 2000, 167, 75, 40, and 10 kDa. Sieving coefficient versus dextran average molecular weight is given in Fig. 6.1.

The pore size distribution of the nanoporous carbon (NPC) membrane was determined by liquid–liquid porosimetry using isopropanol and a mixture of isopropanol–water as the wetting and intrusion fluid, respectively. Figure 6.2 shows the flux versus pressure data. The pore size distribution obtained from the data is shown in Fig. 6.3. The carbon membrane has pores that lie primarily between 8 and 18 nm. The mean pore radius is 12.3 nm with a standard deviation of 2.3 nm.

A. F. Ismail et al., *Carbon-based Membranes for Separation Processes*,
DOI 10.1007/978-0-387-78991-0_6,

Table 6.1 Water permeability and sieving coefficient of blue dextran

	Water permeability ($L\ m^{-2}\ h^{-1}\ bar^{-1}$)	Sieving coefficient of blue dextran
Stainless steel (SS) support	8550	1.0
Modified SS support	2550	1.0
First coat carbon membrane	1670	0.7
Second coat carbon membrane	240	0.2
Third coat carbon membrane	41	0.0076
Additional coat	21	0.003
100 kDa Biomax[a]	396	0.0035

[a] Shown for comparison

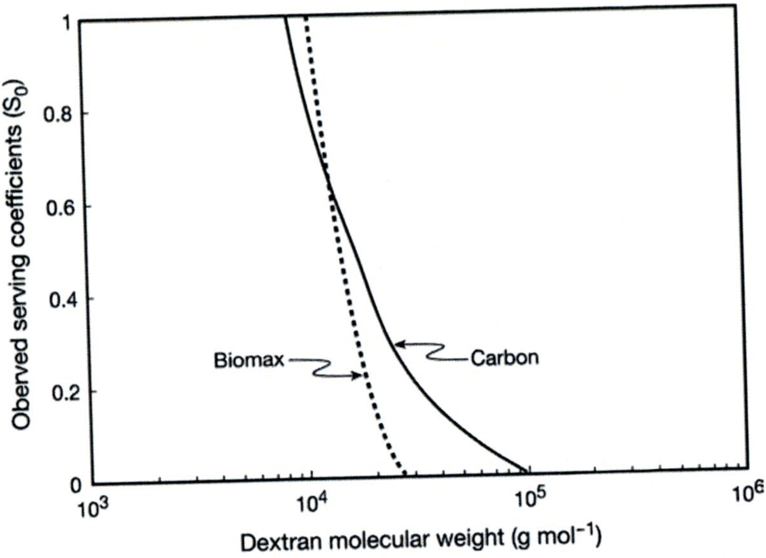

Fig. 6.1 Dexran sieving curves for the nanoporous carbon membrane formed after deposition of three carbon layers and for the Biomax 100 kDa membrane. (From [5])

6.2 Physical Characterization

Some examples of the physical characterizations of the membranes prepared in Chap. 4 are given in the following sections.

6.2.1 Themogravimetric Analysis (TGA)

Thermogravimetric analysis (TGA) is a technique often used in CMSM research. By TGA, weight loss of a substance as a function of increased temperature is measured. It is used to determine at which temperature(s) decomposition takes place.

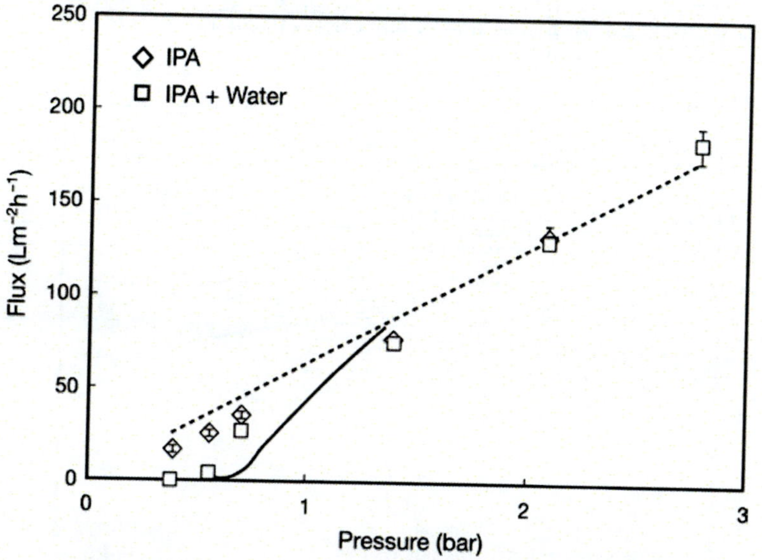

Fig. 6.2 Liquid–liquid porosimetry data for nanoporous carbon membrane formed after deposition of three carbon layers. The flux-pressure curves are shown for the fully intruded (*solid curve*) and iso-propanol wet (*dashed curve*) membranes. (From [5])

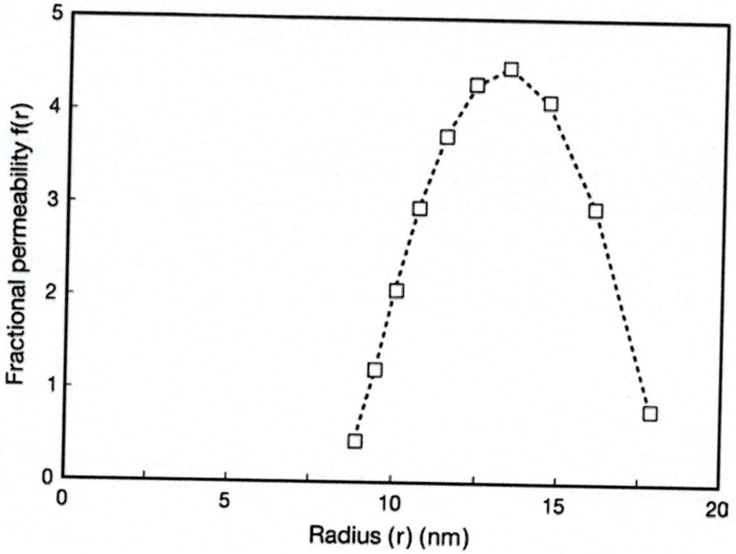

Fig. 6.3 Pore size distribution for the nanoporous carbon determined from liquid–liquid porosimetry. (From [5])

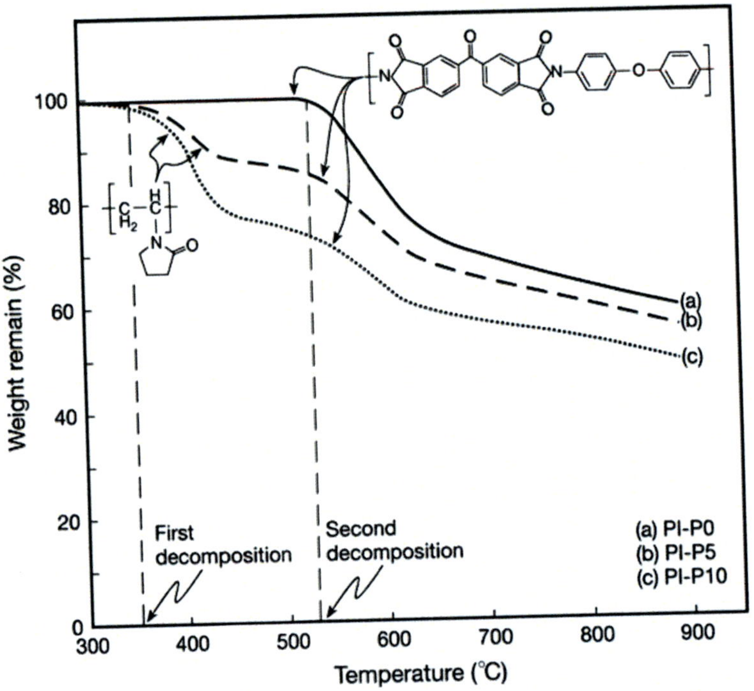

Fig. 6.4 TGA thermograms of PI and PI/PVP precursors: (*a*) PI precursor without PVP blend (PI–P0); (*b*) PI precursor with 5% PVP (PI–P5); (*c*) PI precursor with 10% PVP (PI–P10). (From [6])

Figure 6.4 of Kim et al. [6] shows TGA thermograms of PI, PVP, PI/PVP blend CMSM precursors. PI (PI–P0) is stable up to 550°C. With respect to the PI–PVP blends, the earlier weight loss in the temperature range 370–420°C corresponds to the decomposition of PVP resulting in lower residual weights.

Figure 6.5 demonstrated in David and Ismail's work is another example of characterization by TGA. TGA thermograms of the PAN hollow fiber are shown for hollow fibers (a) stabilized in oxidative atmosphere, (b) stabilized under inert atmosphere and (c) pyrolyzed under oxygen atmosphere. Comparing (a) and (b), it is clear that (a) is thermally more stable particularly in the 300–350°C range that is near the PAN's glass transition temperature (317°C). When pyrolysis is made in oxygen atmosphere at 650°C, the membrane is totally lost. This means that the oxidative stabilization and pyrolysis in an inert atmosphere is necessary. Figure 6.6 shows that two kinds of ladder polymers formed as intermediates during the inert and oxidative stabilization.

Figure 6.7 illustrated by Centeno and Fuertes shows another example of TGA thermograms. The figure shows initial 20 and 44% weight loss by the preoxidation at 150 and 200°C, respectively. But the final weight loss after 600°C did not change very much.

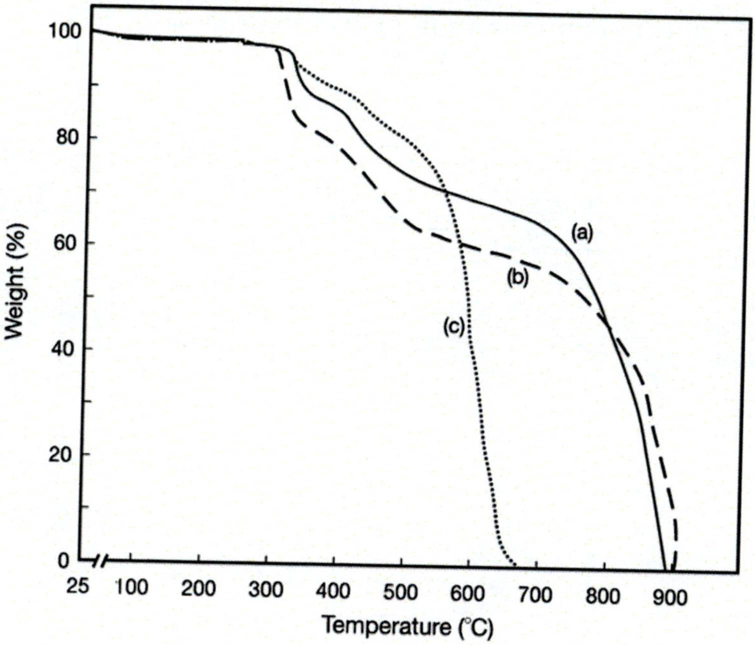

Fig. 6.5 TGA result for the membrane (*a*) stabilized under oxidative atmosphere; (*b*) stabilized under inert atmosphere; (*c*) pyrolyzed under oxygen atmosphere. (From [7])

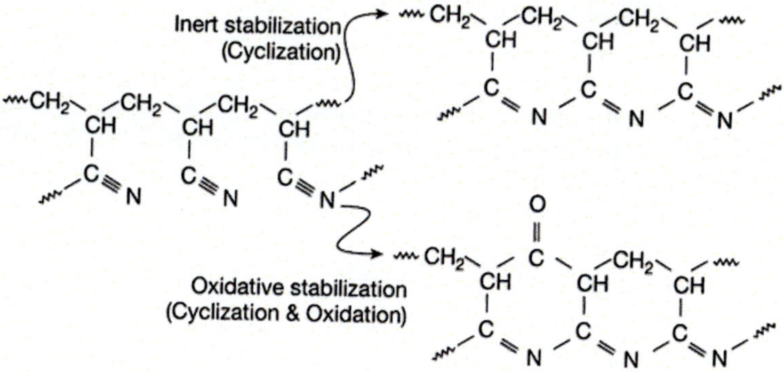

Fig. 6.6 Formation of ladder polymer in inert stabilization and oxidative stabilization. (From [7])

6.2.2 *Wide Angle X-ray Diffraction (WAXD), Scanning Electron Microscopy (SEM), Transmission Electron Microscopy (TEM), Atomic Force Microscopy (AFM)*

It is an X-ray diffraction technique that is often used to determine the crystalline structure of polymers. This technique specifically refers to the analysis of Bragg Peaks scattered to wide angles, which (by Bragg's law) implies that they are caused

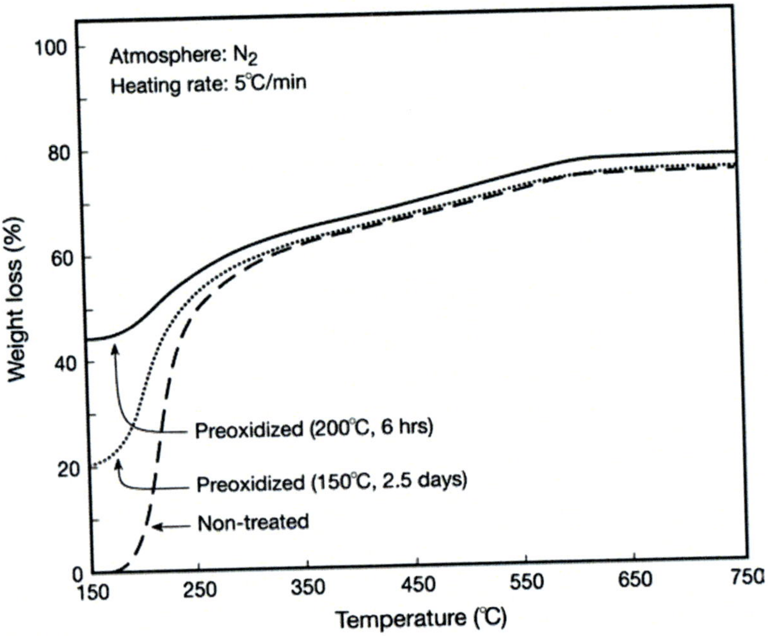

Fig. 6.7 Weight loss of PVDC-PVC materials (untreated and preoxidised) during pyrolysis. (From [8])

by sub-nanometer sized structures. Electron microscopy (EM) is one of the techniques that can be used for membrane characterization. Two basic techniques can be distinguished: scanning electron microscopy (SEM) and transmission electron microscopy (TEM). In particular, some examples of SEM are given in this chapter. The Atomic Force Microscopy (AFM) consists of a cantilever with a sharp tip at its end. The tip is brought into close proximity of a sample surface. The tip scans over the surface of the sample, its position and cantilever deflection are recorded, and a surface image is produced.

Figure 6.8 of Xiao et al. shows the WAXD of the 6FDA-Durene-p-intA copolymer with different p-intA contents. The figure indicates that the 2θ value increases and d-spacing decreases as the content of p-intA increases in the polymeric precursor, because the p-intA is flat and has no bulky substituent group as in Durene.

Figure 6.9 of Kim et al. shows another WAXD pattern of CMSMs prepared from PI/PVP blend precursors with different contents of PVP. The carbonization was made at two temperatures (550 and 700°C). Table 6.2 shows the d-spacing obtained from WAXD results. The d-spacing increases with the increase in PVP content and decreases with the carbonization temperature.

Figure 6.10 of Centeno and Fuertes displays SEM pictures of (a) cross-section of the PVDC-PVC precursor, (b) cross-section of the CMSM, and (c) top view of the fractured CMCS. Figure 6.10b particularly shows that the thickness of the CMS layer is ca 0.8 μm.

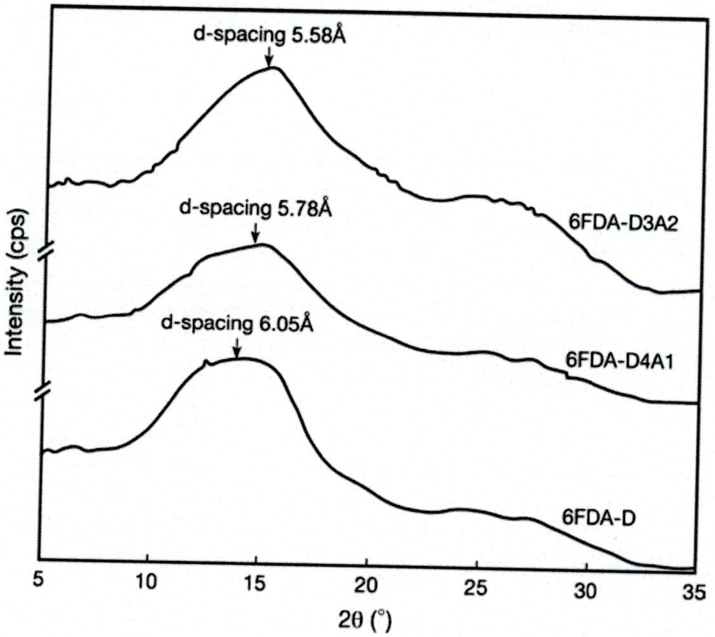

Fig. 6.8 WAXD of 6FDA-Durene-p-intA copolymer; D4A1 (Durene:p-intA=4:1); D3A2 (Durene: p-intA=3:2). (From [9])

The SEM micrographs in Fig. 6.11 show the dense homogeneous structure of the hollow fiber of trimethylsilyl (TMS) substituted PPO precursors and carbon membranes.

Scanning electron microscope (SEM), atomic force microscopic (AFM) and high resolution transmission electron microscope (HRTEM) were used to characterize the SNPCM-11 membrane (supported NPC membranes with the carbon mass of 11 mg on the support) prepared by Shiflett et al. [10]. The SEM picture (Fig. 6.12) shows a defect-free homogeneous surface as evidenced by the high selectivity values. AFM (Fig. 6.13a) images show two regions of carbon. One is lighter (harder) and the other is darker (softer). The darker areas arise from pores filled with water, while the lighter areas are non-porous solid carbon. Lateral dimension of the darker areas was from 2.5 to 4.0 nm. A 3 mm diameter sample of NPC layer was removed from the surface of SNPCM-11 membrane with a sharp razor blade and the sample was subjected to HRTEM. Typically, these carbons are completely amorphous (Fig. 6.14a). However, there were other regions that showed a high degree of ordering. The onion like fullerene features are also presented in Fig. 6.14b.

Asymmetric, flexible hollow fiber carbon membrane with high mechanical strength was produced by thermooxidative stabilization and gradual carbonization of PAN based precursor membrane [11]. SEM showed that the inner surface had the pores of 3–10 μm in diameter penetrating into the hollow fiber wall but not reaching the outer surface. Three membranes A, B, C were produced, the differ-

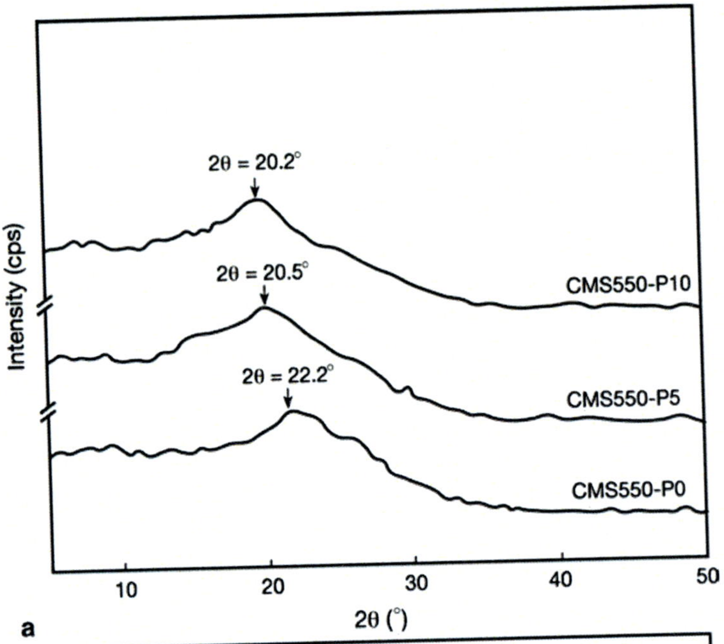

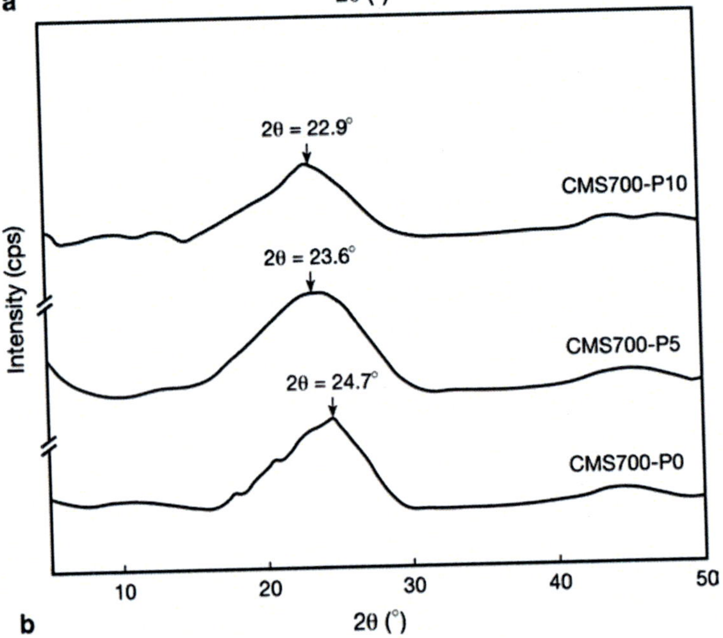

Fig. 6.9 WAXD patterns of the CMS membranes pyrolyzed at (**a**) 550°C and (**b**) 700°C as a function of PVP content. (From [6])

Table 6.2 Average d-spacings of prepared CMS membranes from WAXD

Sample codes[a]	d-spacing (nm)	Sample codes	d-spacing (nm)
CMS550-P0	0.40	CMS700-P0	0.36
CMS550-P5	0.43	CMS700-P5	0.38
CMS550-P10	0.44	CMS700-P10	0.39

[a] CMS carbonization temperature-PVP wt.%

ence being basically the molecular weight of PAN and its concentration in the spinning dope. Membrane A was from a co-polymer of AN-MAC (methyl acrylate, 2%) with intrinsic viscosity of 2.35. Membrane B was from commercial PAN of average Mw = 150,000. Membrane C was from the AN-MAC co-polymer of intrinsic viscosity 3.21. The stabilization was at 260°C in air and carbonization was in nitrogen at 800°C.

Figure 6.15 shows the Scanning Probe Microscopic (SPM) image of the outer surface of membrane A spun from the dope of 10% polymer concentration. The membrane has a relatively smooth surface with pores of average diameter 1 μ. Figure 6.16 shows the SPM image of the membrane B. Membrane B does not contain any pores visible at a nanometer resolution level. The high concentration of the polymer in the spinning dope (20%) prevented the liquid–liquid phase separation from extending to the outer surface of the membrane and the formation of the structure was caused by the gelation process. Figure 6.17 shows the image of membrane C spun from dope with concentration of 8%. High viscosity and low concentration of the dope resulted in a rough outer surface with pores of average diameter 0.4 μm and the maximum 0.65 μm.

6.2.3 *Fourier Transform Infra-Red (FTIR)*

Infrared spectroscopy (IR spectroscopy) is spectroscopy that deals with the infrared region of the electromagnetic spectrum. It covers a range of techniques, the most common being a form of absorption spectroscopy. The modern instruments are FTIR instruments. Fourier transform infrared (FTIR) spectroscopy is a powerful tool for identifying types of chemical bonds in a molecule by producing an IR absorption spectrum that is like a molecular *fingerprint.* Thus, by FTIR technique presence of functional groups in a membrane and/or interactions between components of a membrane can be detected.

Figure 6.18 of Xiao et al. [9] shows the FTIR spectra of 6FDA-Durene-p-intA copolymer. Three IR bands at 1780, 1720 and 1370 cm^{-1} characteristic to polyimides (PIs) are found in all polymer samples. The PIs containing p-intA revealed the IR bands in 1430–1590 cm^{-1} region due to aromatic C=C stretching vibrations. The intensity of these bands increases when aromatic rings are conjugated with C=C or C=C.

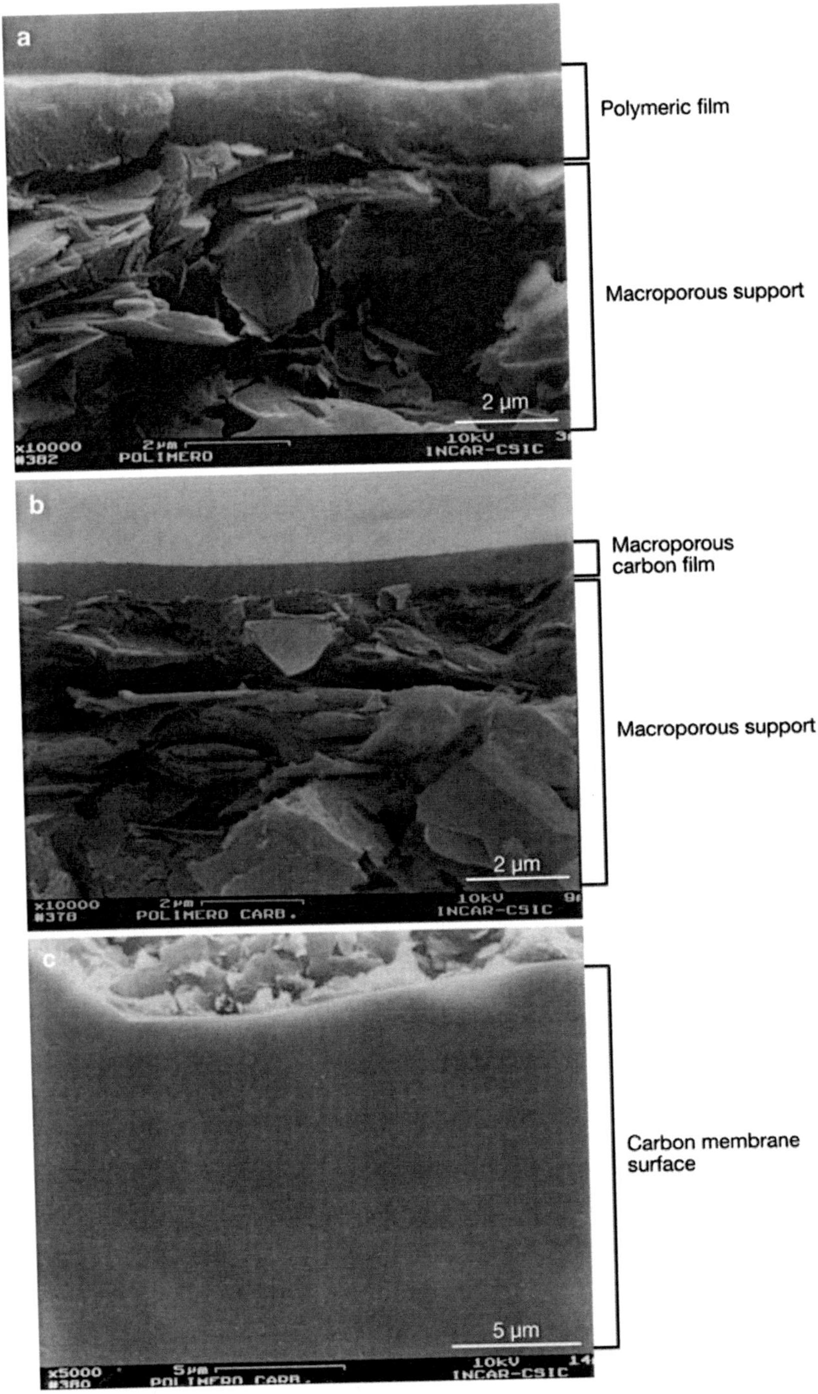

Fig. 6.10 SEM micrographs of membranes; (**a**) cross-section of PVDC-PVC film; (**b**) cross-section of carbon membrane; (**c**) top view of the fractured carbon membrane. (From [8])

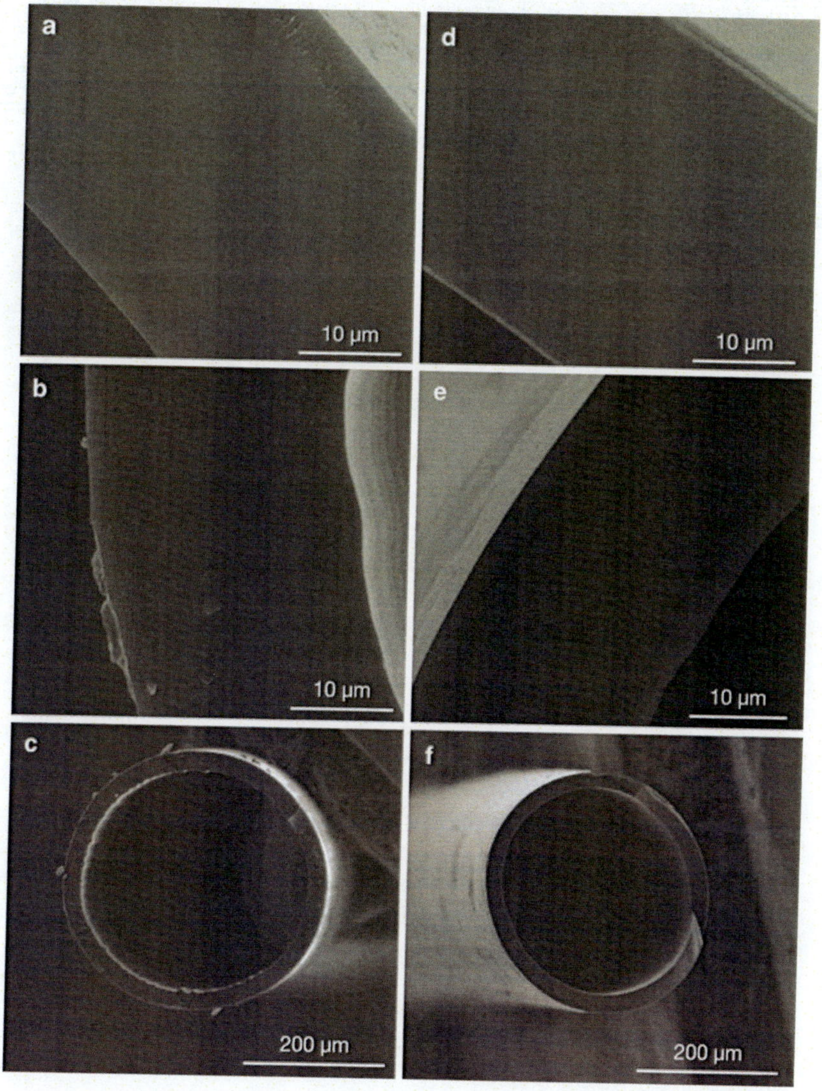

Fig. 6.11 SEM images of the cross-section for (**a**) P(PPO), (**b**, **c**) C(PPO), (**d**) P(TMS80), (**e**, **f**) C(TMS80). Carbon samples pyrolyzed at 923 K. (From [12])

Figure 6.19 of Kim et al. [6] shows FTIR spectra of the PI/PVP blend precursor and CMS membranes. The precursor PI–P10 shows the bands at 1800 cm^{-1} (C=O stretching) and 1380 cm^{-1} (C–N stretching), indicating successful blending of PI and PVP. Upon carbonization at 550°C, these bands largely disappeared and at 700°C, no IR bands were observed.

Figure 6.20 of Yoshimune et al. [12] shows the FTIR spectra of trimethylsilyl (TMS) substituted PPO precursors and carbon membranes. The spectra (a) corre-

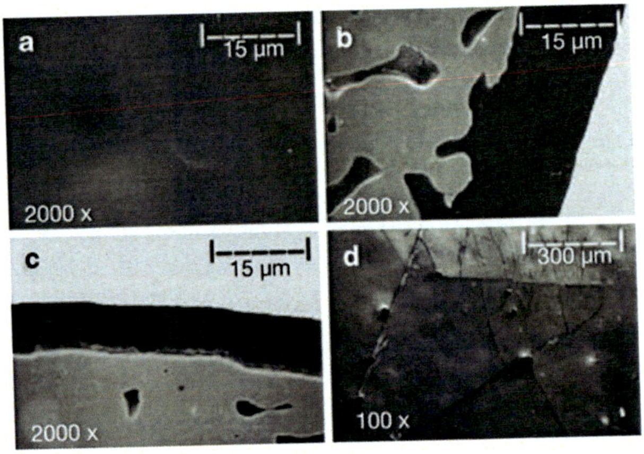

Fig. 6.12 SEM images of (**a**) SNPCM surface, (**b**) radial cross-section, (**c**) axial cross-section, and (**d**) microcracks on SNPCM surface. (From [10])

spond to polymeric PPO and show characteristic bands of PPO. The spectra (b) correspond to PPO where 80% of the repeat unit is substituted by TMS (P(TMS80)). A new band appeared at 1240 cm^{-1} that was attributed to the Si–C stretching vibration of TMS substituent. The spectra (d) correspond to PPO carbonized at 923 K (CPPO). All bands characteristic to PPO disappeared. Finally, spectra (e) are for the carbonized TMS80 (C(TMS80)). A band characteristic to Si–O–Si appeared, indicating that a silica-like species is present in the C(TMS80) membrane.

6.2.4 Adsorption and Sorption Experiments

Nitrogen adsorption is often a very powerful characterization for the pore structure of particles and membranes. However, due to the distribution of ultra-micropores with narrow pore size distribution, N_2 sorption at −196°C becomes unpractical because of extremely slow adsorption rate [13]. The adsorption of supercritical gases seems to be more appropriate to characterize the CMSM pore structures for the following reasons. (1) Supercritical adsorption in micropores is faster than sub-critical adsorption. (2) Supercritical adsorption takes place mostly in micropores while sub-critical adsorption may occur even on the outer surface of the membrane.

A CMSM (referred to as KP800) was prepared by Nguyen et al. via the controlled pyrolysis of Kapton® PI film under vacuum. The highest pyrolysis temperature was 800°C. The membrane showed reasonable sieving effect for gas molecules with different kinetic diameters.

Adsorption desorption isotherms of nitrogen at three temperatures are shown in Fig. 6.21. There is no hysteresis, indicating no capillary condensation due to the absence of mesopores, since the temperature is above supercritical temperature. Data were analyzed by the model proposed by Nguyen and Do [14]. The fitted results are presented in Fig. 6.22a–c, in which pressure is given in normal and logarithmic scale. A narrow pore size distribution is shown in Fig. 6.22c. The majority of the

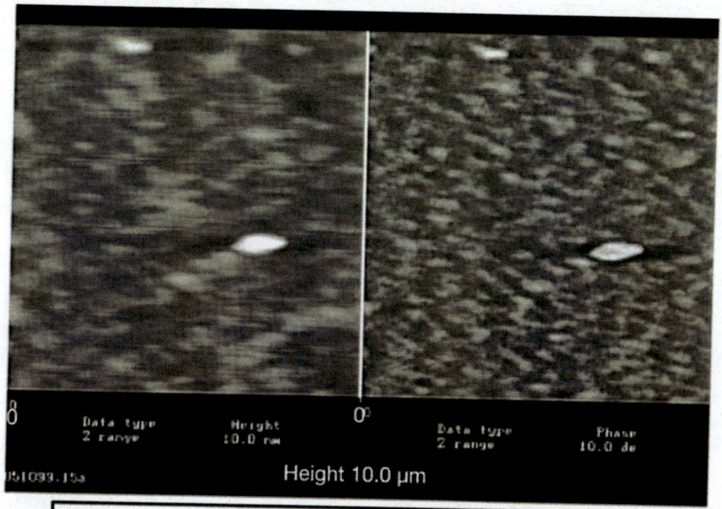

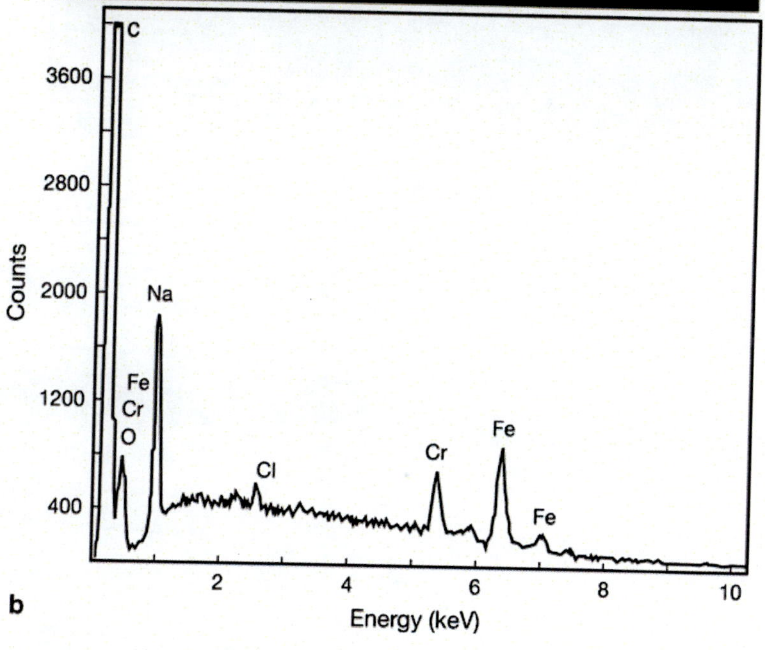

Fig. 6.13 **(a)** AFM surface images (200 nm × 200 nm), **(b)** EDX analysis of SNPCM-11. (From [10])

pores are below 10 Å without the presence of mesopores. This pore size distribution is in accordance with the gas permeation experiments, which showed that CF_4 molecule with kinetic diameter of 4.7 Å virtually could not penetrate the 125 µm thickness CMSM. Adsorption of methane was also tested at 90°C. Fitted data,for pore size and the pore size distribution are shown in Fig. 6.23a–c. The pore size distribution was shifted toward the smaller pore size. Difference between the pore size distributions in Figs. 6.22 and 6.23 is due to the difference in the size (N_2 3.64 Å

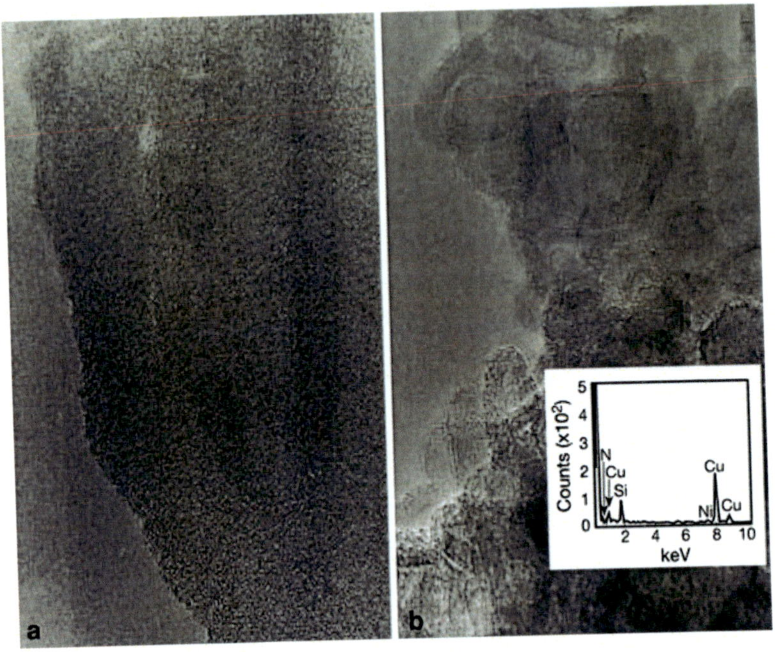

Fig. 6.14 HRTEM images (scale bar represents 12 nm) of SNPCM-11 showing (**a**) a typical amorphous region, and (**b**) a nanostructured region with EDX analysis (*insert*). (From [10])

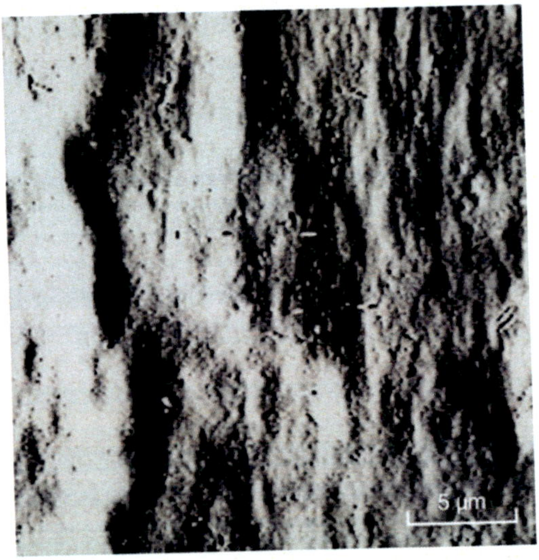

Fig. 6.15 SPM micrograph of the surface of the carbon membrane A. (From [11])

and methane 3.80 Å) and the shape (N_2 dump-bell shape and methane spherical) of the adsorbent molecules.

Another example of using the adsorption of supercritical fluid for the characterization of CMSM is the work done by Katsaros et al. [15]. It was attempted to

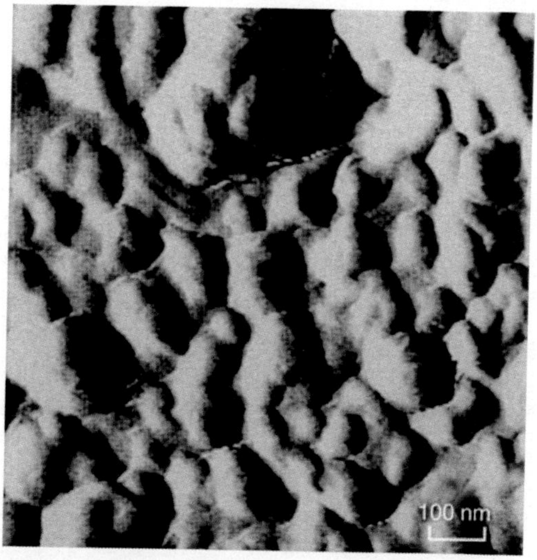

Fig. 6.16 SPM micrograph of the surface of the carbon membrane B. (From [11])

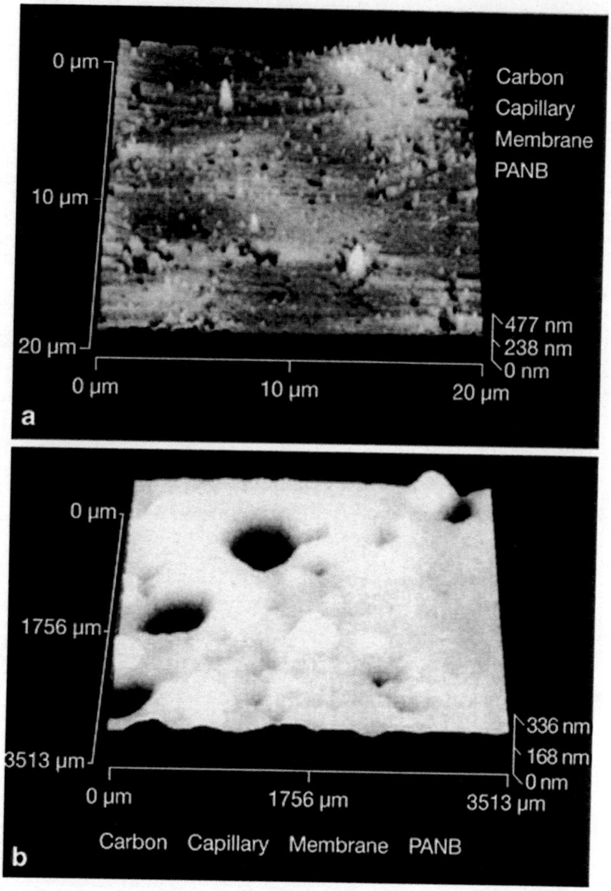

Fig. 6.17 SPM micrograph of the surface of the carbon membrane C: (**a**) 20 µm 3-D image reveals the roughness of the membrane surface; (**b**) pore structure is clear from this 3513 nm × 3513 nm image. (From [11])

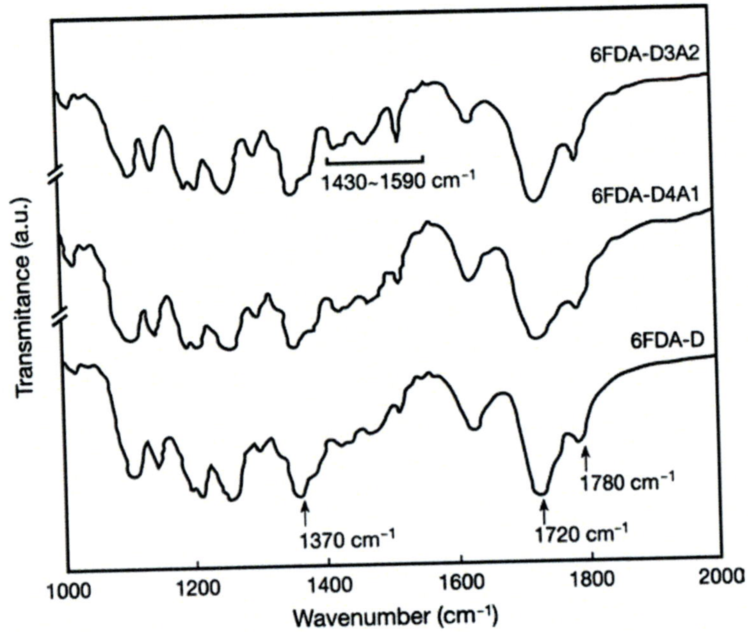

Fig. 6.18 FTIR spectra of copolyimides, 6FDA-Durene-p-intA copolymers. (From [9])

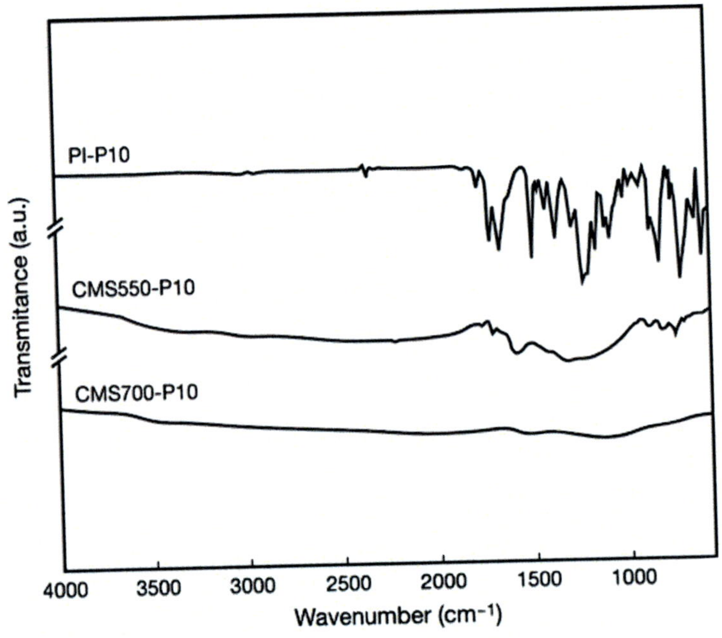

Fig. 6.19 FTIR spectra of: (*a*) PI precursor with 10% PVP (PI–P10); (*b*) PI–P10 carbonized at 550 °C; (*c*) PI–P10 carbonized at 700 °C. (From [6])

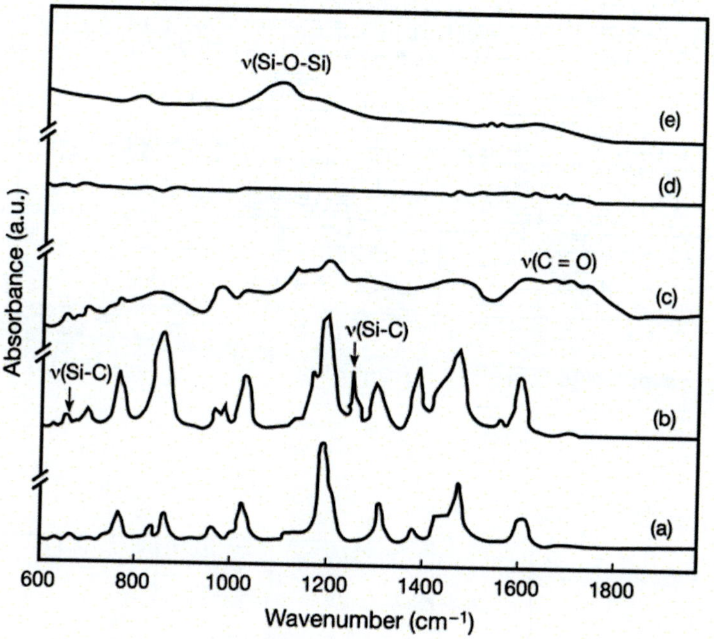

Fig. 6.20 FTIR spectra of (*a*) P(PPO), (*b*) P(TMS80), (*c*) preoxidized TMS80, (*d*) C(PPO), and (*e*) C(TMS80). Carbon samples were pyrolyzed at 923 K. (From [12])

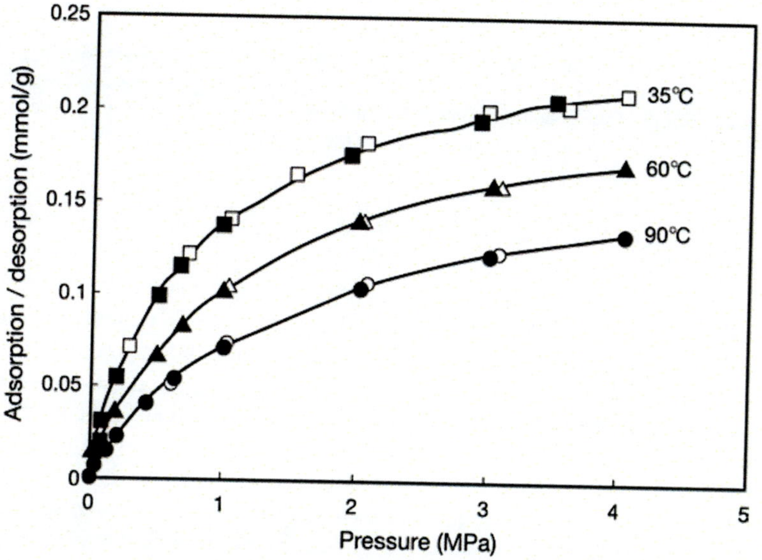

Fig. 6.21 Nitrogen adsorption on KP 800 at different temperatures; *filled symbols*, adsorption; *hollow symbols*, desorption. (From [13])

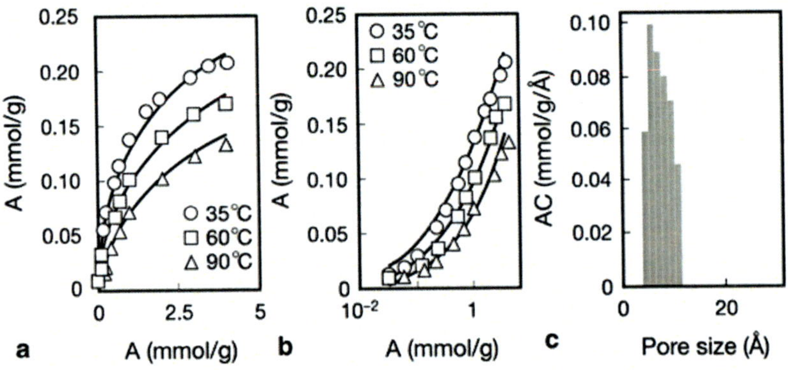

Fig. 6.22 Fitting of nitrogen adsorption isotherms at three temperatures (**a** and **b**) and the derived pore size distribution (**c**); *symbols*, experimental data; *lines*, model fitting. (From [13])

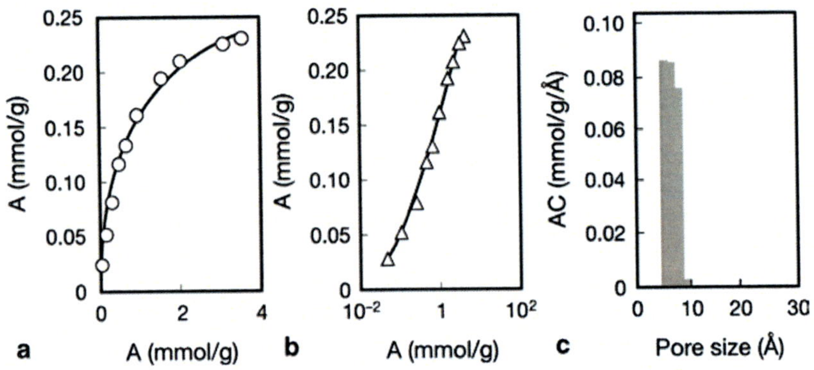

Fig. 6.23 Fitting of methane adsorption isotherm at 90°C (**a** and **b**) and the derived pore size distribution (**c**); *symbols*, experimental data; *lines*, model fitting. (From [13])

use CO_2 as an adsorbent at 194.5 K and more commonly at 308 K, which is at the supercritical conditions.

Carbon membranes were prepared by carbonization and activation of phenol–formaldehyde resin (PFR) [15]. The membranes consist of a cylindrical, macroporous Novolac based carbon substrate and a 40 μm thick microporous Resol based layer.

Commercial granular Novolac J1048 was initially heated at 423 K for the partial cross-linking. Grains were mixed with water and hydroxyethyl cellulose to form dough, which was extruded to form a cylinder of 6 mm outer diameter. Cylinder was then heated at 403°C to complete the cross-linking. Surface layer was produced by spraying with a thin layer of Resol (the same resin as Novolac with a different grain size) mixed with 20% ethanol. After air drying the cylinder was heated in air at 403 K. Carbonization was performed at 1073 K in N_2 atmosphere. The membrane

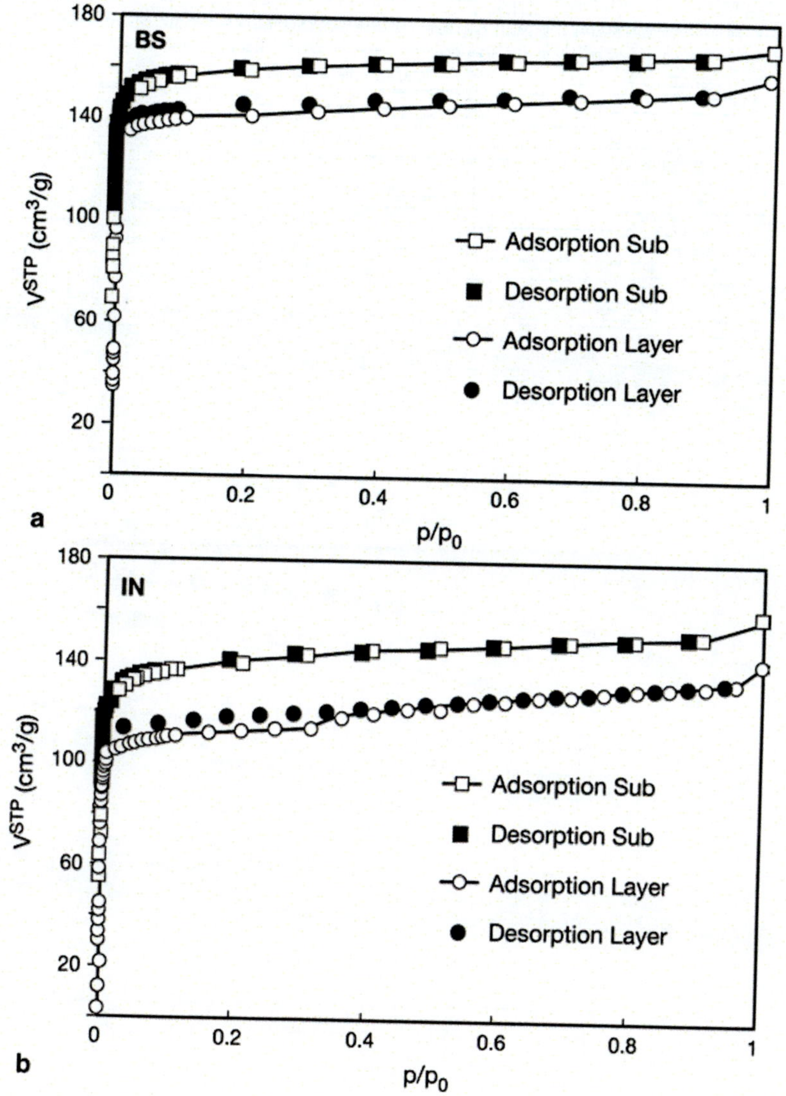

Fig. 6.24 Nitrogen adsorption isotherms at 77 K. (From [15])

was further activated by heating in CO_2 stream at 1073 K. CO_2 stream was either inside the tube (IN) or both sides (BS) of the tube.

Figure 6.24 shows the N_2 adsorption isotherm at 77 K. Adsorption isotherms are shown for both IN and BS treatment and also for both substrate (Sub) and coated layer (Layer). The adsorption isotherm is Type I according to the IUPAC classification and reveals a pure micro-pore structure. The slight gain in uptake at the high pressure end is due to slight condensation of N_2 liquid. One interesting feature of the N_2 adsorption isotherms on both BS and IN layers is the low pressure hysteresis,

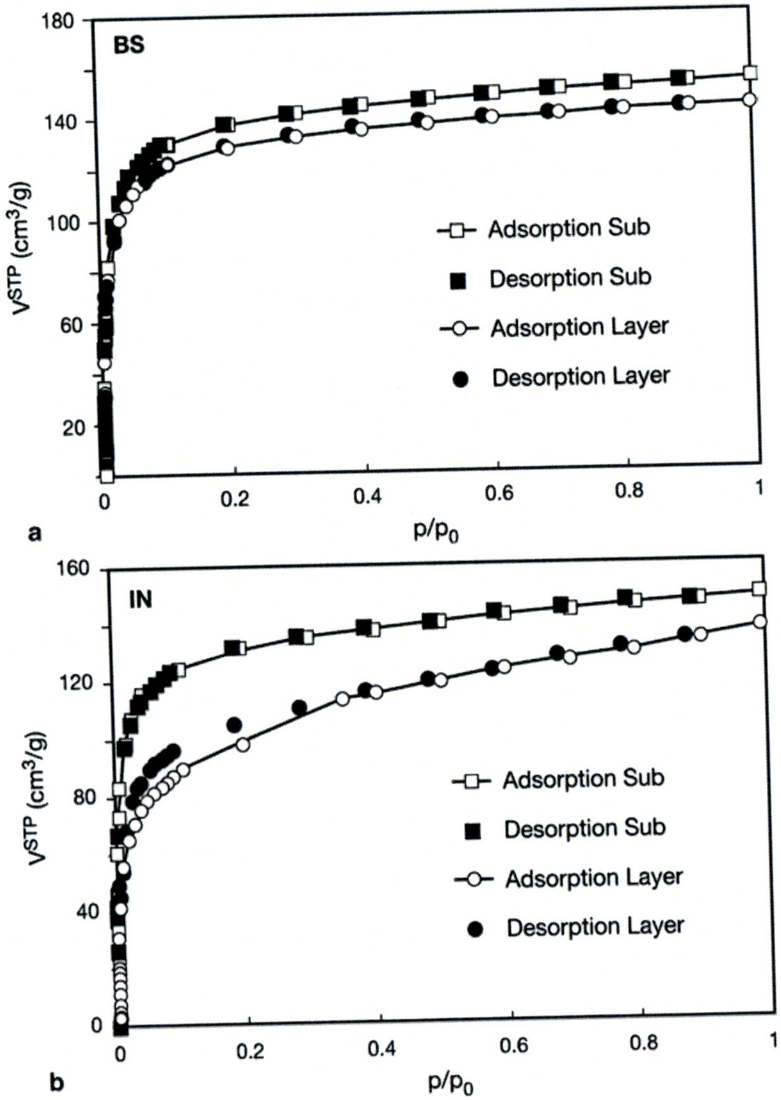

Fig. 6.25 CO_2 adsorption isotherms at 195 K. (From [15])

which is, however, more pronounced in IN layer. This is attributed to kinetic reasons, i.e. the possible existence of constrictions, which hinder the access of nitrogen to certain pores.

CO_2 adsorption at 195 K is shown in Fig. 6.25. Some hysteresis is also found in IN layer even though the temperature was as high as 195 K.

Based on the non local density functional theory (NLDFT) pore analysis, the pore size distributions for slit shaped pores were deduced from the N_2 adsorption isotherms.

The results are given in Fig. 6.26. It is emphasized that for all the samples, the majority of the pore volumes lie below 7.5 Å, thus belonging to ultra-micropore

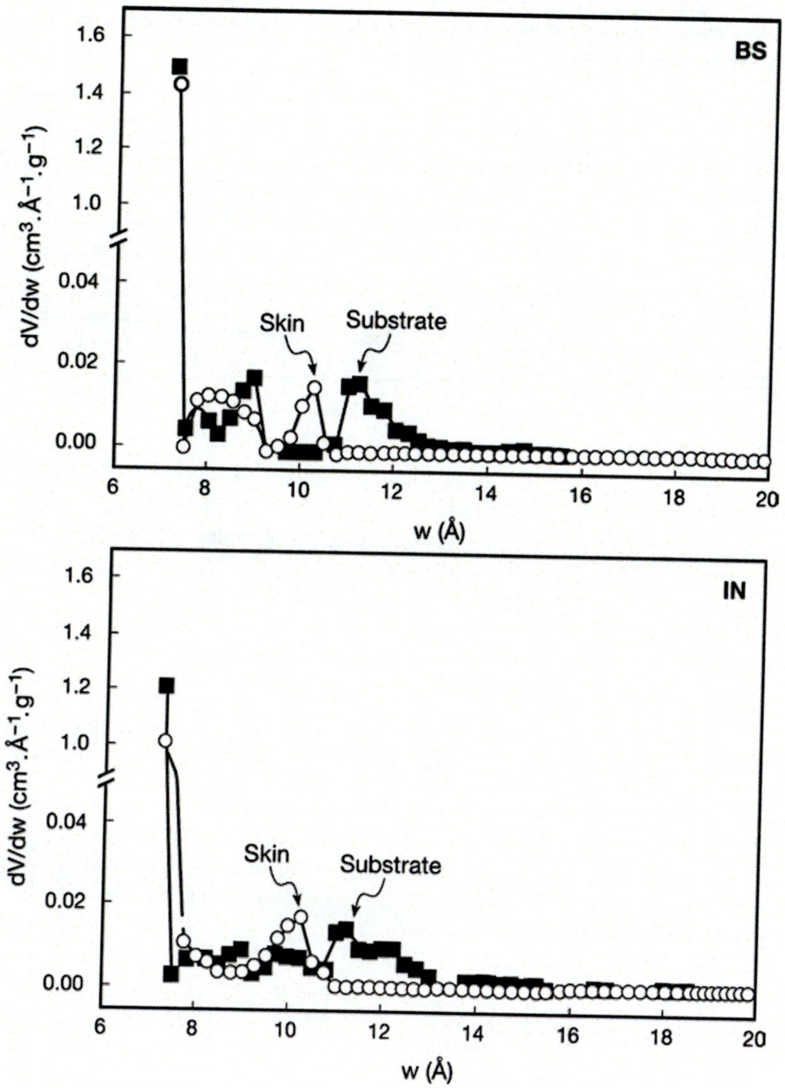

Fig. 6.26 NLDFT pore size distributions for slit pores based nitrogen adsorption at 77 K; (**a**) sample BS, (**b**) sample IN. (From [15])

range. Obtaining the distribution below this pore size was not possible. Beyond 8 Å, the peak appeared at 10 Å for the skins and at 11 Å for the substrates. Probably, N_2 adsorption at 77 K is not adequate to characterize the pore structure of the skin samples due to the existence of constrictions, which render some volume inaccessible to N_2.

Table 6.3 displays the results obtained from the analysis of adsorption isotherms for N_2 and CO_2 by the Dubinin-Radushkevic (DR) method. The pore volumes measured by the carbon dioxide isotherm are consistently higher than the pore volume

Table 6.3 Pore structure characterizations derived from N_2 (77 K) and CO_2 (194.5 K) adsorption isotherm. (Data taken from [15])

	BS-substrate	BS-layer	In-substrate	In-layer
Nitrogen isotherm (77 K)				
V_{pore} (cm^3/g)	0.263	0.245	0.246	0.218
A_{BET} (m^2/g)	639.9	579.5	545.8	444.1
V_{DR} (cm^3/g)	0.266	0.232	0.214	0.181
E_0 (kJ/mol)	25.8	27.6	26.2	27.8
Porosity, ε (%)	35.4	33.5	33.7	30.9
Carbon dioxide isotherm (194.5 K)				
V_{DR} (cm^3/g)	0.284	0.265	0.280	0.226
E_0 (kJ/mol)	25.4	25.9	25.3	26.4

obtained from the nitrogen adsorption isotherm. This is because CO_2 (2.8 Å) is smaller than N_2 (3.0 Å) and more accessible to the smaller pores. Hence, CO_2 is more adequate than N_2 for the pore analysis. However, the hysteresis observed for CO_2 adsorption isotherm at IN layer suggests an extremely constricted pore structure. Probably, the adsorption isotherm should be generated at even higher temperature than the super-critical temperature of CO_2 (308 K).

Shiflett and co-workers coated thin uniform layers of PFA (Dija laboratories) to porous stainless steel tubes by ultrasonic deposition and carbonized at 723 K to form crack-free supported nonporous carbon (NPC) membranes [10]. Methyl chloride porometry was performed with variation in both pressure and temperature. Figure 6.27 shows the results of the measurement and calculation. Applying Hor-

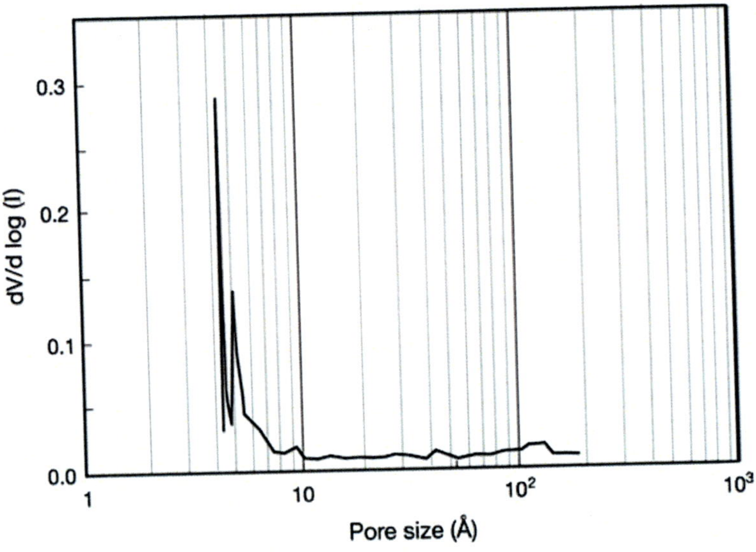

Fig. 6.27 Pore size distribution derived from the adsorption isotherm of methyl chloride for SNPCM. (From [10])

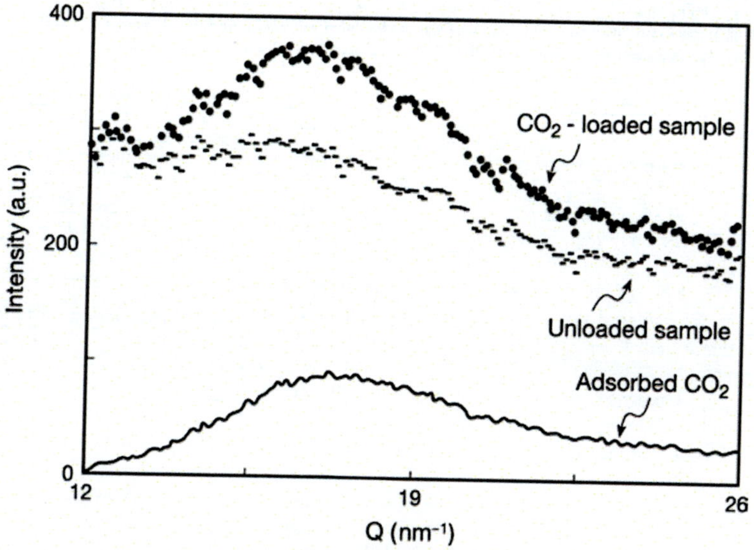

Fig. 6.28 Diffraction patterns of carbon membrane with adsorbed CO_2 at 308 K and 50 bar. (From [16])

vath-Kawazoe model, the pore size distribution was calculated. A multimode pore size distribution with peaks at 4.5, 5.0, 5.5, 6.0 and 6.5 Å was obtained.

6.2.5 Other Methods

An ethanol solution of commercial phenol formaldehyde resin (PFR) was sprayed on a stainless-steel plate [16]. The resin was cross-linked at 150°C and carbonized at 800°C under N_2 environment and further activated at 800°C in CO_2 stream.

Figure 6.28 illustrates the neutron diffraction pattern of the CO_2 loaded and unloaded sample at 50 bar. The subtraction curve is also included. The neutron diffraction pattern of the unloaded sample shows a reflection at $Q=16.4\ nm^{-1}$. The subtraction curve is assigned to the adsorbed CO_2 molecules. The peak at $Q=17.5\ nm^{-1}$ corresponds to the most probable distance between nearest neighbor molecules.

References

1. Strano MS, Zydney AL, Barth H, Wooler G, Agarwal H, Foley HC (2002) Ultrafiltration membrane synthesis by nanoscale templating of porous carbon. J Membr Sci 198 (2): 173-186
2. Strano MS, Agarwal H, Pedrick J, Redman D, Foley HC (2003) Templated pyrolytic carbon: The effect of poly(ethylene glycol) molecular weight on the pore size distribution of poly(fufuryl alcohol)-derived carbon. Carbon 41 (13): 2501-2508

3. Rajagopalan R, Merritt A, Tseytlin A, Foley HC (2006) Modification of macroporous stainless steel supports with silica sub-micron particles for size selective carbon membranes with improved flux. Carbon 44 (10): 2051-2058
4. Merritt A, Rajagopalan R, Foley HC (2007) High performance nanoporous carbon membranes for air separation. Carbon 45 (6): 1267-1278
5. Shah TN, Foley HC, Zydney AL (2007) Development and characterization of nanoporous carbon membranes for protein ultrafiltration. J Membr Sci 295 (1-2): 40-49
6. Kim YY, Park HB, Lee YM (2004) Carbon molecular sieve membranes derived from thermally labile polymer containing blend polymers and their gas separation properties. J Membr Sci 243 (1-2): 9-14
7. David LIB, Ismail AF (2003) Influence of the thermastabilization process and soak time during pyrolysis process on the polyacrylonitrile carbon membrane for O_2/N_2 separation. J Membr Sci 213 (1-2): 285-291
8. Centeno TA, Fuertes AB (2000) Carbon molecular sieve gas separation membranes based on poly(vinylidene chloride-co-vinyl chloride). Carbon 38 (7): 1067-1073
9. Xiao Y, Chung T-S, Guan HM, Guiver MD (2007) Synthesis, cross-linking and carbonization of co-polyamides containing internal acetylene units for gas separation. J Membr Sci 302 (1-2): 254-264
10. Shiflett MB, Pedrick JF, McLean SR, Subramoney S, Foley HC (2006) Characterization of supported nanoporous carbon membranes. Adv Mater 12 (1): 21-25
11. Linkov Y, Sanderson RD, Jacobs EP (1994) Scanning probe microscopy study of carbon membrane surfaces. J Mater Sci Lett 13 (8): 600-601
12. Yoshimune M, Fujiwara I, Haraya K (2007) Carbon molecular sieve membranes derived from trimethylsilyl substituted poly(phenylene oxide) for gas separation. Carbon 45 (3): 553-560
13. Nguyen C, Do DD, Haraya K, Wang K (2003) The structural characterization of carbon molecular sieve membrane (CMSM) via gas adsorption. J Membr Sci 220 (1-2): 177-182
14. Nguyen C, Do DD (1999) Adsorption of supercritical gases in porous media: Determination of micropore size distribution. J Phys Chem B 103 (33): 6900-6908
15. Katsaros FK, Steriotis TA, Ramanos GE, Konstantakou M, Stubos AK, Kanellopoulos NK (2007) Preparation and characterization of gas selective microporous carbon membranes. Microporous Mesoporous Mater 99 (1-2): 181-189
16. Katsaros FK, Steriotis TA, Stefanopoulos KL, Kanellopoulos NK, Mitropoulos AC, Meissner M, Hoser A, (2000) Neutron diffraction study of adsorbed CO_2 on a carbon membrane. Physica B 276-278: 901-902

Chapter 7
Membrane Module Constructions

7.1 Honey Comb Membrane Module by Blue Membranes GmbH

Membrane module construction is seldom refered to in the literature for CMSM. To date, only the laboratory scale tubular or hollow fiber membranes have been described in detail. Industrial scale applications of CMSMs were hampered by the difficulty involved in the large scale membrane production, poor mechanical stability of the membranes and the poor reproducibility of the membrane modules. The assembly of large scale modules were described only in the patents by Soffer and co-workers [1–3]. The only commercial modules are now available from Carbon Membranes Ltd (Israel) with a hollow fiber configuration [4]. Lagorsse and co-workers reported on the development of the honey comb membrane module configuration (HM) designed and developed by Blue Membranes GmbH (Germany) [5].

The success of the membrane separation system depends not only on the development of high performance membrane but also on the design and construction of an efficient and economical membrane module. Currently there are two types of membranes; one is based on flat sheet and the other tubular. Carbon Membrane Ltd. makes use of hollow fiber (small tubular) configuration with the packing density of 2,000 m^2/m^3. The novel CMSM module of Blue Membranes GmbH is based on the concept of "honey comb module" and has a high packing density of 2,500 m^2/m^3. The flat supported membrane is produced first. The support consists of industrial grade paper modified with ceramic fibers. The use of this support minimizes the gap that occurs between the support and the coated polymeric precursor layer due to the differential shrinking rate, thus avoiding the residual stresses from the membrane. The support is coated with a precursor polymer solution using an imprinting technique. As a precursor, blended phenolic resin (PR) (a commercially available resol type-Phenodur PR 515) and an epoxy resin (Beckopox EP309) are used. The PR-derived activated carbon was described by Tennison [6]. Resins were dissolved in methyl ethyl ketone for the coating purposes. A picture of the membrane sheet is given in Fig. 7.1. From the flat precursor coated sheet, a corrugated sheet is ob-

A. F. Ismail et al., *Carbon-based Membranes for Separation Processes*,
DOI 10.1007/978-0-387-78991-0_7, © Springer Science+Business Media, LLC 2011

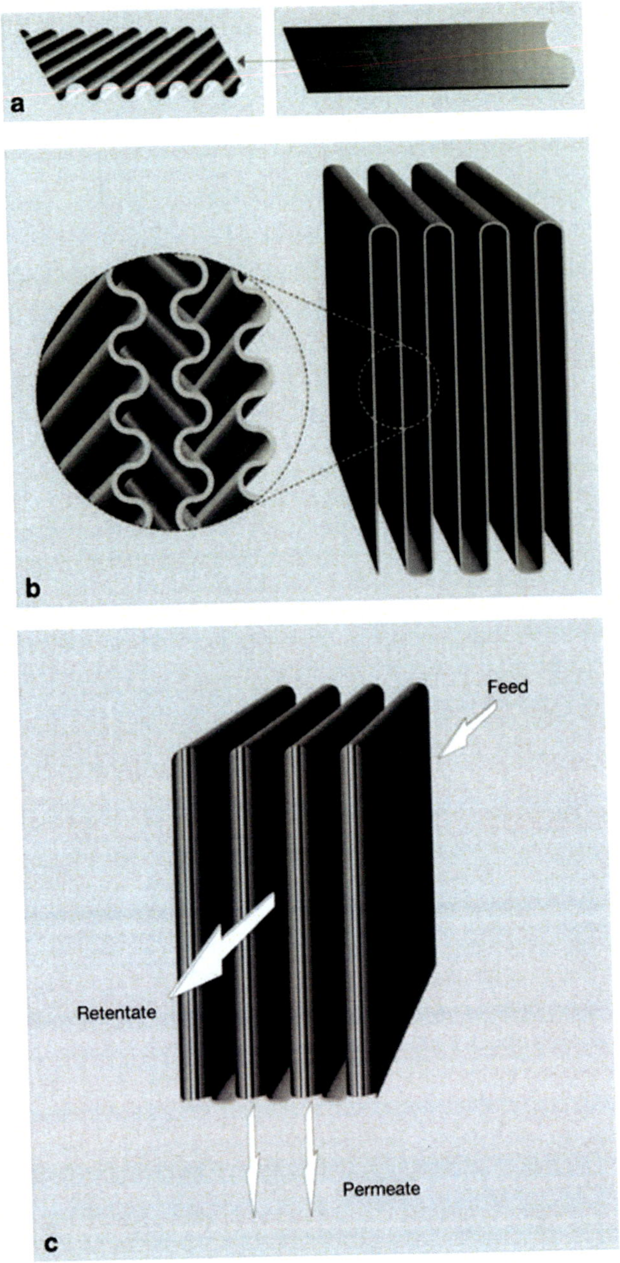

Fig. 7.1 Schematic of the honeycomb CMSM module preparation procedure; (**a**) flat corrugated precursor coated sheet, (**b**) pleated precursor coated sheet, (**c**) membrane module. (From [5])

tained using a stamping procedure (Fig. 7.1a). The wavy pattern is oriented diagonally. The sheet is further pleated (Fig. 7.1b). By overlapping the pleated corrugated sheet, flow channels in the cross-flow direction is obtained without the need for the turbulence promoter. The permeate and the feed sides become independent from each other by sealing, at the edges, those alternate sets of layers on which the polymeric film layers face each other (Fig. 7.1c).

This creates a pocket-like chambers which constitute the permeate side of the module. During the subsequent pyrolysis the pore size is tailored by controlling the temperature and the duration of the process. Typically, the membrane is heated at a heating rate of 1 K/min up to 1,053 K. The membrane is held at the maximum temperature for 3 h, before the membrane is cooled down to the room temperature. Then, the fine tuning of the molecular sieving properties is done by CVD followed by an activation step. The membrane is heated in a quartz tube furnace under inert atmosphere of nitrogen up to 923 K at a rate of 10 K/min. At this temperature the membrane is brought into contact with propylene for 1 min. After purging the system with nitrogen the temperature is brought down to the room temperature. Then, the pore opening by controlled oxidation will follow, by heating from 523 to 623 K in air stream for a period of time ranging 1–3 h. This creates pore size distribution that is narrower than the one before CVD. The CVD together with oxidation procedure enables the performance enhancement of the membrane as well as the repair of the defects. After the heat treatment the module is sealed into the housing. This entire procedure is summarized in Fig. 7.2. Figure 7.3 shows the cross-sectional SEM images of the membrane. Corrugated shape is shown in Fig. 7.3a. The presence of two distinctive layers, one meso-/macro-porous support and the other the selective dense layer is quite obvious. Total thickness of the membrane is 50–100 μm, where as the carbon layer thickness is 10–30 μm.

Some of the experimental results from the single gas permeation test using the membrane fabricated are given in Table 7.1.

Pyrolysis conditions are the same. For the final activation step, sample A was activated for a time period two times longer than B. A1 and B1 were submitted to the highest temperature whereas B3 and A2 were submitted to the lowest temperature during the activation step.

Comparable literature data were also reported in [5]. The adsorption isotherms data at 303 K are given in Fig. 7.4 for various gases. Figure 7.5 shows the adsorption isotherm for N_2 and CO_2 at three different temperatures. H_2, O_2 and N_2 were fitted to Langmuir equation and CO_2 to UNILAN equation. The parameters associated in each equation are given in Table 7.2.

It is interesting to note that no adsorption of SF_6 was detected up to 5 bar. It is attributed to the size of SF_6 which is larger than the pore size.

Nguyen and Do's method [7] to evaluate micro-/meso-pore size distribution was applied in this work. According to the method the Langmuir adsorption isotherm is used to describe the behavior of the adsorbent molecules in the pores. Different isotherms are considered for different pore sizes. Hence, the adsorption isotherm parameters are the function of the pore radius. Thus, the fractional coverage, θ, becomes [5].

$$\theta(r) = \frac{q(r)}{q_m(r)} = \frac{b_{poe}(r)p_{pore}(r)}{1 + b_{pore}(r)p_{pore}(r)} \tag{7.1}$$

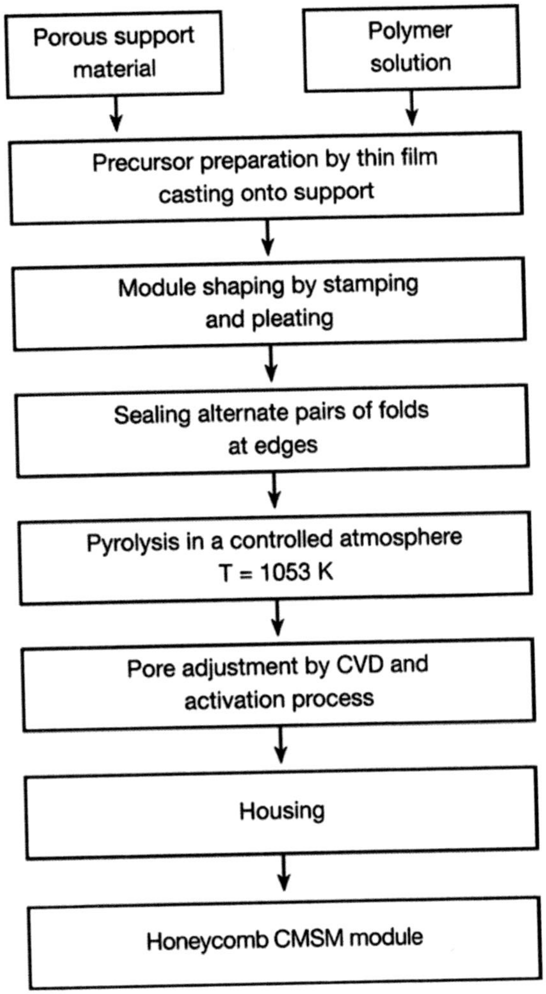

Fig. 7.2 Steps involved in the production of Blue Membranes' honeycomb CMSM module. (From [5])

where parameters $b_{pore}(r)$ is estimated as a function of heat of adsorption inside the pore and is a function of temperature; the pressure inside the pore is related to the pressure outside the pore and the potential energy of the molecules of the gas phase inside the pore. Slit shape is assumed for the pore. The potential in the pore is the sum of the potential energy due to the two opposite pore walls. The potential of a molecule relative to a flat surface is calculated by the Steele equation (or Steele potential);

$$H(z) = 2\pi \rho_s \varepsilon_{sf} {\sigma_{sf}}^2 \Delta \left[\frac{2}{5}\left(\frac{\sigma_{sf}}{z}\right)^{10} - \left(\frac{\sigma_{sf}}{z}\right)^4 - \frac{{\sigma_{sf}}^4}{3\Delta(0.61\Delta + z)^3} \right] \tag{7.2}$$

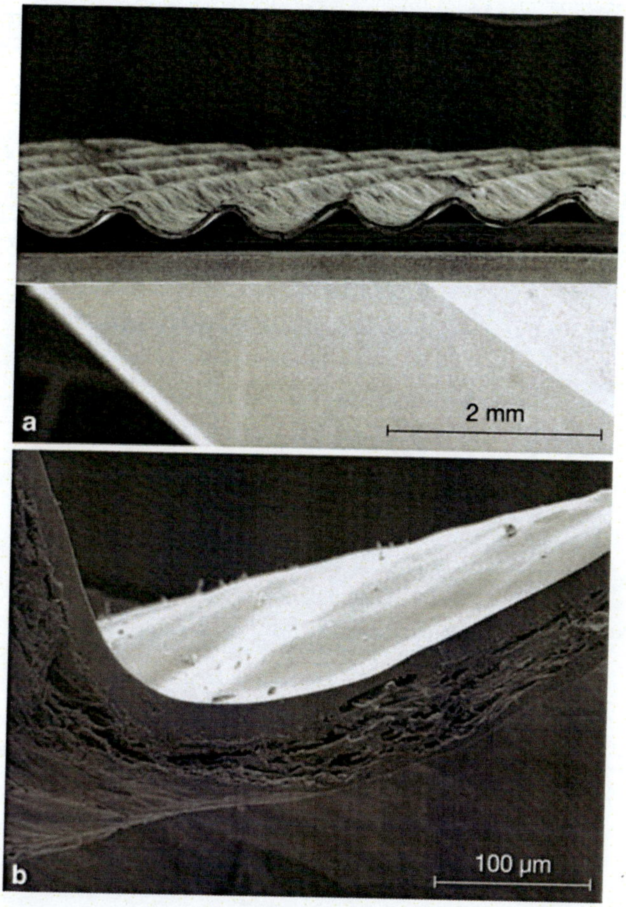

Fig. 7.3 Cross-sectional SEM pictures of the honeycomb CMSM; (**a**) corrugated membrane, (**b**) selective layer on top of the support layer. (From [5])

Table 7.1 Experimental results from single gas permeation experiments

Membrane sample	Permeance ($\times 10^{-8}$ m^3STP/m^2 kPa s)				Ideal selectivity			
	H_2	CO_2	O_2	N_2	H_2/N_2	H_2/CO_2	O_2/N_2	CO_2/N_2
A1	162	36.5	6.47	0.78	208	4.2	8.3	40.5
A2	126	21.6	4.75	0.56	227	5.8	8.6	39.0
B1	157	59.4	10.6	1.00	157	2.6	10.6	59.4
B2	142	39.8	7.30	0.61	233	3.6	12.0	65.1
B3	84.8	23.8	4.39	0.33	254	3.6	13.2	71.3

Feed pressure 2 bar, permeate pressure < 1 bar, at 303 K

where z is the distance relative to the flat surface, ρ_s, density of graphite, Δ, distance between two successive graphite layer, ε_{sf}, interaction energy solid-fluid, and σ_{sf}, the geometrical parameter of interaction solid-fluid (collision diameter). The last parameters are known to be Lennard-Jones potential parameters. Then the surface coverage at pressure p becomes

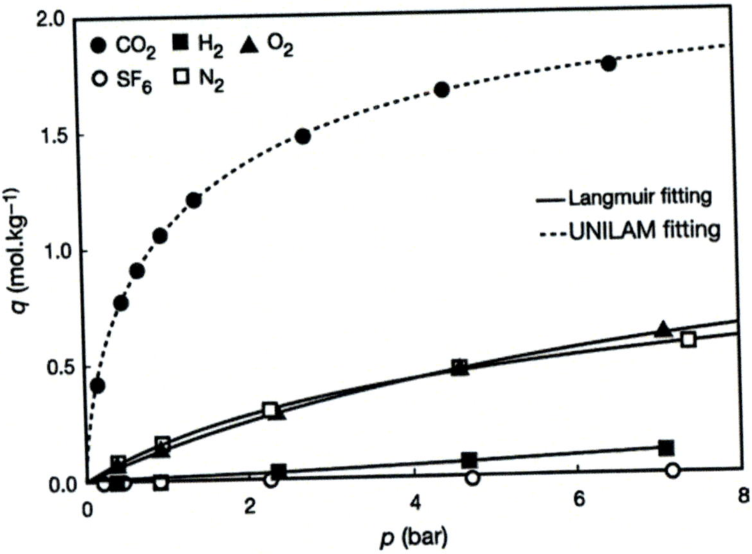

Fig. 7.4 Adsorption equilibrium isotherm on sample B2 at 303 K; (○) SF_6, (■) H_2, (▲) O_2, (□) N_2, (•) CO_2; the *solid* and *dashed lines* correspond to Langmuir and UNILAN equations, respectively. (From [5])

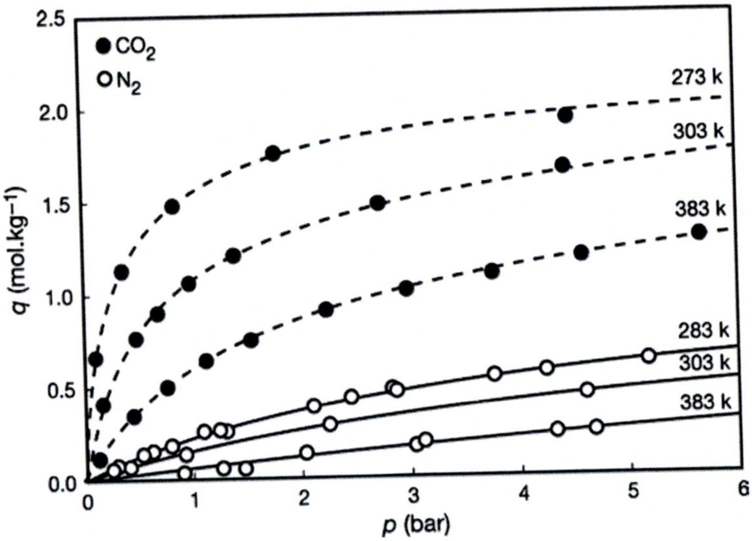

Fig. 7.5 Adsorption equilibrium isotherms on sample B2 at 273, 303 and 383 K for CO_2 (•) and at 283, 303 and 383 K for N_2 (○); the *solid* and *dashed lines* correspond to Langmuir and UNILAN equations, respectively. (From [5])

Table 7.2 Parameters associated with adsorption isotherms (membrane sample B2)

Gas	Temperature (K)	Langmuir parameters		
		b (bar)	q_m (mol/kg)	
H_2	303	0.002	6.184	
O_2	303	0.119	1.313	
N_2	283	0.285	1.058	
	303	0.29	0.937	
	338	0.057	1.219	
		UNILAN		
		b (bar)	q_m (mol/kg)	S
CO_2	273	2.854	2.192	1.802
	303	0.861	2.236	2.089
	338	0.359	1.976	1.654

$$\theta_{\exp}(p) = \int_{r\,\min}^{r\,\max} \theta_{calc}(p,r) f(r) dr \tag{7.3}$$

where $f(r)$ is the pore size distribution function and is

$$\sum_{1}^{n} f(r_i) = 1. \tag{7.4}$$

Figure 7.6 gives the pore size distribution obtained by the above method. It shows a bimodal distribution, 91.6% in the range of 0.60–0.68 nm and about 7.3% in the

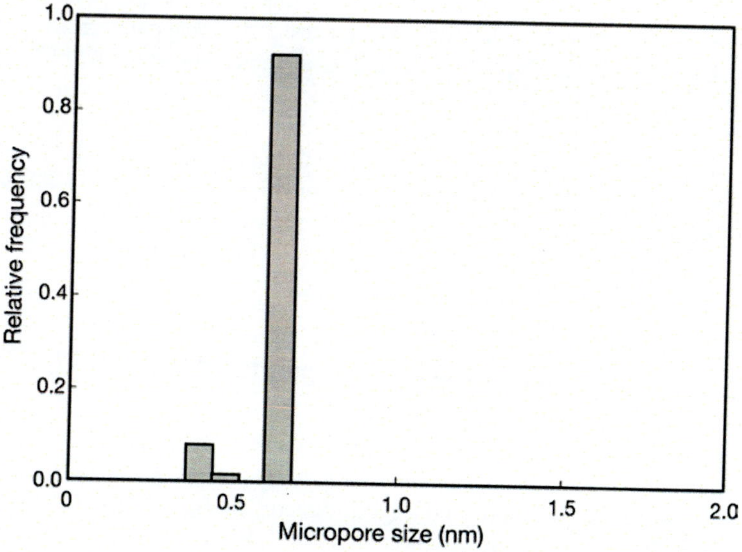

Fig. 7.6 Micropore size distribution for the sample B2. (From [5])

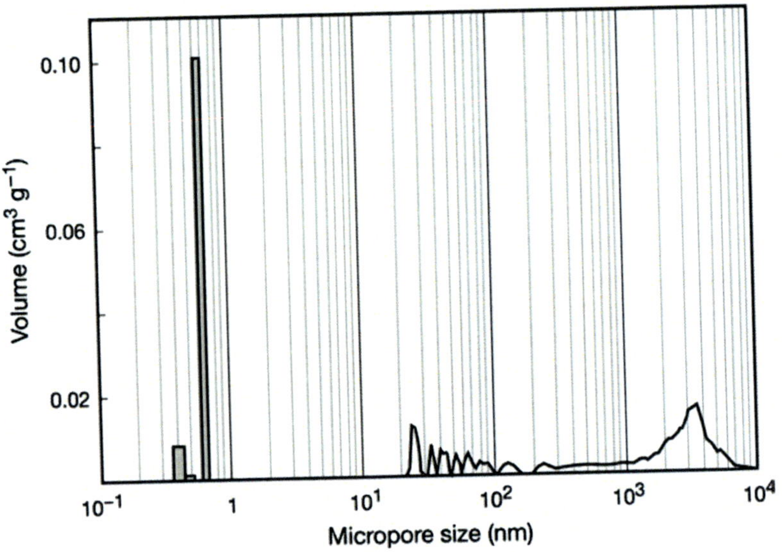

Fig. 7.7 Overall pore size distribution for the sample B2. (From [5])

range of 0.36–0.44 nm. An effective average size is 0.62 nm. This value is slightly higher than 0.51 nm obtained by the Stoeckli method. It should be noted that the pore size was calculated from the center of the carbon atom at the wall on one side to the center of the carbon atom at the wall on the other side. This binodal distribution supports the pore system proposed for CMSM; i.e., an ultra-microporous network comprised of "wide" segments ("cavities") randomly interrupted by short constrictions [8]. The latter provides the molecular sieving performance (selectivity), but since they constitute a small fraction of a spieces' diffusional path across the membrane, relatively high permeabilites are still obtainable. Adsorptions of SF_6 and Xe were also examined. While there was no noticeable SF_6 adsorption up to 5 bars, there was also very little Xe adsorption. Since the molecular sizes of SF_6 and Xe are 0.502 and 0.39 nm, respectively, these molecules are sieved by the 0.36–0.44 nm pores. Mercury porosimetry provided the information on the pore size distribution of the support material. Combining the ND method and the mercury porosimetry, the pore size distribution in the entire range will be like Fig. 7.7.

7.2 Capillary Type CMSM Developed By Haraya et al.

A method of preparing a capillary type CMS membrane was described in detail by Haraya et al. [9]. The schematic of the casting equipment is shown in Fig. 7.8. A 1.6 wt% solution of polyamic acid (PA) in dimethyl acetamide (DMAC) was used for the casting dope. A polytetrafluoroethylene (PTFE) tube (e.d. 1.8 mm) traveled through the polymer dope in a container at a speed of 3 cm/min, coming out through

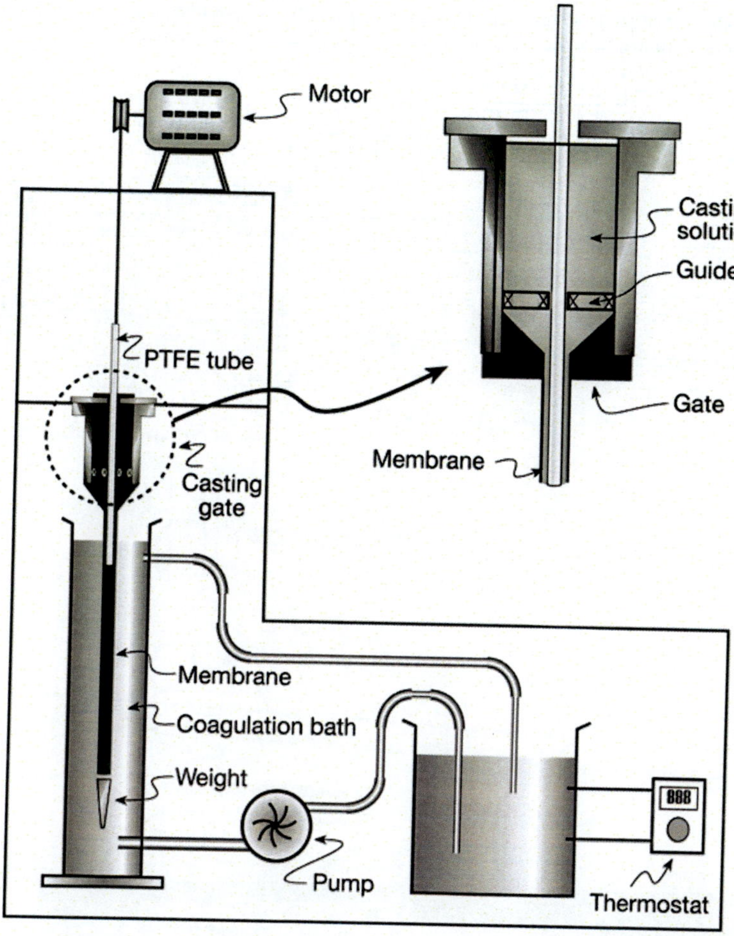

Fig. 7.8 Schematic illustration of the apparatus used for the casting of tubular CMS membrane. (From [9])

a gate at the bottom of the container. There was a gap of 500 μm between the radius of the PTFE tube and that of the circular gate, by which the thickness of the coated layer was determined. The coated layer together with the PTFE tube went into the gelation bath, water or ethanol, after a 2 s exposure to the air. The PA layer on the PTFE capillary tube stayed in the gelation bath for 10 h before being washed by water and dried for 1 day. Then, the PTFE tube was removed. The PA capillary so obtained was further dried under vacuum before being imidized to Kapton type PI by the following two heating steps; i.e. at 473 K for 30 min and at 673 K for 1 h under vacuum. The PI membrane thus obtained was pyrolyzed under vacuum by heating at 20 K/min to 1,223 K. The membrane was held at the latter temperature for 1 h. The reaction taking place in this fabrication procedure is schematically shown in Scheme 7.1. Figure 7.9 shows the asymmetric structure of the capillary

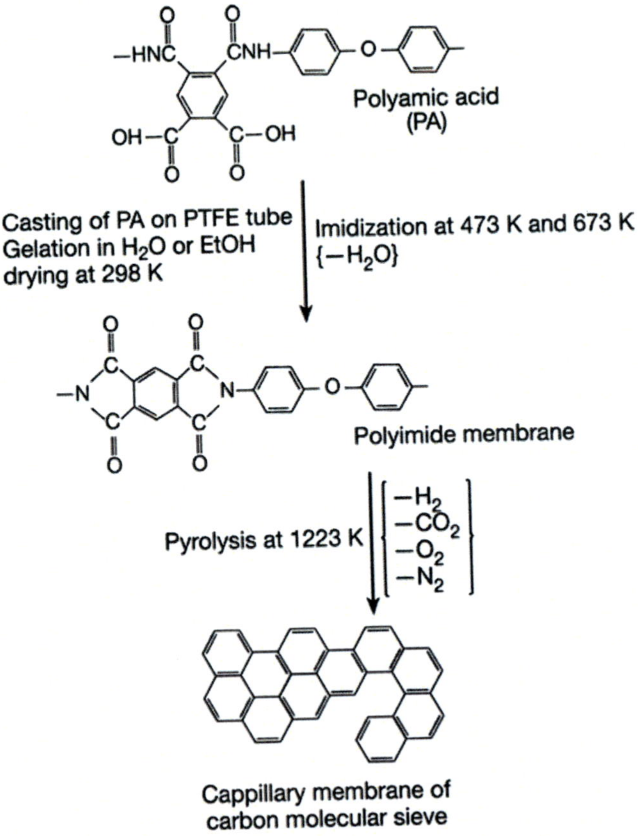

Scheme 7.1 Chemical reaction taking place at each step of membrane fabrication. (From [9])

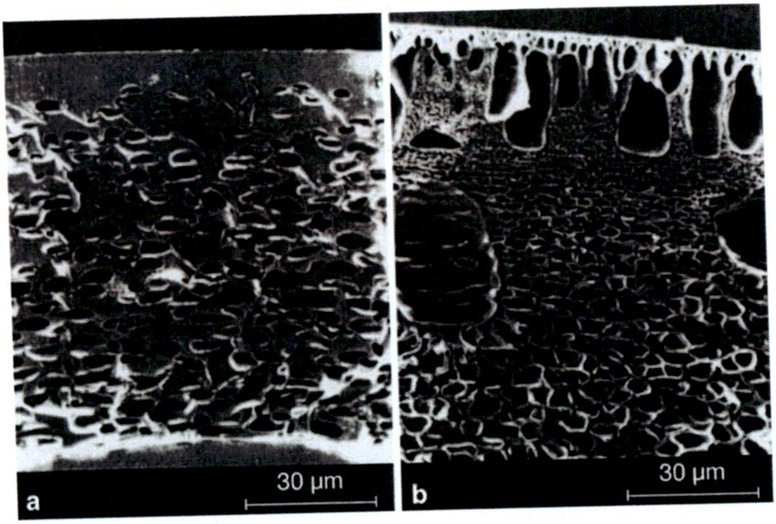

Fig. 7.9 Scanning electron microscopic images of the prepared CMT membrane; (**a**) ethanol coagulation, and (**b**) water coagulation

Table 7.3 Permeance and ideal separation factor of the asymmetric capillary CMS membranes

	Coagulant, ethanol		Coagulant, water	
	308 K[a]	373 K[a]	308 K[a]	373 K[a]
P_{H_2} (GPU)[b]	0.74	1.17	39.8	48.6
H_2/N_2[c]	1080	430	95.9	83.9
P_{He} (GPU)[b]	0.43	0.69	16.4	23.7
He/N_2[c]	630	250	39.5	40.9
P_{CO_2} (GPU)[b]	0.046	0.065	8.50	8.13
CO_2/N_2[c]	68	24	20.5	14.0
P_{O_2} (GPU)[b]	0.012	0.022	2.63	2.93
O_2/N_2[c]	17	8	6.34	5.06
P_{N_2} (GPU)[b]	0.00068	0.0027	0.415	0.579

[a] Temperature at which gas permeation was measured
[b] Gas permeance
[c] Ideal separation factor

CMSM so obtained. The structure formed at the gelation step of PA is believed to be maintained in the following steps of imidization and pyrolysis. Thus, when the coagulation bath is water, instantaneous demixing of the polymer occurs, resulting in a thin dense film and large pores in the support layer. Alternatively, when ethanol is used as the coagulation media, the demixing is delayed leading to thicker dense layer and smaller pores in the porous sublayer. This morphological change makes a large difference in the performance of the capillary CMS membrane. The performance data are given in Table 7.3.

On top of the macropores displayed by SEM the capillary of the CMS membranes have ultra-micro-pores whose dimensions are nearly the same as the flat sheet membrane.

References

1. Soffer A, Koresh J, Saggy S (1987) Separation device. US Patent 4,685,940
2. Soffer A, Saguee S, Golub D, Cohen H, Azariah M (1996) Selective clogging of failed fibers. US Patent 5,575,963
3. Soffer A, Azariah M, Amar A, Cohen H, Golub D, Saguee S, Tobias H (1997) Method of improving the selectivity of carbon membranes by chemical vapour depositions. US Patent 5,695,818
4. Lagorsse S, Magalhäes FD, Mendes A (2004) Carbon molecular sieve membranes: Sorption, kinetic and structural characterization. J Membr Sci 241 (2): 275-287
5. Lagorsse S, Leite A, Magalhães FD, Bischofberger N, Rathenow J, Mendes A (2005) Novel carbon molecular honeycomb membrane module: Configuration and membrane characterization. Carbon 43 (4): 809-819
6. Tennison SR (1998) Phenolic-resin-derived activated carbons. Appl Catal A Gen 173 (2): 289-311

7. Nguyen C, Do DD (1999) Adsorption of supercritical gases in porous media: Determination of micropore size distribution. J Phys Chem B 103 (33): 6900-6908
8. Koresh J, Soffer A (1987) The carbon molecular sieve membrane. General properties and the permeability of CH_4/H_2 mixture. Sep Sci Technol 22 (2-3): 973-982
9. Haraya K, Suda H, Yanagishita H, Matsuda S (1995) Asymmetric capillary membrane of a carbon molecular sieve. J Chem Soc Chem Commun: 1781-1782

Chapter 8
Other Carbon-based Membranes

8.1 Carbon Nanotubes Membrane

Since Iijima identified carbon nanotubes (CNTs) in 1991, CNTs have been investigated in various fields and become extremely desirable for a wide range of applications. CNTs, with diameters in nanometer scale and a smooth surface may offer a very unique molecular transport through their pores. In fact, several studies in recent years suggest that the water transport through single-walled carbon nanotubes (SWNT) would become much faster than the transport rate that the continuum hydrodynamic theory would predict. This was attributed by Molecular Dynamic (MD) simulation to the smoothness of the nano-tube wall [1, 2].

Recently Holt et al. [3] developed CNT membranes whose pore sizes were smaller than 2 nm by using a microelectro-mechanical system (MEMS)-compatible fabrication process.

Dense, vertically aligned double-walled carbon nanotubes (DWNTs) were grown on the surface of a silicon tip by catalytic chemical vapor deposition (Step 3 in Fig. 8.1a, b). It was followed by conformal encapsulation of the nanotubes by a hard, low pressure chemical vapour deposited silicon nitride (Si_3N_4) matrix (Sep 4 in Fig. 8.1a, b). There is no gap between the nanotubes according to the SEM image. Silicon tip is then etched to enlarge the support pore (Step 5 in Fig. 8.1a). Silicone nitride is etched to expose carbon nanotubes and catalyst nano-particles are also removed by Ar ion milling (Step 6 in Fig. 8.1a). Finally, the nanotubes are uncapped by reactive ion etching (Step 7 in Fig. 8.1a). According to the TEM image (Fig. 8.2c–e) silicon nitride coats the DWNTs conformally and there is no gap between the nanotube and silicon-nitride.

Size exclusion experiments revealed that gold particles of the size less than 1.3 nm could pass the membrane pore freely while gold particles of the size more than 2 nm was completely rejected by the pore. Therefore the pore size was estimated to be between 1.3 and 2.0 nm. This is further supported by the TEM image analysis given in Fig. 8.2b which shows the average value of the inner diameter is 1.6 nm. The TEM image shown in Fig. 8.2c also reveals that the holes that go through the entire cross-section of the membrane have the sizes of the inner diameter of DWNTs. From these results, it was concluded that the transport through

A. F. Ismail et al., *Carbon-based Membranes for Separation Processes*,
DOI 10.1007/978-0-387-78991-0_8,

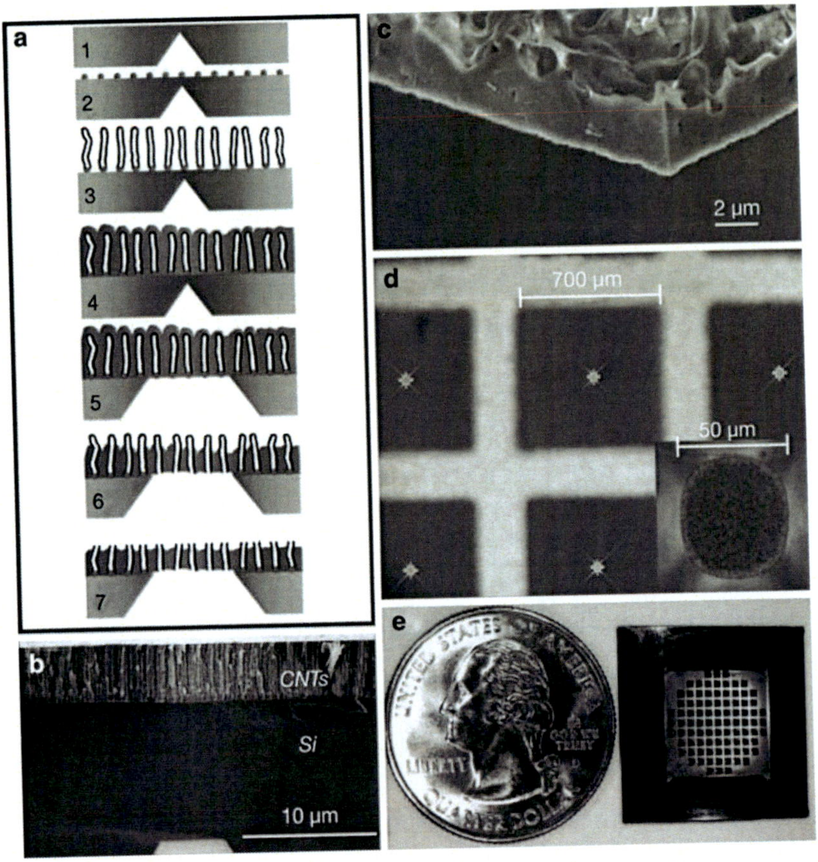

Fig. 8.1 (**a**) Schematic of the fabrication process. *Step 1*: microscale pit formation (by KOH etching). *Step 2*: catalyst deposition/annealing. *Step 3*: nanotube growth. *Step 4*: gas filling with low-pressure chemical vapor-deposited Si_3N_4. *Step 5*: membrane area definition (by XeF_2 isotropic Si etching). *Step 6*: silicon nitride etch to expose nanotubes and remove catalyst nano-particles (by Ar ion milling); the membrane is still impermeable at this step. *Step 7*: nanotube uncapping (reactive ion etching); the membrane begins to exhibit gas permeability at this step. (**b**) SEM cross-section of the as-grown DWNT's (CNTs). (**c**) SEM cross-section of the membrane, illustrating the excellent gap filling by silicon nitride. (**d**) Photograph of the open membrane areas: insert shows a close-up of one membrane. (**e**) Photograph of the membrane chip that contains 89 open windows; each window is 50 μm in diameter. (From [3])

the CNT membrane occurs only through the inner pores of the CNTs spanning the membrane.

The results of the gas permeation experiments are as follows. Since the Knudsen number, λ/d, where λ is the mean free path and d is the pore diameter, is 10–70 for the membrane and for the tested gases, the gas flow through the membrane pore should be in Knudsen flow regime. The flux through the membrane can then be calculated from the pore size, pore length and the transmembrane pressure difference. Surprisingly, the experimental permeation data revealed that they are at least

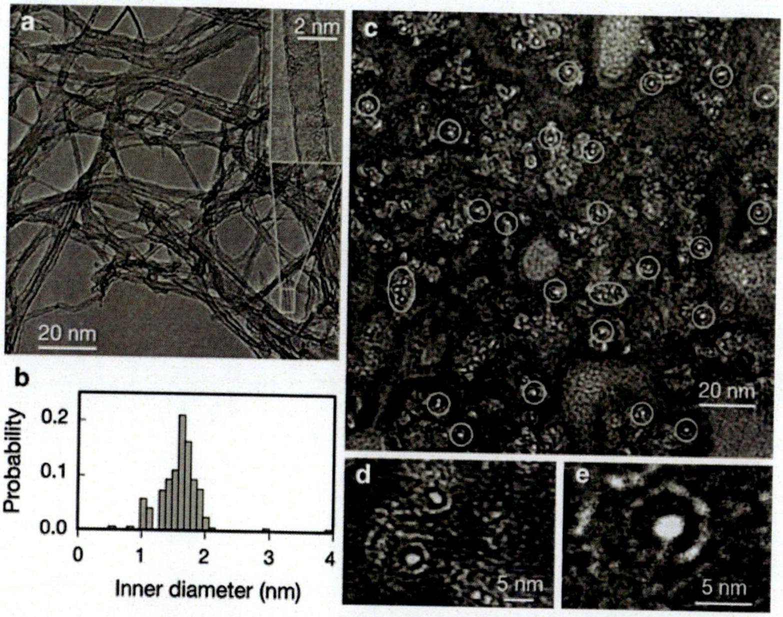

Fig. 8.2 (**a**) TEM images of as-grown nanotubes, prepared by removing them from the silicon substrate and dispersing them in DMF. The majority of the CNTs are double-walled, as indicated in the high resolution insert. (**b**) Pore size distribution, derived from TEM measurements of the inner diameter of 391 individual CNTs, reveals an average pore size of 1.6 nm. The average outer diameter of these DWNTs is estimated to be 2.3 nm. (**c**–**e**) Plan-view TEM images of CNT membrane taken with the beam parallel to the nanotube axis. In **c** the nanotube membrane shows continuous nitride coating on the scales examined in this image (ca. 0.2 by 0.2 μm). No microcracks or microvoids can be seen. The bright white spots (*circled*) are CNT pores, which can be identified by the surrounding ring-shaped coating of silicone nitride. For clarity, not all visible nanotubes were circled. The density of CNTs is measured to be ca 2.5×10^{11} cm^{-2} from several similar TEM images. In **d** and **e**, HRTEM images of selected areas from **c** show conformal coating of silicon nitride. The bright white spots in the images have the same inner diameter as the CNTs. (From [3])

one or two orders of magnitude higher than the calculated values. Table 8.1 shows the range of the enhancement over the calculated value based on Knudsen model. On the other hand, for the polycarbonate membrane the experimental and calculated values agreed in an order of magnitude. The observed enhancement over the

Table 8.1 The pore structure of the carbon nanotube membrane and the enhancement of flows over the calculated values based on Knudsen and hydrodynamic flow model

Membrane	Pore diameter (nm)	Pore density (cm^{-2})	Thickness (μm)	Enhancement over Knudsen model	Enhancement over hydrodynamic model	Minimum slip length
DWNT1	1.3–2.0	0.25×10^{12}	2.0	40–120	1500–8400	380–1400
DWNT2	1.3–2.0	0.25×10^{12}	3.0	20–80	680–3800	170–600
DWNT3	1.3–2.0	0.25×10^{12}	2.8	16–60	560–3100	140–500
Polycarbonate	15	6×10^{8}	6.0	2.1	3.7	5.1

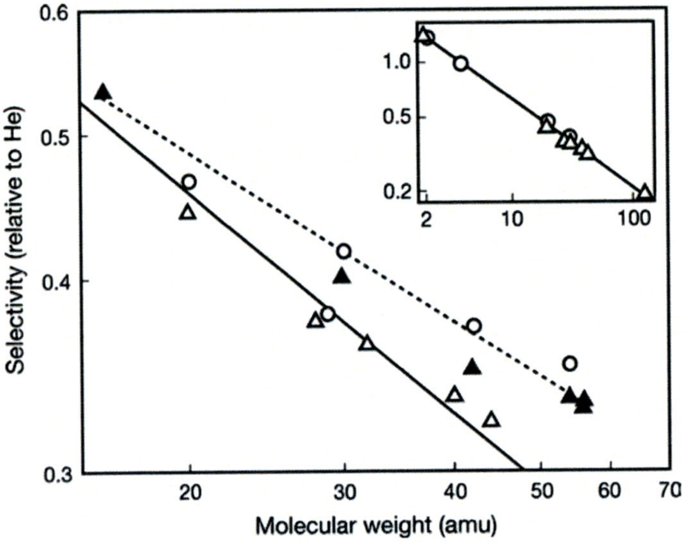

Fig. 8.3 Gas selectivity (indicated as permeability relative to He) data for sub 2 nm DWNT (*triangles*) and MWNT (*circles*) membranes. Open symbols denote nonhydrocarbon gases (H_2, He, Ne, N_2, O_2, Ar, CO_2, Xe); *solid symbols* denote hydrocarbon gases (CH_4, C_2H_6, C_3H_6, C_4H_6, C_4H_8). This *solid line* is a power-law fit of the nonhydrocarbon gas selectivity data, showing a scaling predicted by the Knudsen diffusion model (exponent of -0.49 ± 0.01). The *dashed line* is power-law fit of hydrocarbon gas data, showing a deviation from the Knudsen model (exponent of -0.37 ± 0.02). The insert shows the full mass range of the nonhydrocarbon gas data, again illustrating agreement with the Knudsen model scaling. (From [3])

Knudsen flow is most likely due to the smoothness of the nanotube surface. This was predicted by the molecular dynamics simulation of the gas flow through SWNT [4–7]. In atomically smooth pores, the gas-wall collision may change from diffuse (Knudsen model) to a combination of specular and diffuse collisions, leading to faster gas transport. Regarding the selectivity between gases, the experimental data of the nonhydrocarbon gases (open symbols on Fig. 8.3) followed the inverse square root of molecular weight rule applicable for the molecular flow regime, including the diffusion flow regime. The selectivity of hydrocarbon over helium was above the line corresponding to the inverse square root relationship, probably due to the preferential interaction of hydrocarbons with the carbon nanotubes' pore walls. The hydrocarbon transport enhancement most likely occurs by the surface diffusion or a solubility diffusivity mechanism.

Regarding the transport of the liquid water, the flow rate observed by the experiment can not be explained by the continuum flow models. The experimental permeation is more than three orders of magnitude faster than the value predicted by the Poiseuille equation. In such a small pore whose diameter is about seven water molecules, the velocity profile, inherent for the continuum theory, seems difficult to apply. The experimentally observed water flux compares well with the value predicted by the molecular dynamic (MD) model [2]. The simulation predicts the flow of 12 water molecules per 1 nm^2 of cross-sectional area per ns. The experimental data, when extrapolated to the pressure drop used in the simulation, correspond to 10–40 water molecules per nm^2 per ns. According to the simulation model, the high water flow rate in the nanoscale pore is due to the formation of water-wires in the confined space.

However, the small difference in the pore diameter may influence the mode of water transport in a narrow pore channel considerably. Therefore, it is difficult to conclude if the flow pattern proposed by the molecular dynamic simulation is responsible for the water flow enhancement or the flow enhancement is due to the presence of frictionless surface.

Kim et al. presented a simple, fast and practical method to vertically align CNTs on a porous support using a combination of self-assembly and filtration method. The authors claimed that this method can be easily scaled up to large surface areas that may allow the fabrication of membranes for practical applications [8].

CNTs were first amine-functionalized and dispersed in tetrahydrofuran (THF) (Step 1 in Fig. 8.4a) and then the solution was filtered through a hydrophobic polytetrafluoroethylene (PTFE) filter of 0.2 μm pore size (Step 2 in Fig. 8.4a). Then, the CNT/PTFE membrane was spin-coated with a dilute polysulfone (PS) solution (Step 3 in Fig. 8.4a). Figure 8.4b shows the SEM picture of the CNT on top of the PTFE support after the step 2. It is shown that most of the fibers are standing although some are not aligned fully to the vertical direction to the support filter surface. The alignment is supposed to be caused by the shear force of the solvent flowing through the pore of the PTFE support in combination with the long range electrical repulsive forces working between the CNTs and between the CNT and the support filter surface. Figure 8.4c shows that the dilute PS solution penetrates into and fill the space between the vertically oriented carbon nanotubes. As well, for most of the CNTs the upper end of the nanotubes are slightly above the surface of the spin coated polymer. Thickness of the CNT/polymer layer is about 600 nm. The composite PS/CNT/PTFE membrane was further coated in some cases with polydimethylsiloxane (PDMS). The structure is shown in Fig. 8.4d. Cross-sectional image of the CNTs embedded in the PS matrix was taken by HRTEM, as shown in Fig. 8.5a. From the number of the bright spots, which represent the carbon nanofiber membranes, the pore densities were evaluated to be $(7.0 \pm 1.75) \times 10^{10}/\text{cm}^2$. In Fig. 8.5b, a bundle of SWNTs (upper part) and an individual SWNT (lower part), each encapsulated in the surrounding graphite sheet, are shown. The diameter of the CNT bundle is 4 nm. Figure 8.5c shows the individual SWNT encapsulated by an additional graphite layer. The diameter of the individual SWNT is ca 1.5 nm.

The gas permeation of the prepared composite membrane was tested using He as the permeant. As shown in Fig. 8.6a, the He permeance was independent of the pressure, which shows that the gas transport through the membrane pore is governed by the Knudsen flow. This is an evidence that there is no viscous flow occurring in the membrane pores and the membranes do not have any large pinholes through which viscous flow would occur. The results displayed in Fig. 8.6b demonstrate that the permeance vs. gas molecular weight follow an inverse square root rule, which is also the evidence that the gas transport in the pore is occurring

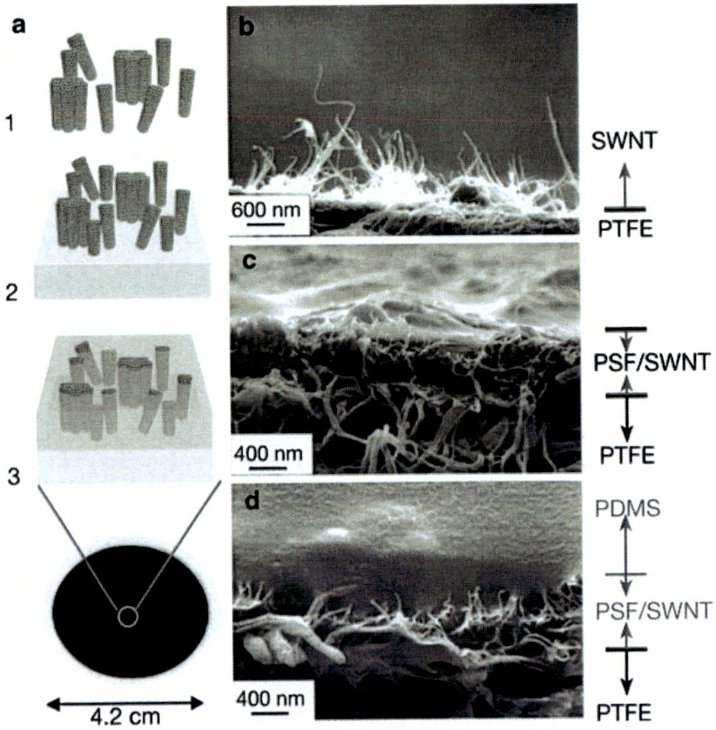

Fig. 8.4 CNT nanocomposite membrane process. (**a**) Schematic membrane fabrication process. *Step 1*: The functionalized CNTs are dispersed in THF solution. *Step 2*: The CNTs/THF solution is filtered through 0.2 μm pore size hydrophobic polytetrafluoroethylene (PTFE) membrane filter. Step 3: The CNTs/PTFE membrane is spin coated with a dilute PS solution. Some nanotube tips are embedded in polymer matrix. (**b**) Side-view SEM image of CNTs standing vertically on a membrane filter. (**c**) Side–view SEM image of aligned nanotube/PS nanocomposite membrane after spin-coating. Polymer coating is so thin that some CNT tips are exposed on top of the surface. (**d**) Side-view SEM image of aligned nanotube/PS/PDMS composite membrane with a protective PDMS coating of 4 μm. (From [8])

by the Knudsen diffusion mechanism. Looking at Fig. 8.5b, the solid line shows the theoretical value of the Knudsen flow. The composite CNT/PS showed, while maintaining the inverse square root of molecular weight rule, nearly an order of magnitude higher permeability than the theoretical values. The enhancement of the permeability is in accordance with the earlier work by Holt et al. [3]. On the other hand, when the composite CNT/PS membrane is coated with PDMS, the permeability data are below the theoretical Knudsen values. This is because a resistance from the PDMS layer was added. Figure 8.6c shows the single gas selectivities for gas/He gas pairs. Again, the data largely satisfy the inverse square root molecular weight rule presented by the solid line. The exceptionally high values of selectivity for CO_2/He gas pair, particularly for the PDMS coated membrane is due to the high solubility of CO_2 into the PDMS layer.

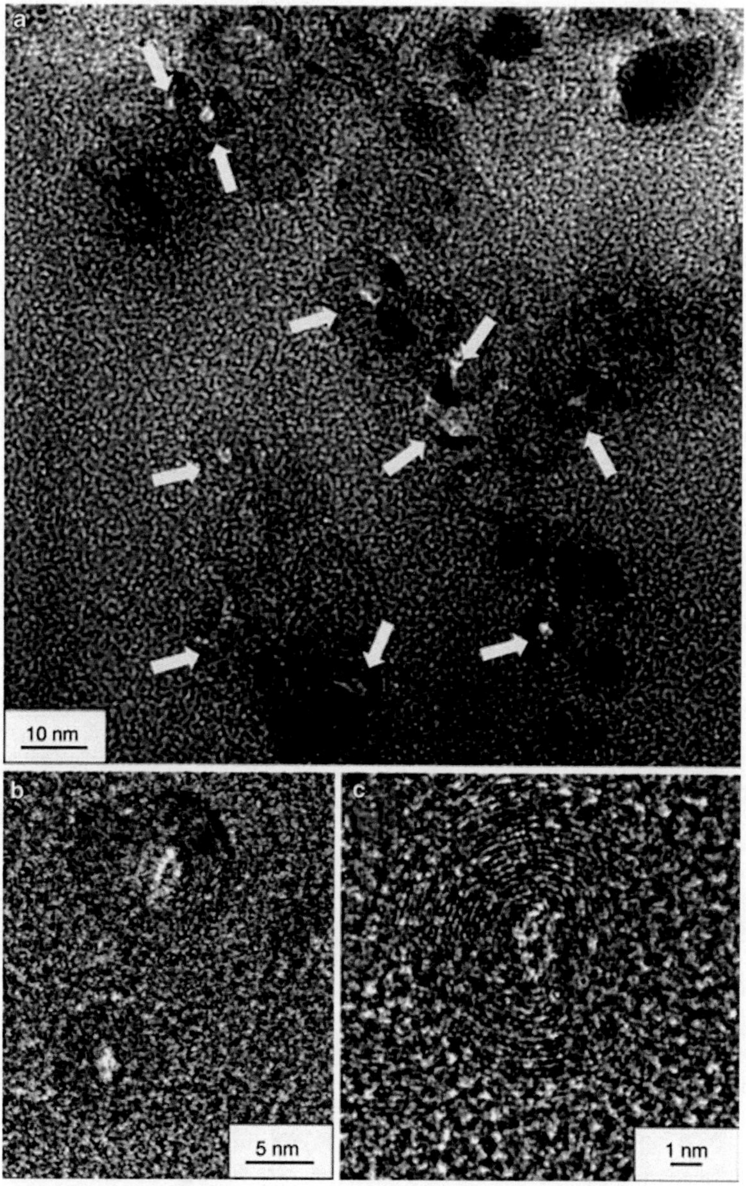

Fig. 8.5 HRTEM images in plan-view orientation of CNTs in a polymer mixture. The TEM specimen has been coated with a thin amorphous carbon film prior to the TEM experiment to prevent specimen from charging. In **a**, bright-white spots (indicated by *arrows*) represent open nanotube pores. (**b**) and (**c**) HRTEM images at higher magnification reveal the structure of these pores. In **b**, an encapsulated SWNT bundle is shown in the upper part of the image, and an encapsulated individual SWNT is in the lower part of the image. This SWNT bundle has an inner diameter of 4 nm. (**c**) The individual SWNT with an inner diameter of ca 1.5 nm is clearly encapsulated by additional graphite layers. The area density of SWNTs was measured to be ca ($7.0 \pm 1.75 \times 10^{10}$/cm^2) from several HRTEM images. (From [8])

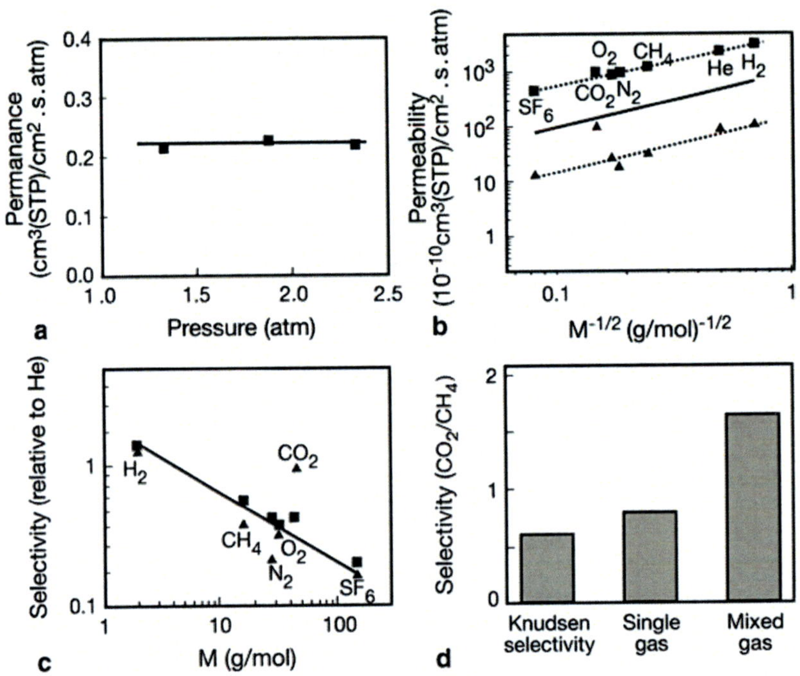

Fig. 8.6 Gas transport properties of CNT nanocomposite membrane. Gas transport properties of CNT/PS/PDMS membrane (*triangle*). CNTs/PS membranes (*square*), and Knudsen diffusion model (*solid line*). (**a**) Effect of the pressure drop on the permeance of helium through CNTs/PS membrane. (**b**) Single-gas permeability as a function of the inverse square root of the molecular weight of the penetrant. (**c**) Single gas selectivity with respect to He calculated from singe-gas permeability data. (**d**) Mixed-gas selectivity (CO_2/CH_4) of CNTs/PS membrane. The composition of gas mixture was CO_2:CH_4 = 1:1. The feed pressure was 50 psi, and the pressure differential across the membrane was maintained by drawing a vacuum on the permeate side. Operating temperature was maintained at 308 K. (From [8])

The permeation of gas mixtures through porous membranes provides more stringent test of the transport mechanism than the single gas permeation. If the transport occurs by the Knudsen diffusion, the selectivity based on the single gas experiments and that for the gas mixture should be equal.

Separation of binary gas mixtures (CO_2/CH_4) was then attempted both experimentally and theoretically. In the theoretical calculation, atomistic simulations were made for the transport of CO_2/CH_4 mixtures in SWNTs. By separately computing adsorption isotherms and loading-dependent transport diffusion coefficients, gas transport was calculated. A broad range of feed pressures, compositions and transmembrane pressure drops were explored in the calculation and CO_2/CH_4 selectivity ranged from 2 to 10. The experimental and the theoretically calculated results are compared in Fig. 8.7 for a feed pressure of 50 psig and the temperature of 308 K. Vacuum condition was maintained on the permeate side. Obviously, the predicted values were far greater than the experimental ones. However, both prediction and experiment show selectivity values higher than those obtained from the Knudsen

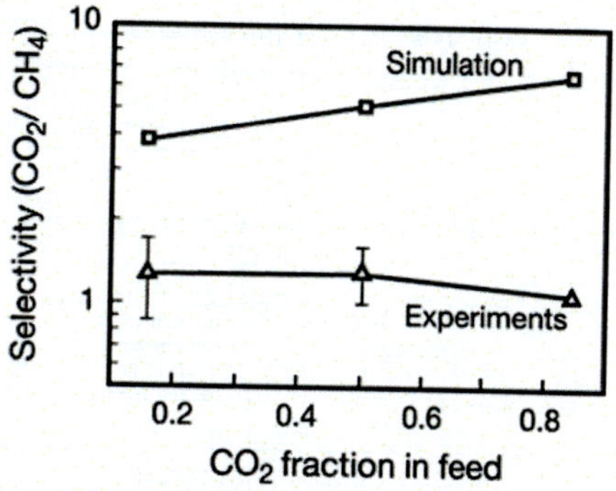

Fig. 8.7 Mixed gas selectivity CO_2/CH_4 of CNTs/PS membrane at different feed compositions. Both experimental and simulation assumed a feed pressure of 50 psi and a temperature of 308 K. The pressure difference across membrane was maintained by drawing a vacuum on the permeate side. (From [8])

mechanism, since according to the latter mechanism the selectivity should be less than unity. Molecular diffusion in SWTNs differs from the Knudsen diffusion because both CO_2 and CH_4 molecules adsorb on the CNT wall. Individual collisions between molecules and the CNT wall do not completely theorize the molecular momentum. More specifically, CO_2 is more strongly adsorbed to the CNT wall, which excludes the adsorption of CH_4 molecules. Furthermore, the slow diffusion of CO_2 reduces the mobility of CH_4.

The discrepancy between the experimental and theoretical values may have occurred for the following reasons: (1) The calculation is based on (10, 10) CNTs. The actual pore sizes may be larger than the assumed pore size. (2) The assumption was made that the intrapore diffusion dominates the membrane transport. In reality, the transport at the pore entrance and the pore exit may also contribute. Since the effect of the transport at the pore entrance and the pore exit is stronger for more strongly adsorbed molecules, this may be the reason why the experimental selectivity is lower than the calculated values. (3) Functionalization of the pore entrance is not considered in the calculation. In fact, the membrane used for the experiment has the property of zwitter ion at the pore mouth.

Briefly, the authors of this paper claim that they have presented a method to facilitate the orientation of CNTs on the porous support (PTFE) and the method can easily be adopted to the large-scale membrane production. The PS/CNT/PTFE composite membrane showed the same fast gas transport observed by Holt et al. Their work also showed for the first time the transport of the binary gas mixtures through the CNTs based membrane.

Another paper in which the potential of large scale application of CNT membranes has been done by Mi et al. [9]. Unlike zeolite membrane it is impossible to synthesize CNT membranes with the nanotubes intergrown together without inter-tube gaps due to the structural features of SWNT and MWNT and their growth mechanism. Hence a special device is necessary to fill the gap between the nanotubes.

The macroporous α-alumina support disks (20 mm in diameter and 2 mm in thickness, with average pore diameter of 300 nm and porosity of 27%) were prepared by pressing α-alumina powder (particle diameter≥2 μm) at a pressure of 242 MPa, followed by sintering in air at 1280 °C for 2 h. The α-alumina disks were polished on one surface with a sand paper and cleaned in an ultrasonic acetone bath for 30 min before being used for CNT growth. An atmospheric CVD reactor made of a quartz tube with an alumina substrate and a catalyst precursor zone was heated by a two–zone furnace. The substrate in the deposition zone was heated to a desired CVD temperature, typically 800 °C, at a heating rate of 10 °C/min. About 0.1 g of ferrocene was placed in the catalyst precursor zone (800 °C) ahead of the substrate. Carrier gas at the flow rate of 100 cm^3/min was passed to carry the catalyst precursor to the substrate for about 2 min. Ferrocene vapor was decomposed and iron nano-particles were formed and deposited on the alumina support surface. A mixture of acetylene and nitrogen (1:10) was introduced to the CVD reactor at a flow rate of 10 cm^3/min for about 10 min. Black carbon was formed on the alumina surface. The reactor was cooled down to room temperature under nitrogen. The inter-tube gaps of the CNT array on the alumina support was filled with 30 wt% solution of polystyrene (PSt) in toluene by spin coating. After coating, the CNT membrane was dried at 60 °C for 3 days. The PSt overlayer was removed by sanding with a 2,000 mesh sand paper. The sanding was continued until the black CNT appeared from under the white PSt overlayer. The composite membrane (CNTs + support layer) was immersed in a concentrated nitric acid (68%) and refluxed for about 1 day to ensure complete removal of the catalyst particles on the bottom of CNT. Figure 8.8 shows the cross-sectional image of a porous alumina disk coated with CNT array. From Fig. 8.8a CNTs are about 10 μm in length grown vertically from the surface of the α-alumina disk. Figure 8.8b is a magnified view of CNT. They are more or less parallel to each other. Figure 8.8c shows the structure of a single CNT. The CNT is of MWNT with a pore diameter of about 6.3 nm. Figure 8.9 shows the SEM image after filling the iner-tube space with PSt. It shows that the CNT layer filled with PSt is about 20 μm in thickness (Fig. 8.9a). PSt layer did not penetrate into the support alumina layer. CNTs are not visible from the top (Fig. 8.9b) because it is covered by PSt overlayer. The cross-sectional and top view of the composite membrane after sanding off the PSt over-layer is shown in Fig. 8.10a, b. The thickness of the CNT layer filled with PSt is now reduced to 10 μm. The top layer view clearly shows the openings of the CNTs. Figure 8.10c shows the end section of a CNT removed from the membrane after sanding and acid treatment.

Table 8.2 shows the permeance data obtained at various stages of CNT membrane fabrication. It should be noted that the membrane after PSt filling is practically gas tight, meaning the inter-tubular space was filled completely without defect. Helium permeance of the finally obtained CNT membrane (after acid treatment) has permeance two orders of magnitude lower than the porous alumina support, meaning 99% of resistance for the gas permeation comes from the CNT layer. The area density of the CNTs was estimated from Fig. 8.8b to be 1.87×10^9 CNTs/cm^2. Figure 8.10b gave a similar value of 1.94×10^9. Tortuosity factor was also estimated as 1.26. Membrane thickness is 10 μm as already mentioned.

The permeation data for helium, hydrogen and nitrogen gas was analyzed by a transport theory in which the Knudsen and viscous flow are linearly combined by $F=F_K+\beta P_m$. F is the total permeance, F_K is the permeance due to the Knudsen flow and βP_m is the permeance due to viscous flow, which increases linearly with the mean transmembrane pressure P_m. The permeance of the porous alumina support showed a linear dependence on the mean transmembrane pressure due to the large

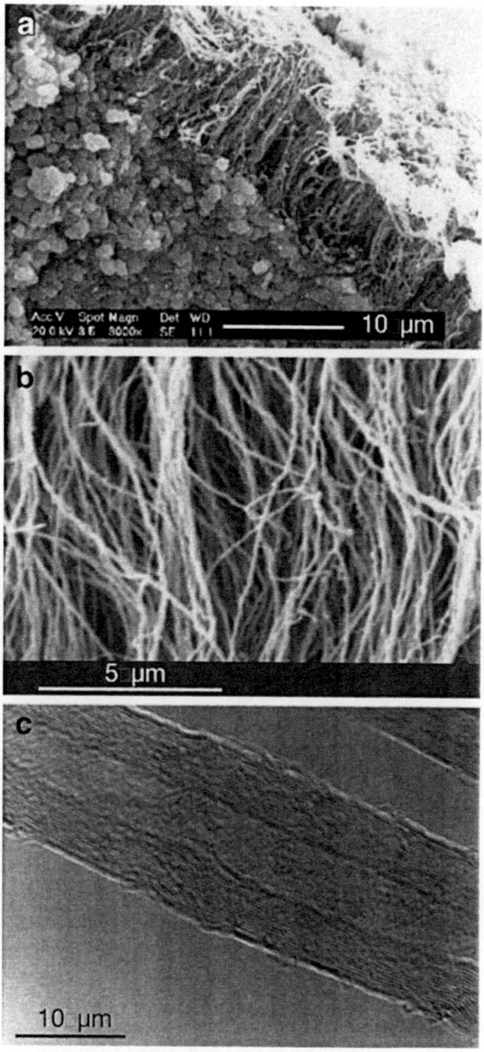

Fig. 8.8 SEM cross-sectional image of macroporous alumina disk coated with aligned CNT array; (**a**) enlarged view of CNT array, (**b**) HRTEM image of a CNT grown on alumina, (**c**) image of a MWNT

pore size that allows the viscous flow. On the other hand, permeance for the CNT composite membrane did not depend on the transmembrane mean pressure, meaning that the Knudsen flow is the dominant mechanism for the gas transport. For the Knudsen flow, the permeance can be written as

$$F_K = \left(\frac{\varepsilon}{\tau}\right)\left(\frac{1}{L}\right)\left(\frac{1}{RT}\right)D. \tag{8.1}$$

If the Knudsen mechanism truly governs the transport in the pore diffusivity D in Eq. (8.1) should be replaced by Knudsen diffusivity

$$D_K = \frac{2}{3}\left(\frac{8RT}{\pi M}\right)^{1/2} r. \tag{8.2}$$

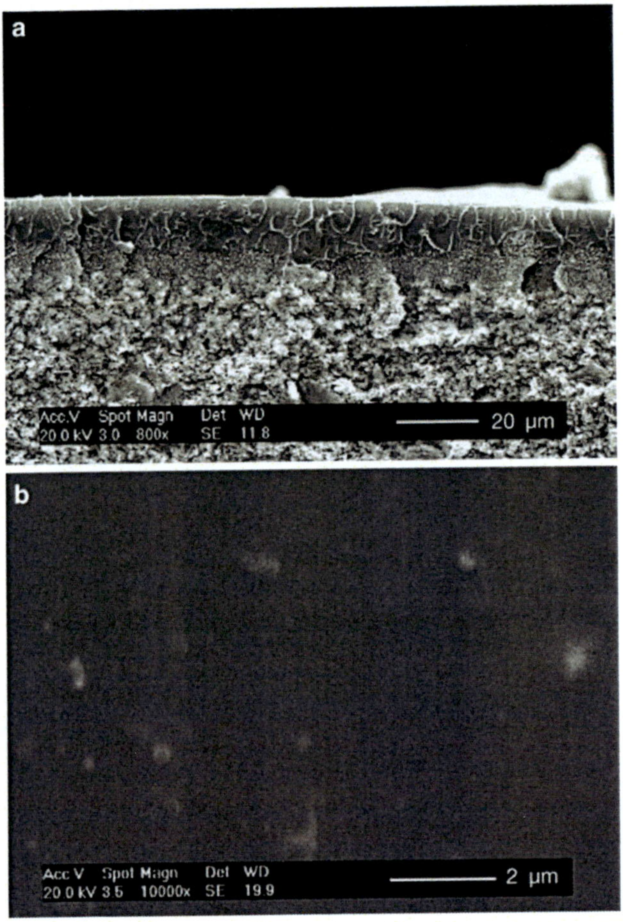

Fig. 8.9 SEM image of supported CNT membrane with inter-tube gaps filled by PSt; (**a**) cross-section, (**b**) the top surface

The gas permeance data were analyzed by the above two equations and the results are given in Table 8.3. As can be seen, the experimental diffusivity data is about 4 times as large as the calculated Knudsen diffusivity.

Finally, in Table 8.4, enhancement of diffusivity obtained by different researchers is summarized.

Clearly, the enhancement is observed as the pore diameter decreases. According to the Knudsen mechanism, the gas transport in a pore is governed by the collision between the gas molecule and the pore wall. For small CNT pores with atomically smooth pore wall, the gas-wall interaction is governed by the combination of specular and diffuse collisions. This explains the larger enhancement of diffusivity over the Knudsen diffusivity for the smaller pores.

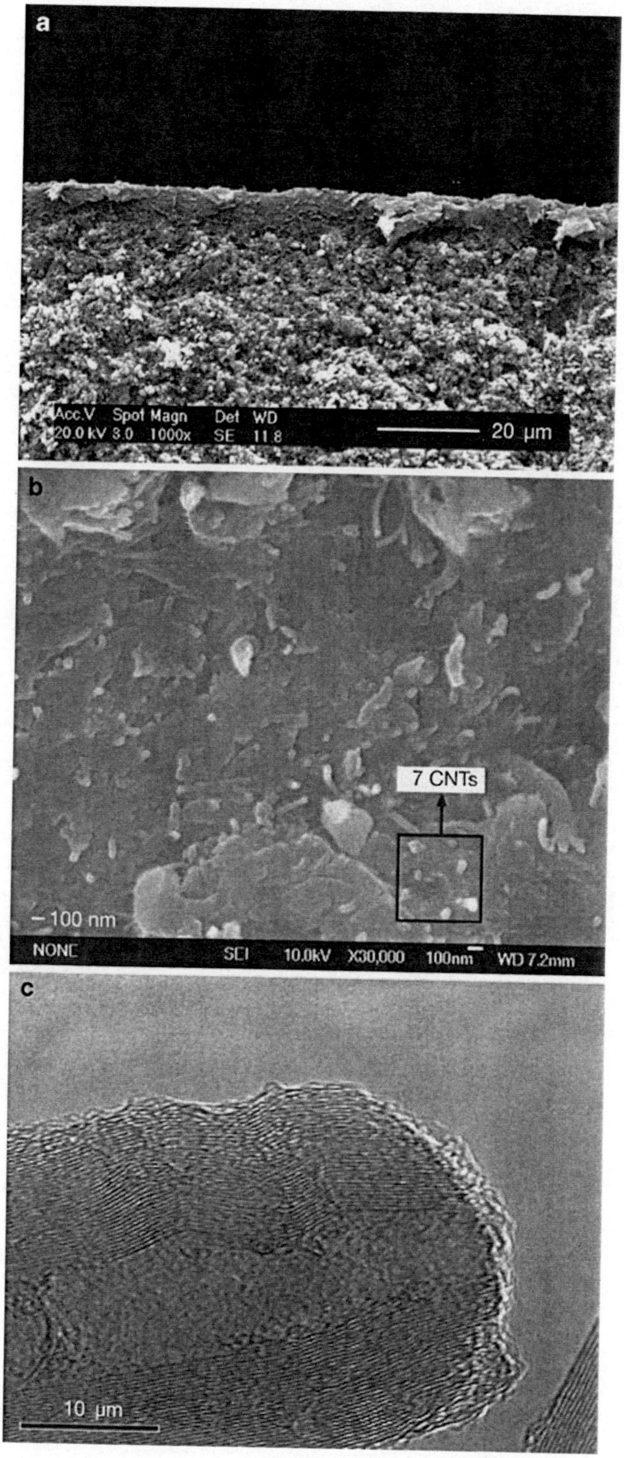

Fig. 8.10 SEM and field emission SEM images of supported CNT membrane with inter-tube gaps filled by PSt and after excess PSt overlayer was removed by mechanical sanding; (**a**) cross-section, (**b**) top surface, (**c**) HRTEM image of the end of a CNT

Table 8.2 Helium permeance (30 °C, and feed pressure 6×10^5 Pa) for CNT membrane at various stages

Preparation stage	Permeance (mol/m^2 s Pa)
Porous alumina support	3.62×10^{-6}
Support coated with CNT arrays	3.11×10^{-6}
After filling with PSt	$< 1.11 \times 10^{-12}$
After sanding	5.23×10^{-9}
After acid treatment	6.92×10^{-8}

Table 8.3 Comparison of measured and theoretical Knudsen diffusivity in the CNT membrane (at 30 °C)

Gas	F, permeance ($\times 10^8$ mol/m^2 s Pa)	D_K[a] (cm^2/s)	D[b] (cm^2/s)	D/D_K
H_2	12.2	0.0376	0.150	4.0
He	6.92	0.0266	0.085	3.1
N_2	3.21	0.0100	0.039	3.9

[a] Calculated from Eq. (8.2) using the pore diameter of 6.3 nm
[b] Calculated assuming $F = F_K$ for the composite CNT membrane

Table 8.4 Enhancement in diffusivity over Knudsen diffusivity for CNT membranes of different pore sizes

	Hinds et al. [10]	Mi et al. [9]	Holt et al. [3]
CNT ore diameter (nm)	7.5	6.5	1.6
Knudsen number	~1	1.2–2.2	10–70
Enhancement over Knudsen diffusivity	~1	~4	16–120

8.2 Molecular Dynamics Simulation

As shown in the foregoing section, the experimental results from CNT membranes have revealed orders of magnitude higher flow rates as compared to the conventional Knudsen flow (gas) or viscous flow (liquid) mechanism in the pores of the CNT membranes. Indeed, this high flow rate through the pore of the CNTs was anticipated by the theoretical calculation based on molecular dynamics (MD) even before the experimental data were obtained. Hence, a brief introduction to MD simulation is made in this section. A paper of Furukawa and Nitta is cited first to understand the nonequilibrium molecular dynamics (NEMD) simulation semi-quantitatively, since, even though the paper deals with various pore shapes, complicated simulation procedure is described clearly. Furukawa and Nitta's work is then followed by Wu et al.'s work where they attempted to calculate the performance of nanoporous carbon (NPC) membrane in the separation of H_2/CO mixture based on an approach similar to Furukawa and Nitta. Chen and Sholl also attempted to predict the selectivity in CH_4/H_2 separation by carbon nanotubes by equilibrium molecular dynamics (EMD). Their calculation shows theoretically that the diffusivity of the CNT membrane should be orders of magnitude higher than the zeolite membrane of a

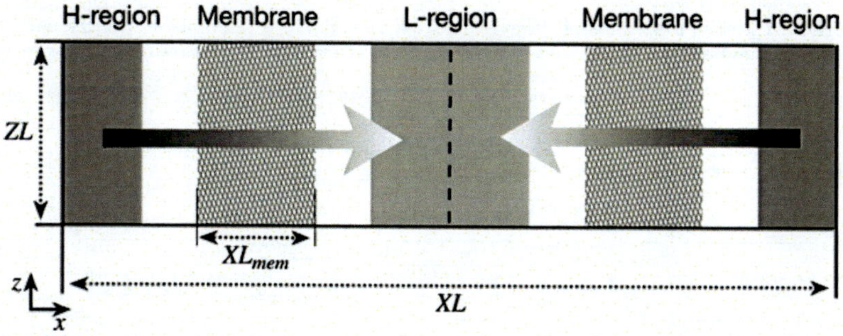

Fig. 8.11 Schematic diagram of a simulation cell for the μVT-NEMD method. (From [11])

similar structure. Finally, the work of Verweij et al. demonstrates that the unusually high transport rate in the CNT membrane can be explained by their new conceptual models.

Furukawa and Nitta [11] applied the MD simulations for the permeation of pure and mixed gases across carbon membranes with three different pore shapes: the diamond pore (DP), zigzag path (ZP) and straight path (SP), each composed of micro-graphite crystalline. The gases, CH_4 and C_2H_6, were used for the simulation. Even though the real carbon membrane may be composed of parallel stacked carbon sheets (microcrystalline) and amorphous carbon domains, the study on the gas permeation through non-slit shaped pores is very limited. This study includes both slit and non-slit shaped pores. Figure 8.11 presents schematically the simulation cells for the μ*VT*-NEMD method used in their work. There are two boxes placed in mirror image in *x*-direction. Each symmetric box has three regions. Two are density control; H-region (high density) and L-region (low density) and one is free of control M-region which is placed between the H- and L- region. For each simulation, the density in the H-region, ρ_H, is maintained to be that of the feed gas and the density in the L-region is maintained at zero, corresponding to the vacuum. The difference in the gas density between the H- and L- region is the driving force for the gas permeation through the M-region which represents the membrane. The transition and rotational velocities are given to each inserted gas molecules randomly based on the Gaussian distribution around an average velocity corresponding to the specified temperature. Molecules spontaneously move from H- to L- region via leap-frog algorithm and a non-equilibrium steady state is obtained at the M-region. The MD time step was 2.0 fs. The first 1,000,000 steps were discarded and the next 15,000,000 steps were collected for the ensemble averaging. The spherical Lennard-Jones (LJ) interaction was used for CH_4 molecule while two interaction sites, each representing one carbon atom, were considered for the C_2H_6 molecule. All interactions between the gas molecules and between the gas molecule and the carbon atoms of the membrane were expressed by the 12–6 LJ potential function

$$\phi_{ij}(r) = 4\varepsilon_{ij}\left[\left(\frac{\sigma_{ij}}{r}\right)^{12} - \left(\frac{\sigma_{ij}}{r}\right)^{6}\right]. \tag{8.3}$$

Table 8.5 Lennard-Jones potential parameters

Site or atom	σ (nm)	ε (kJ/mol)	Bond length (nm)
CH_4	0.3730	1.230	
CH_3	0.3775	0.866	
C (graphite)	0.3400	0.233	
CH_3–CH_3			0.153

Where r is the distance between the two interaction sites i and j. ε_{ij} and σ_{ij} are the LJ interaction parameters for the ij pair. All the parameters involved in the LJ potential functions are listed in Table 8.5.

The cross-parameters for the pair ij are calculated by the Lorentz–Bertherot rule. The cut-off distance for the intermolecular interaction was set at 3.5 σ_{CH3}.

The simulation cell is divided into subcells. The density of the component k (either methane or ethane) in the Lth subcell is calculated by

$$\rho_k^{(L)} = \frac{\langle N_k{}^{(L)} \rangle}{V_{SC} N_A}, \quad (V_{SC} = XL_{SC} \times YL \times ZL) \tag{8.4}$$

Where superscript (L) denotes subcell number L, $N_k^{(L)}$ is the number of the molecules of the component k in the Lth subcell, V_{SC} is the volume of a subcell and XL_{SC} is the distance in the x-direction of a subcell. The permeation flux ($J_k^{(L)}$) of the component k in the Lth subcell is calculated by

$$J_k^{(L)} = \langle \rho_k \bar{u}_{x,k} \rangle^{(L)} \tag{8.5}$$

Where ρ_k is the instantaneous density of the component k and $\bar{u}_{x,k}$ the instantaneous average molecular velocity of the component k in the x-direction. In addition to the (non-equilibrium molecular dynamics) NEMD simulations, equilibrium simulations were also conducted. Equilibrium (between H- and M-region) densities of CH_4 and C_2H_6 were obtained throughout the three membrane elements by setting the densities of the L-region equal to that of the H-region.

Three different pore shapes were considered; i.e. diamond path (DP), a zigzag path (ZP) and a straight path (SP). Figure 8.12a–c shows the cross-sectional view of each pore shape.

DP (A) has two different pore mouths; one a large (pore a) and the other a small mouth (pore b). ZP (B) has zigzag shaped pores whose sizes (diameters) are all the same at the pore entry. SP (C) has straight pores which can be called slit-shaped pores. The minimum pore width (W_p) is set equal to 0.5 nm for all three types of pores. The membrane thickness, given in Fig. 8.12 as XL_{mem} was set equal to 6.3, 4.6 and 5.2 nm for DP, ZP and SP membranes. These values are based on the arrangement of micro-graphite crystallites (shown as a nodule in the figure) whose unit size was fixed at 2.7 × 3.4 × 1.3 nm for the DP and ZP membranes.

Figure 8.13a–c shows the potential surfaces for a CH_4 molecule for the three types of pore shapes, where the darker the shade, the deeper is the potential energy. In Fig. 8.13a, the potential energy near the pore wall is deeper than in the middle of

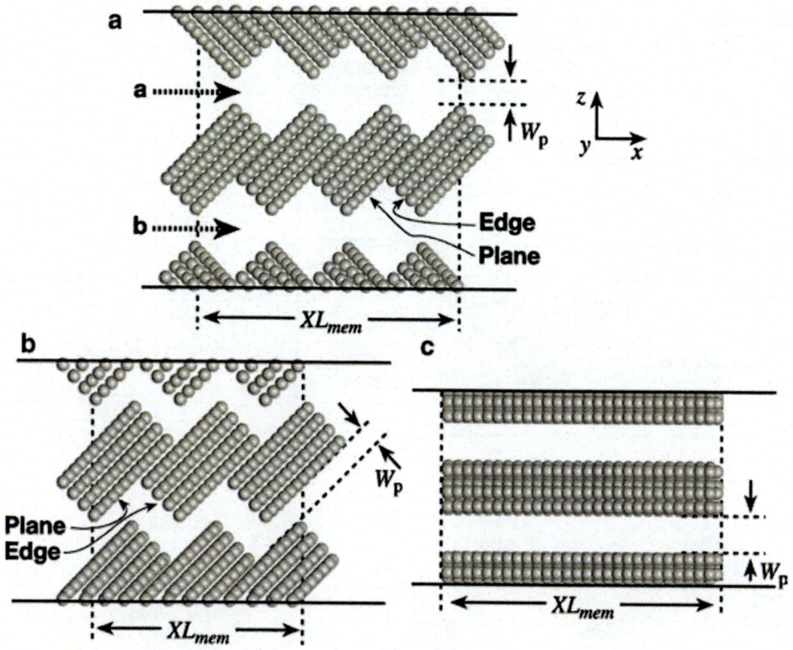

Fig. 8.12 Three membrane pore shapes; (**a**) *diamond* path (DP), (**b**) *zigzag* path (ZP), (**c**) *straight* path (SP). (From [11])

the pore. As a result, when the CH_4 molecule moves toward the x-direction, the molecule should overcome a potential barrier. The deepest potential energy in the DP cage is −15 kJ/mol while the deepest potential at the window is −6 kJ/mol. Hence the largest potential barrier is 9 kJ/mol.

In Fig. 8.13b, the potential energy near the surface of micro-graphite edge (see Fig. 8.12b) is shallower than the surface of the micro-graphite plane due to the lower density of the carbon atoms, which causes the potential barrier. The largest potential barrier is calculated to be 8 kJ/mol and slightly smaller than that of the DP membrane. In Fig. 8.13c, the potential energy in the pore is uniform and −16 kJ/mol. The potential barrier for the molecule to move from inside the pore to the outside (vacuum) of the pore is 16 kJ/mol.

The simulation was made under the following conditions. Either pure CH_4 gas, or pure C_2H_6 gas or equimolar mixture of CH_4 and C_2H_6 was considered for the simulation. Temperature was 300 K and pressure in H region was 0.5 MPa for pure gas. For the gas mixture the partial pressure of each component was set equal to 0.5 MPa.

Figure 8.14a–c show the snapshots for the mixed gas permeation. Methane molecules are shown by gray circles while ethane molecules are shown by black circles. One obvious thing is that more molecules are gathering at the micro-graphite planes (along the gray lines) while less molecules are gathering at the surface made by the micro-graphite edges (at the edges of the gray lines). It is also noted hat C_2H_6 molecules are more enriched in the pore than CH_4 molecules.

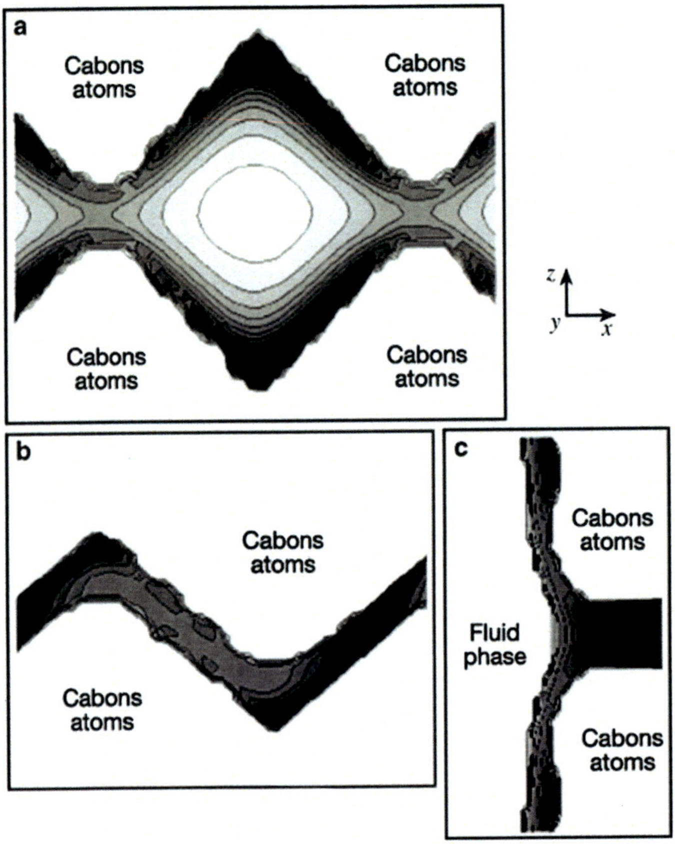

Fig. 8.13 Potential energy surfaces for CH_4 molecule; (**a**) DP, (**b**) ZP, (**c**) SP. (From [11])

Figure 8.15a, b displays the density profiles in the M-region (region shadowed gray) for (a) pure and (b) mixed gas permeation through the pore A of the DP carbon membrane (see Fig. 8.12a). There are three peaks that correspond to the deepest potential energy regions shown in Fig. 8.13a. The peaks for C_2H_6 (broken line) is higher than CH_4 (solid line) due to the higher adsorption of C_2H_6. The peak height decreases toward the moving direction of the gas because of the resistance for the gas permeation. In the case of the mix gas permeation, the peak height of CH_4 is much lower than that of the pure gas permeation. The peak height of C_2H_6 is almost unchanged. This is due to the competitive adsorption between CH_4 and C_2H_6 gases. Figure 8.16a, b show the density profiles of pore b of DP carbon membrane. Since the permeation resistance is greater in pore b, the peak heights decrease more quickly than in pore a.

Density profiles for ZP are depicted in Fig. 8.17a, b. The shape of the density profiles is more complicated than DP. Two density valleys inside the membrane (shadowed region) correspond to pore surfaces composed of the micrographite edges. The density of both CH_4 and C_2H_6 decreases suddenly in the last channel connecting to the membrane exit. Hence, molecules move to the external surface of the membrane with very little potential barrier.

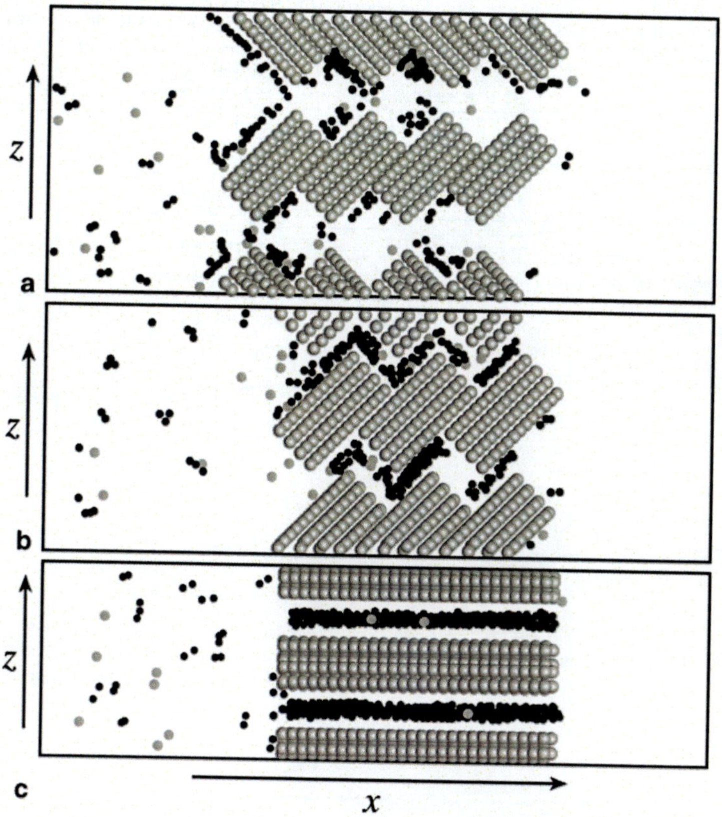

Fig. 8.14 Snapshots for the molar density of CH_4 (*gray*) and C_2H_4 (*black*) in the pore; (**a**) DP, (**b**) ZP, (**c**) SP (Feed gas composition $CH_4/C_2H_6 = 1/1$). (From [11])

Figure 8.18a, b presents the density profile in SP. The most distinguishable feature of SP is that the density profile is very flat inside the membrane. This is because the potential surface in the pore is very flat and the potential barrier for the gas permeation exists only at the membrane exit. In the mixed gas permeation, the density of CH_4 decreases toward the membrane exit while that of C_2H_6 increases. This observation can be explained by the two characteristics of the SP membrane, i.e. different potential barriers for CH_4 and C_2H_6 at the membrane exit and the high mobility of the molecules inside the membrane.

The simulation value of the permeation flux of the component k, J_k, is given by averaging $J_k^{(L)}$ which is calculated by Eq. (8.5), over the subcells in M-region.

The permeation flux of component k is usually given by

$$J_k = \frac{\hat{P}_k}{XL_{mem}}(P_{k,H} - P_{k,L}) \tag{8.6}$$

Where $\hat{P}_k$, $P_{k,H}$ and $P_{k,L}$ are the permeability, the pressure on the feed side and the pressure on the permeate side, respectively.

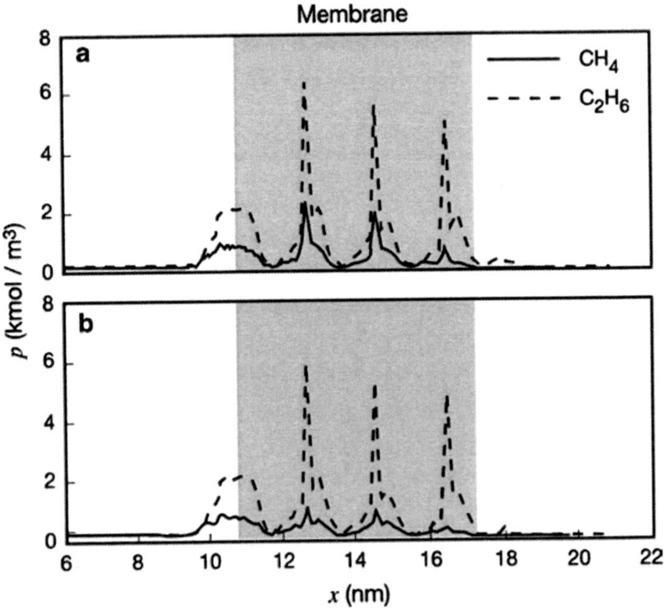

Fig. 8.15 Density profiles of CH_4 and C_2H_6 in the DP with the pore mouth A; (**a**) pure gas system, (**b**) mixed gas system ($\Delta P = 0.5$ MPa for pure gas, $\Delta P_k = 0.5$ MPa for mixed gas, temperature 300 K). (From [11])

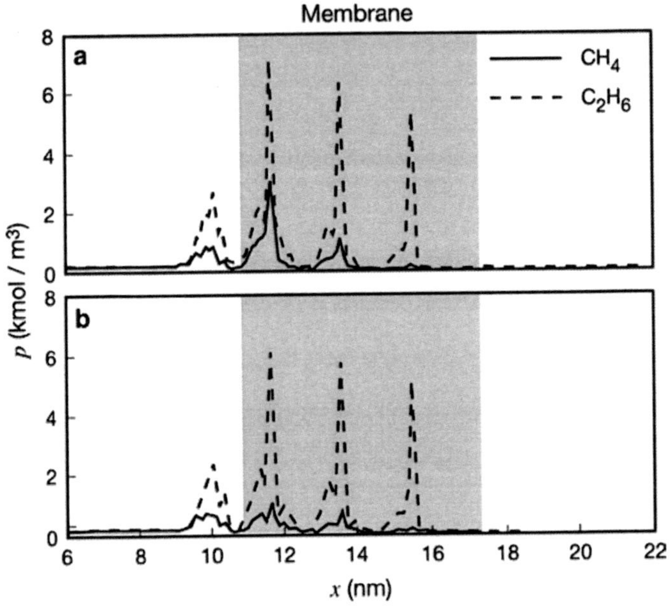

Fig. 8.16 Density profiles of CH_4 and C_2H_6 in the DP with the pore mouth B; (**a**) pure gas system, (**b**) mixed gas system ($\Delta P = 0.5$ MPa for pure gas, $\Delta P_k = 0.5$ MPa for mixed gas, temperature 300 K). (From [11])

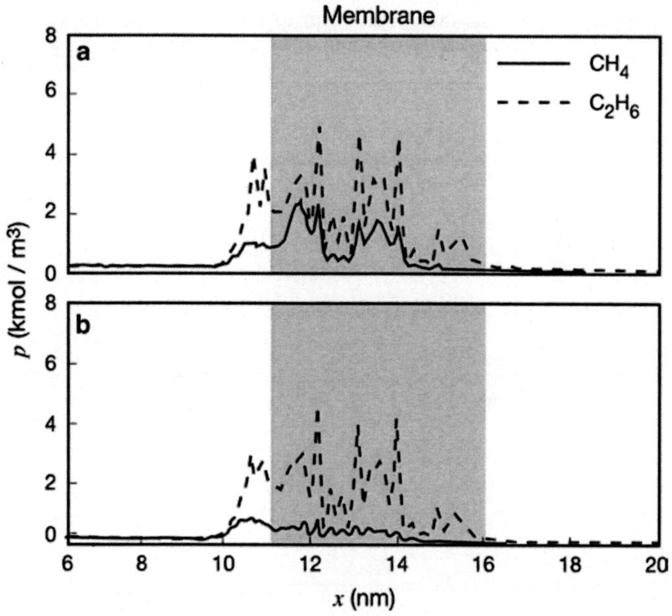

Fig. 8.17 Density profiles of CH_4 and C_2H_6 in the ZP; (**a**) pure gas system, (**b**) mixed gas system ($\Delta P = 0.5$ MPa for pure gas, $\Delta P_k = 0.5$ MPa for mixed gas, temperature 300 K). (From [11])

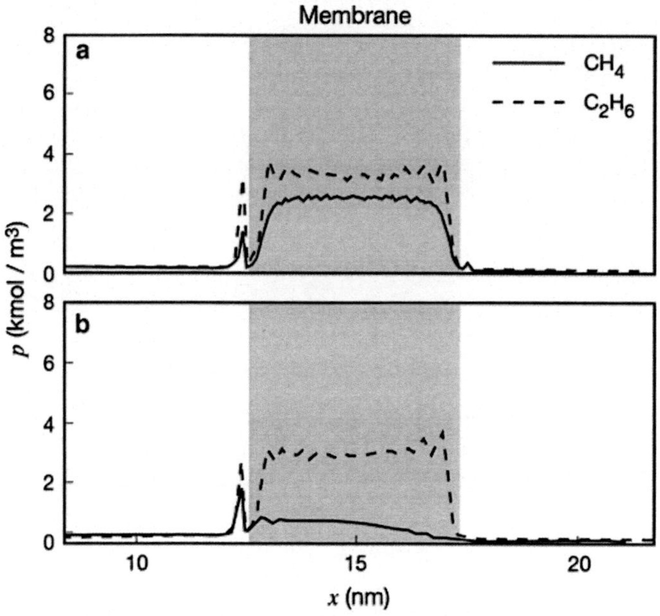

Fig. 8.18 Density profiles of CH_4 and C_2H_6 in the SP; (**a**) pure gas system, (**b**) mixed gas system ($\Delta P = 0.5$ MPa for pure gas, $\Delta P_k = 0.5$ MPa for mixed gas, temperature 300 K). (From [11])

An alternative expression for the permeation rate may be the permeation resistance, R_k, based on the difference of densities in the membrane $q_{k,H} - q_{k,L}$ as a driving force, i.e.

$$J_k = \frac{1}{R_k}(q_{k,H} - q_{k,L}) \tag{8.7}$$

Where $q_{k,H}$ and $q_{k,L}$ are the molar densities of the component k inside the membrane in equilibrium with the feed gas at $P_{k,H}$ and the permeate gas at $P_{k,L}$, respectively. q_k is calculated by

$$q_k = \frac{N_{k,mem}/N_A}{V_{mem}} \quad (V_{mem} = XL_{mem} \times YL \times ZL) \tag{8.8}$$

Where $N_{k,mem}$ is the number of molecules of component k inside the membrane, N_A the Avogadro number and V_{mem} is the membrane volume. The resistance R_k is related to the permeability, $\hat{P}_k$, using adsorption equilibrium constant $K_k\,(q_k/P_k)$ as

$$R_k = \frac{K_k XL_{mem}}{\hat{P}_k} \tag{8.9}$$

Table 8.6 summarizes the simulation results for J, q, R and $\hat{P}$ for pure gas permeation along with the ideal separation factor $(\hat{P}_2/\hat{P}_1)$. q for C_2H_6 is larger than CH_4 because of a stronger C_2H_6 adsorption. They are however almost equal for SP membrane. The order in q values is DP<ZP<SP since the potential energy becomes deeper in the same order. R values are always larger for C_2H_6, which is attributed to the heavier mass and the larger potential barrier for C_2H_6 than CH_4. Interestingly, the permeability ratio is more than unity for DP and ZP but it is less than unity for SP membrane. This occurs as a result of the interplay of equilibrium constant K and resistance R.

Table 8.7 shows J, q and R for the mixed gas permeation along with the separation factors. Two separation factors, α_{12} and $\alpha_{12}^{(eq)}$ are given. The former is defined as

$$\alpha_{21} \equiv \frac{J_2/J_1}{y_2/y_1} \tag{8.10}$$

Table 8.6 Permeation flux, J, equilibrium density, q, resistance, R, permeability, $\hat{P}$, and ideal separation factor $(\hat{P}_2/\hat{P}_1)$ for single gas permeation

Membrane	Gas	J, kmol/m² s	q, kmol/m³	R, s/m	$\hat{P}$, 10^{-12} mol m/m² s Pa	$\hat{P}_2/\hat{P}_1$
DP	CH_4	6.8	1.0	0.15	86	1.1
	C_2H_6	7.4	2.0	0.27	93	
ZP	CH_4	7.0	1.8	0.26	65	1.3
	C_2H_6	9.0	3.0	0.33	83	
SP	CH_4	16	3.2	0.21	160	0.63
	C_2H_6	10	3.5	0.35	100	

ΔP=0.5 MPa and T=300 K

Table 8.7 Permeation flux, *J*, equilibrium density, *q*, resistance, *R*, separation factor and the ratio of resistances for mixed gas (CH_4/C_2H_6 50/50) permeation

Membrane	Gas	*J*, kmol/m² s	*q*, kmol/m³	*R*, s/m	α_{21}	$A_{21}^{(eq)}$	R_2/R_1
DP	CH_4	4.4	0.62	0.14	1.5	3.2	2.1
	C_2H_6	6.6	1.9	0.29			
ZP	CH_4	3.7	0.56	0.15	2.4	4.3	2.0
	C_2H_6	8.8	2.6	0.30			
SP	CH_4	4.5	0.39	0.087	2.2	8.3	3.8
	C_2H_6	10	3.3	0.33			

$\Delta P = 0.5$ MPa and $T = 300$ K

The latter separation factor is defined as

$$\alpha_{21}^{(eq)} \equiv \frac{q_2/q_1}{y_2/y_1} = \frac{K_2}{K_1} \tag{8.11}$$

Therefore, the latter corresponds to the sorption selectivity.

From Table 8.7, it is found that for each membrane the *q* value for CH_4 is much smaller than C_2H_6. This is due to the competitive adsorption of C_2H_6. To consider the change in the *R* value of CH_4 from the pure gas (Table 8.6) to the mixed gas (Table 8.7), it is affected by two factors, i.e. one is the increase in *R* due to the increase in the total density in the pore (total density effect) and the other is the decrease in *R* due to the decrease in the effective potential barrier for CH_4 (effective potential barrier effect), since the strong adsorption regions are occupied by C_2H_6 molecules.

Comparing Tables 8.6 and 8.7, *R* values for CH_4 are almost the same for DP membrane. This means the two factors are compensating each other for this membrane. On the other hand, for the ZP and SP membranes, the *R* values for CH_4 in Table 8.7 are considerably lower than those in Table 8.6. This means that the second effect is dominant for the latter two membranes.

In all three membranes the separation factor α_{21} is always smaller than $\alpha_{21}^{(eq)}$. This is because the relationship between α_{21} and $\alpha_{21}^{(eq)}$ is

$$\alpha_{21} = \frac{\alpha_{21}^{(eq)}}{R_2/R_1} \tag{8.12}$$

and R_2/R_1 is more than unity as shown in Table 8.7. α_{21} is however larger than ideal separation factor $\hat{P}_2/\hat{P}_1$ shown in Table 8.6.

Recently, Wu et al. used a similar NEMD method to simulate the permeation and separation of H_2/CO in nanoporous carbon membrane [12]. The schematic diagram of the simulation cell adopted by Wu et al. is given in Fig. 8.19. The difference between Wu et al. [12] and Furukawa and Nitta [11] is that there are buffer zones between the H and M and M and L zones. The schematic representation of the slit pore is as shown in Fig. 8.20. It corresponds to Furukawa and Nitta's SP membrane.

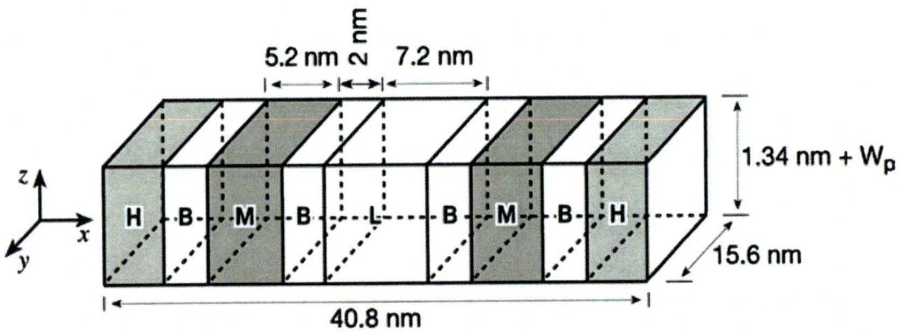

Fig. 8.19 Schematic diagram of the simulation cell. The length of the cell in the *x*-, *y*- and *z*- directions are 40.8, 15.6 and 1.34 + W_p (see Fig. 8.20) nm. The lengths of the H-, B-, M- and L- regions are 7.6, 2, 5.2 and 3.6 nm, respectively. (From [12])

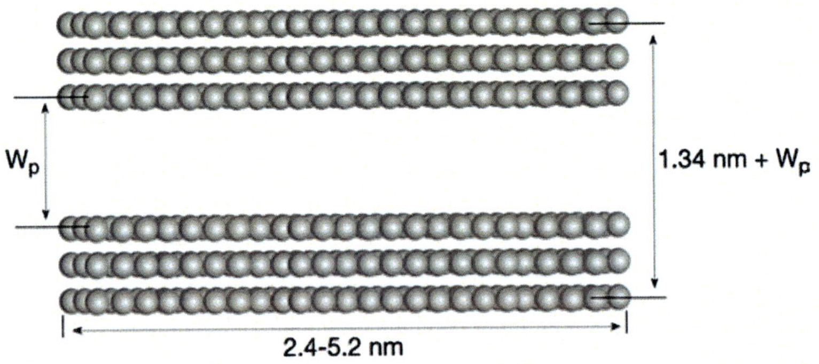

Fig. 8.20 Schematic representation of a slit pore W_p is the pore width. (From [12])

Table 8.8 Lenard-Jones interaction parameters used by Wu et al.

	σ (nm)	ε/k_0 (K)
Carbon	0.340	28.0
H_2	0.2960	34.2
CO	0.3763	100.2

The pore width, W_p, was changed in a range of 0.50–2.01 nm. The Lennard-Jones interaction parameter for this system is given in Table 8.8.

The simulation results are as follows.

When the gas pressure and the temperature are 0.5 MPa and 300 K, respectively, and the pore width is 1.01 nm, the density profile of H_2 and CO is given in Fig. 8.21a. In Fig. 8.21b indicates the buffer zone. It is found that the density of CO inside the membrane is kept almost constant at a value of 840 mol/m^3 while the density of hydrogen decreases from 47.3 to 25.4 mol/m^3 from the pore inlet (left) to the pore outlet (right). This gradient indicates that there is a resistance to the permeation inside the membrane. It is interpreted that the resistance for CO molecules to exit the pore mouth is much greater than that for the H_2 molecules. There are two peaks

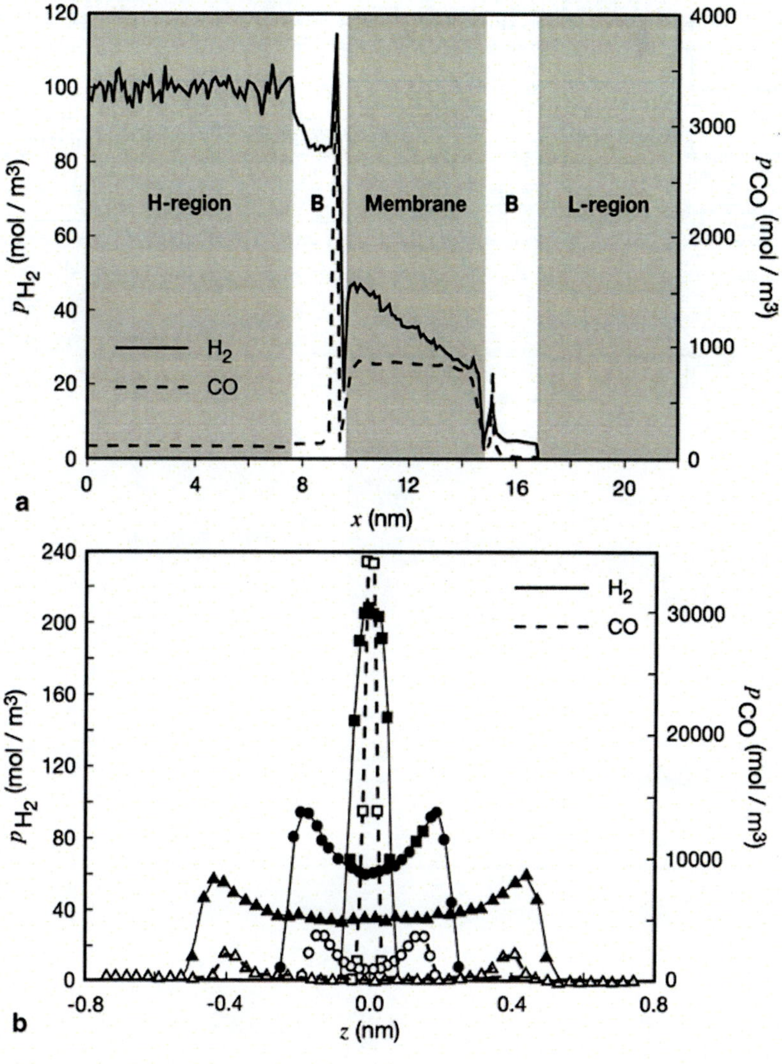

Fig. 8.21 Density profile of H_2 and CO corresponding to the H_2 mole fraction in the feed, temperature and feed pressure 0.5, 300 K and 0.5 MPa, respectively; (**a**) Density profile in the x-direction. W_p is 1.01 nm. B represents the buffer zone. (**b**) Density profile in th z-direction for three W_p values, i.e. (■, □) 0.67, (●, ○) 1.01 and (▲, Δ) 1.51 nm, respectively. *Solid lines* and *filled symbols* for H_2, *dotted lines* and *open symbols* for CO. (From [12])

near the entrance and exit of the pore for both CO and H_2, which correspond to the adsorbed layers. The density peak values 3,500 and 750 mol/m^3, at the entrance and exit, respectively, for CO are much greater than those for H_2, i.e. 115 and 10 mol/m^3. This is due to the much stronger adsorption of CO, at the pore entrance and exit, than H_2. In Fig. 8.21b density profiles of CO and H_2 are given for the z-direction for different pore widths (0.67, 1.01 and 1.51 nm). The density peak heights of CO are generally two orders of magnitude higher than those of H_2, reflecting the much stronger adsorption of CO than H_2. When the pore width is 0.67 nm, only one peak appears both for CO and H_2. However, when the pore width is 1.01 or 1.51 nm two peaks are formed, which means that the molecules are adsorbed near the two inner surfaces of the slit-like pores. The peaks of CO are much broader than H_2, meaning that the CO molecules occupy most of the adsorption sites, while the hydrogen molecules appear mostly in the center of the pores.

In Fig. 8.22 the flux and the separation factor are plotted versus pore width. Because of the size of the CO molecule (radius = 0.34 nm) CO can hardly permeate

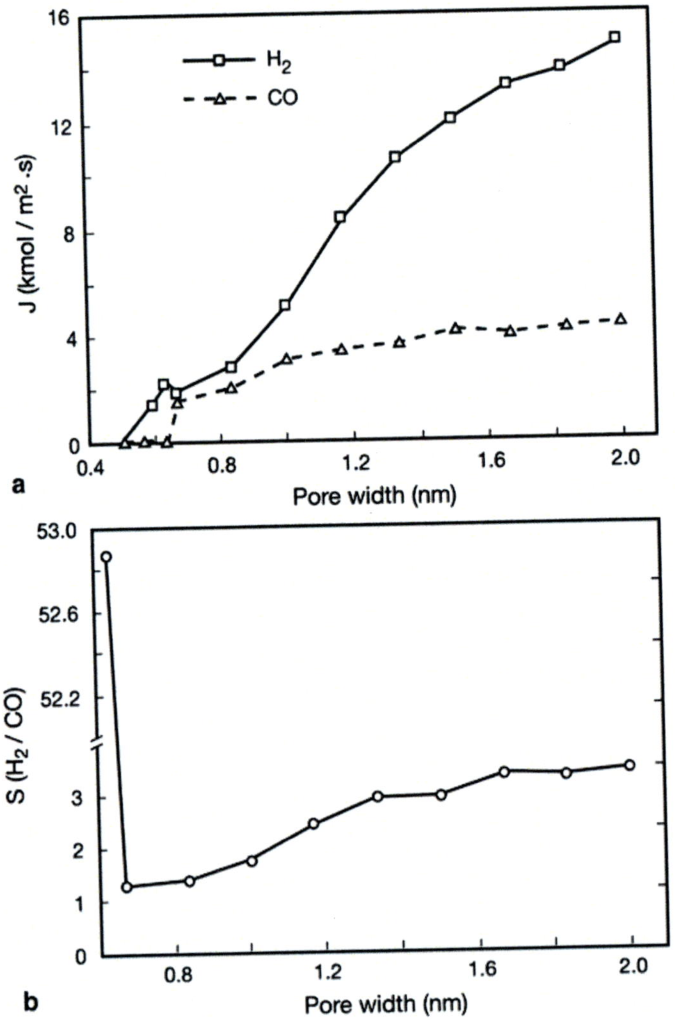

Fig. 8.22 Effect of pore width, W_p, on the flux (**a**) and separation factor (**b**); The mole fraction of H_2 in the feed, temperature and feed pressure are 0.5, 300 K and 0.5 MPa, respectively. (From [12])

through the pore whose size is smaller than 0.64 nm, while hydrogen molecules can. Hence, the separation factor shows a maximum of 52.88 at the pore size of 0.64 nm. When the pore size becomes larger than 0.64 nm both the flux and the separation factor increase with the pore width. The separation factor increases since the flux increase of H_2 is faster than that of CO.

In the pore size range of 0.64–0.84 nm, H_2 flux decreases first and then slowly increases. This is because in this pore size range the adsorbed CO molecules are tightly packed. H_2 molecules have difficulty to enter the pore, and when they enter, they move very slowly, even though the speed of the movement is faster than CO molecules. When the pore size is more than 0.84 nm, H_2 molecules begin to move faster than before in the center of the pore where CO adsorption is less.

The effects of temperature on the flux and the separation factor were studied at the H_2 mole fraction, the pore width and the gas pressure of 0.5, 1.01 nm and 1.0 MPa. The results are shown in Fig. 8.23. From the figure, H_2 flux and the separation factor increase while the flux of CO decreases slightly. This is due to the interplay of the faster molecular movement and weaker molecular adsorption at the higher temperature. As a result more H_2 molecules enter the pore and the H_2 flux increases. It is interesting to note that both the H_2 flux and separation factor reach a plateau, because of the competition between the adsorption and permeation of H_2 and CO molecules.

Figure 8.24 shows the effect of the feed gas pressure. The fluxes of both gases increase due to an increase in the driving force for the gas permeation. The separation factor decreases with the increase in the pressure because more CO is adsorbed at higher gas pressures. Subsequently, the entry of H_2 molecules into the pore becomes more difficult. (From [12]).

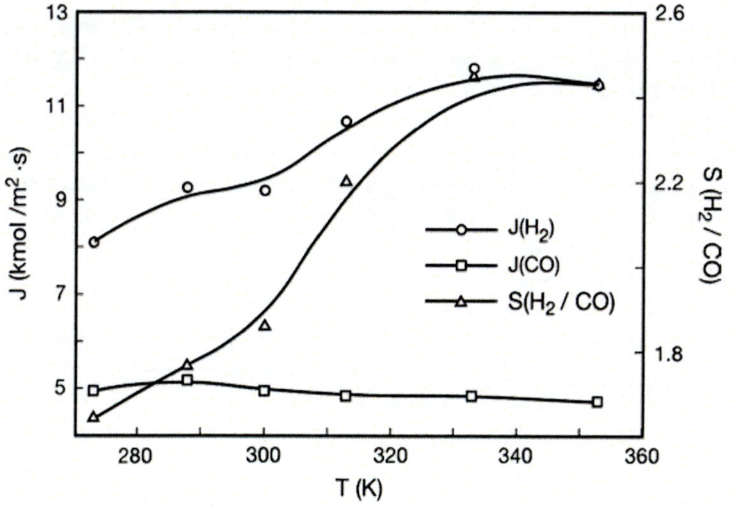

Fig. 8.23 Effect of temperature on the flux and the separation factor; The mole fraction of H_2 in the feed, the pore width, W_p, and the feed pressure are 0.5, 1.01 nm and 1.0 MPa, respectively. (From [12])

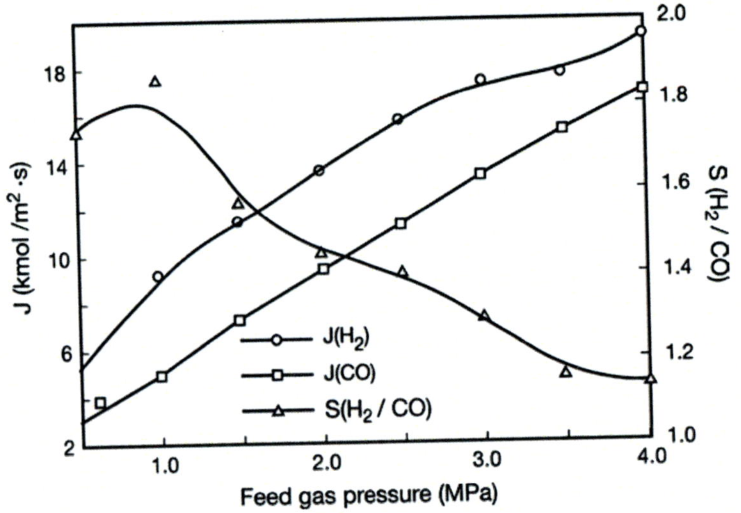

Fig. 8.24 Effect of feed gas pressure on the flux and the separation factor; The mole fraction of H_2 in the feed, the pore width, W_p, and the temperature are 0.5, 1.01 nm and 300 K, respectively. (From [12])

Figure 8.25 depicts the effect of the membrane thickness, ranging from 2.4 to 5.2 nm, on the fluxes and the separation factor of the membrane. The simulation was made for the two different pore resumes, i.e. 0.64 and 1.01 nm. Interestingly, the separation factor increases for the slit size of 0.64 nm, since for the slit size of 0.64 nm the sieving effect dominates and CO hardly can enter the pore. Hence, the separation factor increases as the membrane thickness increases. On the other hand, for the slit size of 1.01 nm, H_2 mainly permeates through the space between the two density peaks of CO. The resistance for this flow of H_2 increases as the membrane thickness increases.

Chen and Sholl presented a detailed model for the permeation of CH_4/H_2 mixtures through membranes constructed from closely packed bundles of single walled carbon nanotubes [13]. Combination of atomically detailed and continuum models that has proven effective in previous treatments of mixture permeation through zeolite membranes was applied.

The gas transport is assumed to occur by adsorption and diffusion through the interior of (10, 10) SWNTs, which have a diameter of 1.356 nm and are representative in many preparation procedures. CH_4 and H_2 are considered to be spherical species that interact with each C atom in a nanotube according to the Lennard-Jones potential function. These potentials are assumed to be applicable for the planar graphite surface. CH_4-CH_4 and H_2–H_2 interactions are also represented by Lennard-Jones function with standard values for Lennard-Jones parameters for pure species. For the CH_4-H_2 interaction the parameters were evaluated by the Lorentz–Berthelot mixing rule.

All adsorption and diffusion were computed for defect-free nanotubes with infinite extent to the axial direction. Another important assumption made is the mole-

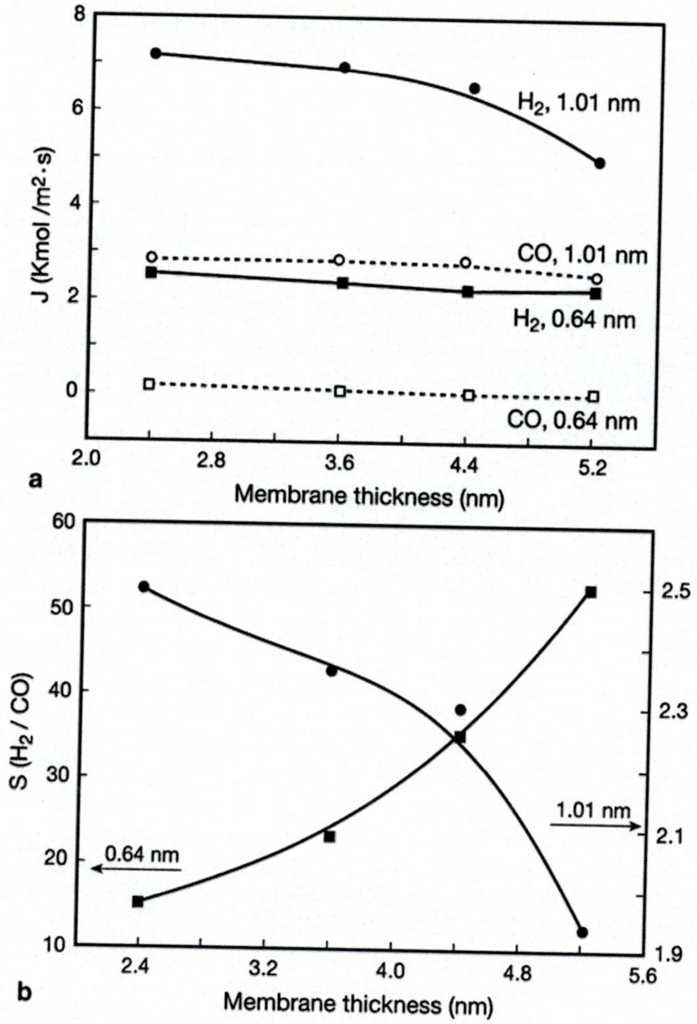

Fig. 8.25 Effect of membrane thickness on the flux (**a**) and the separation factor (**b**); The mole fraction in the feed, temperature and the feed gas presser are 0.5, 300 K and 0.5 MPa, respectively. (From [12])

cule-nanotube wall collision. It was assumed that the CNT wall remained rigid and momentum was not transferred to/from the colliding molecule from the nanotube wall. For the computation of adsorption isotherm, the Grand Canonical Monte Carlo (GCMC) method was adopted. The simulation was performed for the combination of CH_4 and H_2 fugacities given as state points in Fig. 8.26.

The GCMC data was fitted by the following equations for the binary system of 1 (CH_4) and 2 (H_2).

$$n_1 = \frac{a_1 p_1}{a_2 p_1 + a_3 p_2 + a_4} + \frac{a_5 p_1}{a_6 p_1 + a_7 p_2 + a_8} \tag{8.13}$$

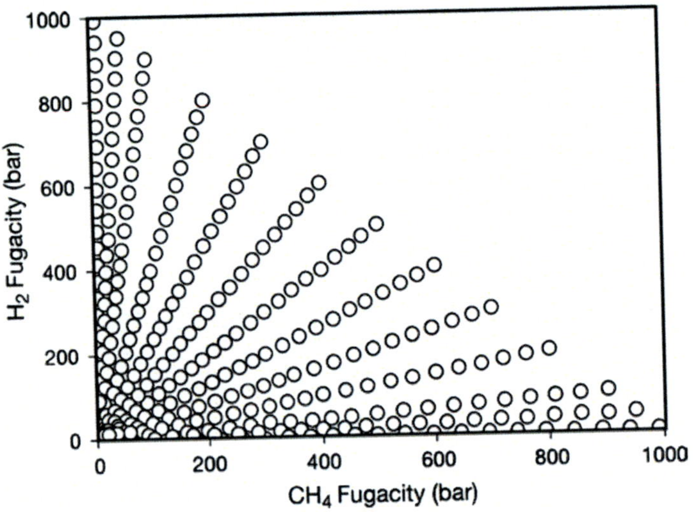

Fig. 8.26 Partial pressures of bulk phase for which GCMC simulations were performed of binary CH_4/H_2 adsorption at room temperature in (10, 10) SWNTs. (From [13])

$$n_2 = \frac{b_1 p_2}{b_2 p_1 + b_3 p_2 + b_4} + \frac{b_5 p_2}{b_6 p_1 + b_7 p_2 + b_8} \tag{8.14}$$

Where p_i denotes the partial pressure of species i and a and b are fitting parameters for which the numerical values are given in Table 8.9. The n has the unit of molecules/nm axial length of nanotube.

Figure 8.27 shows the agreement of the adsorption isotherm calculated from Eqs. (8.13) and (8.14) with the GCMC simulation results for the equi-molar CH_4/H_2 mixtures.

The flux, J, computed by the molecular dynamic model was also applied to calculate the Fickian diffusion coefficients, D, and the Onsager coefficient, L, by using the matrices

Table 8.9 The numerical parametrs in fitting the binary adsorption of CH_4/H_2 mixtures in (10, 10) SWNTs

Parameter	Value	Parameter	Value
a_1	15.19	b_1	0.228
a_2	3.098	b_2	0.206
a_3	0.064	b_3	0.014
a_4	7.455	b_4	14.33
a_5	1.896	b_5	0.680
a_6	0.340	b_6	13.94
a_7	0.029	b_7	0.061
a_8	5.252	b_8	9.255

a_1 a_5 b_1 b_5 have dimensions of molecules/nm, a_4 a_8 b_4 b_8 have dimensions of bar. The rest of the parameters are dimensionless

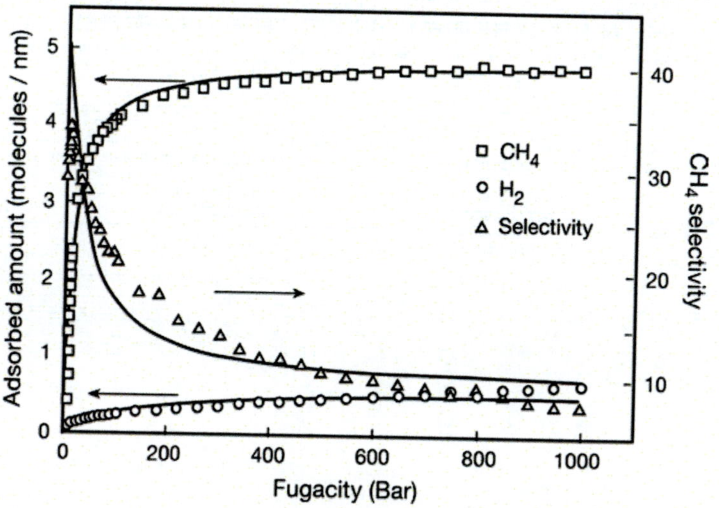

Fig. 8.27 The adsorption isotherm at room temperature for an equimolar mixture of CH_4/H_2. Symbols are from GCMC simulation while curves are for the fitting equations

$$\begin{matrix} J_1 \\ J_2 \end{matrix} = \begin{matrix} D_{11} & D_{12} \\ D_{21} & D_{22} \end{matrix} \begin{matrix} \nabla c_1 \\ \nabla c_2 \end{matrix} \tag{8.15}$$

Where c is the concentration of the gas

$$\begin{matrix} J_1 \\ J_2 \end{matrix} = \begin{matrix} L_{11} & L_{12} \\ L_{21} & L_{22} \end{matrix} \begin{matrix} \nabla \mu_1 \\ \nabla \mu_2 \end{matrix} \tag{8.16}$$

Where μ is the chemical potential.

It should be noted that the both Fickian diffusivity and the Onsager coefficient are the functions of the concentration of each gaseous component, 1 and 2. Furthermore, $D_{12} \neq D_{21}$ while $L_{12} = L_{21}$.

For this calculation equilibrium molecular dynamics (EMD) was used. EMD calculation was done for the state points given in Fig. 8.28 as adsorption phase composition of each gaseous species. Results are given in Figs. 8.29 and 8.30 by the solid symbols together with the bars indicating the uncertainties. One thing immediately obvious from Fig. 8.30 is that the Fickian diffusivity values are in a range of 0.1–1.0 cm^2/s which is extremely large compared to the values obtained for zeolites (10^{-4} cm^2/s). Diffusivities in the polymeric materials are even lower than in zeolite. This anomalous behaviour of molecular diffusion in CNTs was already mentioned in the foregoing section. The origin of these rapid transport rates is the smoothness of the potential energy surface defined by CNTs. It should be noted that this argument applies only for the Fickian diffusion coefficient defined by Eq. (8.15), that is associated with the net mass transport. The motions of individual tagged molecules are not necessarily very high.

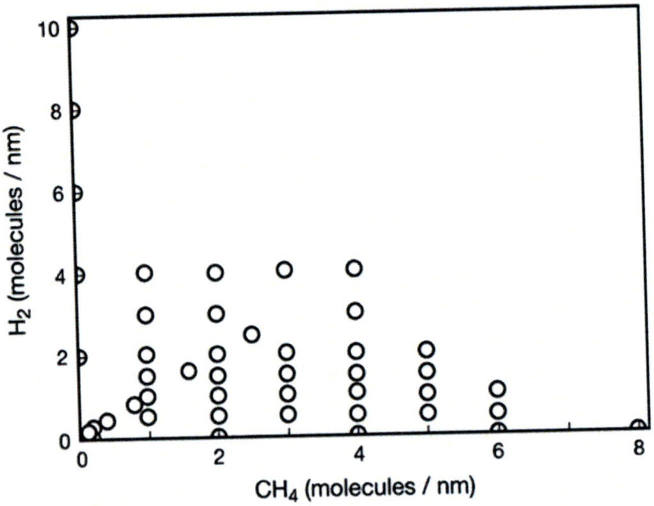

Fig. 8.28 Adsorbed phase concentrations for which EMD simulations were performed to compute binary Onsager transport coefficients in (10, 10) SWNTs at room temperature. (From [13])

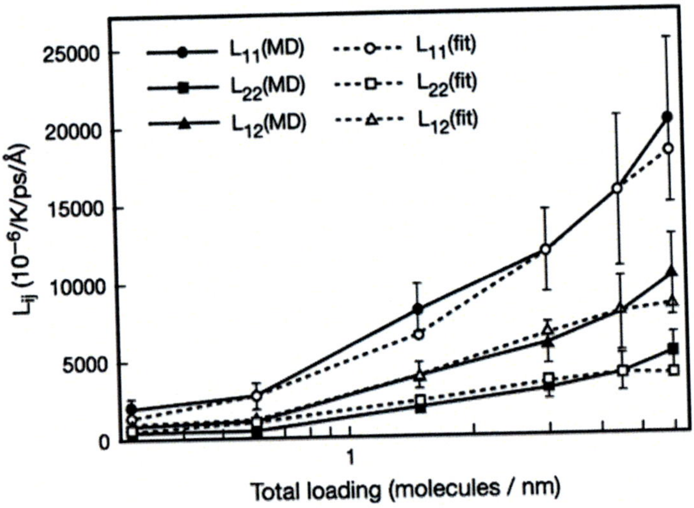

Fig. 8.29 The Onsager coefficients for adsorbed mixtures with a ratio of CH_4:H_2=2:1. *Filled symbols* and *solid lines* show the results of EMD, while the *open symbols* and *dashed lines* correspond to the fitting equations. (From [13])

The diagonal Onsager coefficients were further fitted by the following empirical expressions

$$L_{11} = d_1 a_1{}^{(b_1-n_1)^2/e_1} a_2{}^{(b_2-n_2)^2/e_2} n_1 \tag{8.17}$$

$$L_{22} = d_2 a_3{}^{(b_3-n_1)^2/e_3} a_4{}^{(b_4-n_2)^2/e_4} n_2 \tag{8.18}$$

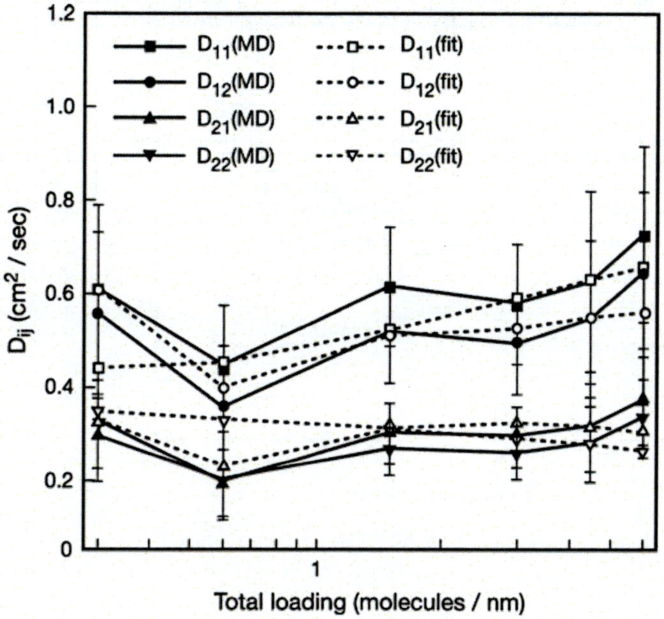

Fig. 8.30 The Fickian diffusivities for adsorbed mixtures with a ratio of CH_4:H_2 = 2:1. *Filled symbols* and *solid lines* show the results of EMD, while the *open symbols* and *dashed lines* correspond to the fitting equations. (From [13])

Furthermore, the off-diagonal Onsager coefficients were obtained by

$$L_{12} = L_{21} = \sqrt{L_{11}L_{22}} \tag{8.19}$$

The numerical values for the parameters involved in Eqs. (8.17) and (8.18) are summarized in Table 8.10.

By using the parameters listed in Tables 8.9 and 8.10, simulation of the membrane performance was conducted. A bundle of (10, 10) SWNTs with the packing density observed experimentally was considered for the simulation. Transmem-

Table 8.10 Numerical values used in calculation of Onsager transport coefficients

Parameter	Value	Parameter	Value
a_1	0.651	a_3	0.816
a_2	1.104	A_4	0.841
b_1	79.20	B_3	20.99
b_2	−1.206	B_4	9.571
e_1	999.7	E_3	15.85
e_2	3.296	E_4	22.32
d_1	557.3	D_2	18.56

b_1 b_2 b_3 b_4 have dimension of molecules/nm, e_1 e_3 e_3 e_4 have dimension of nm^2 /molecules^2, d_1 d_2 have dimension of 10^5 mol^{-1}K^{-1} ps^{-1}. The rest of parameters are dimensionless

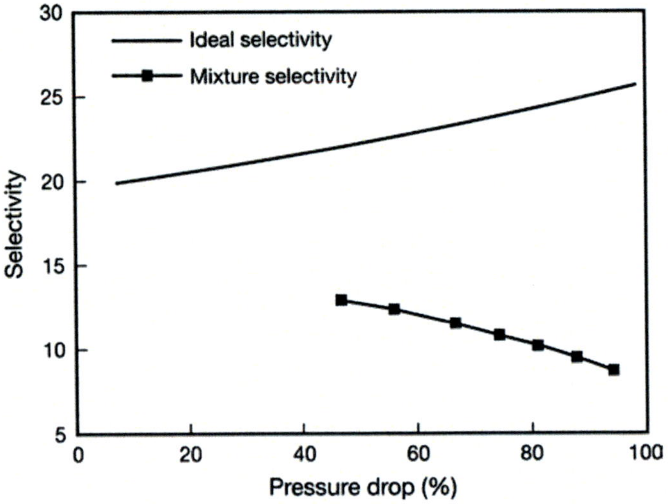

Fig. 8.31 Comparison of the binary selectivity and the ideal selectivity for a (10, 10) SWNT with a equi-molar feed as a function of % pressure drop when the feed pressure is fixed to 296 kPa. Selectivity means CH_4 over H_2

brane transport was assumed to occur only through the CNTs, not in the interstitial region between the CNTs. The adsorption isotherm was assumed to be established instantly at the both ends of the CNTs. This assumption is valid for a sufficiently thick membrane.

The gas mixtures on the feed and permeate side were assumed to be well mixed. The thickness of the membrane was taken to be 10 μm. Figure 8.31 shows the results of the simulation for the equi-molar CH_4-H_2 mixture when the feed pressure is kept constant at 296 kPa, while the permeate pressure is changed. Figure 8.31 also shows the ideal selectivity for the same membrane determined from the ratio of single component fluxes when each single component has a feed pressure of 148 kPa.

The selectivity of the binary mixture is significantly lower than the single component system particularly at the higher percentage of pressure drop. This is primarily due to the contribution of the off-diagonal component of the gas transport. For example, when the pressure drop is 240 kPa across the membrane, about 90% of CH_4 transport is caused by the CH_4 concentration gradient across the membrane while only about 20% of H_2 transport is caused by its concentration gradient. In other words, a large amount of H_2 is dragged by the faster transport of CH_4.

Figure 8.32 shows the calculated results for selectivity and CH_4 flux as a function of CH_4 mole fraction in feed. Pressure drop was maintained at 138 kPa while feed pressure was changed. Selectivity decreases with an increase in the CH_4 mole fraction but is quite insensitive to the feed pressure. The CH_4 flux increases as the CH_4 mole fraction in the feed increases. Let us compare with a similar simulation data for 100 μm silicate at the feed pressure of 292 kPa and permeate pressure of

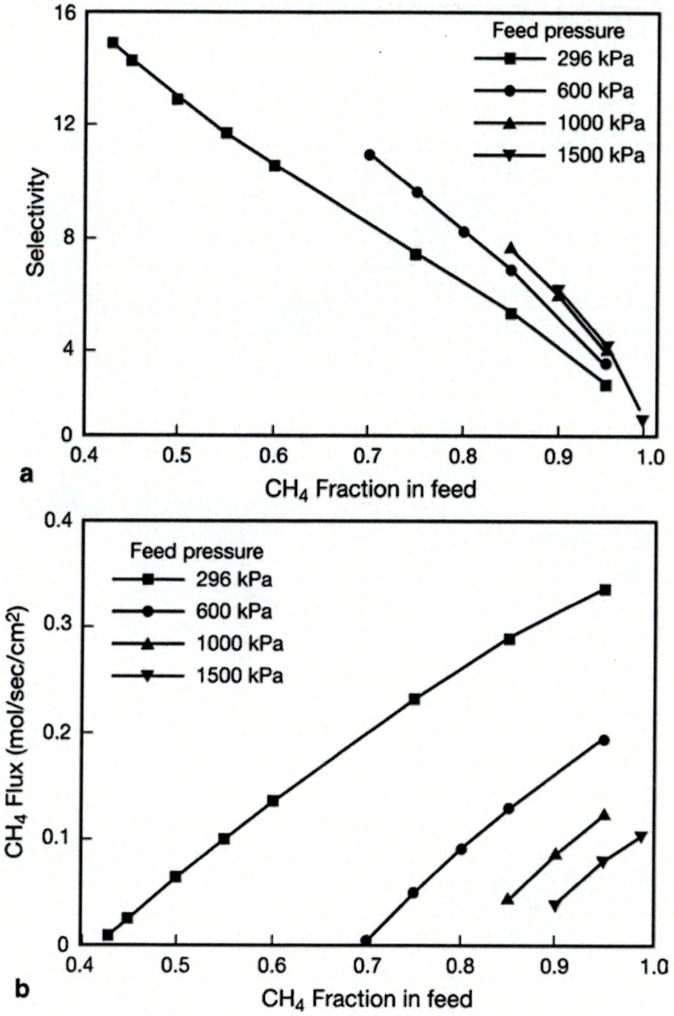

Fig. 8.32 Calculated results for the selectivity and CH_4 flux as functions of CH_4 mole fraction in feed. The transmembrane pressure drop was fixed to 138 kPa. The feed pressure was changed. (From [14])

138 kPa for 50/50 CH_4/H_2 feed mixture. Simulated CH_4 flux for the silicate membrane is ca 10^{-5} mol/s cm^2. Corresponding simulation value for SWNT membrane for 10 μm thickness is from Fig. 8.32 ca. 0.06 mol/s cm^2. Even for the thickness of 100 μm, the CH_4 flux is orders of magnitude higher than the comparable zeolite membrane. This is the direct consequence of the rapid diffusion of molecules inside the CNTs. The effect of the transmembrane pressure drop is given in Fig. 8.33. Feed pressure was kept constant at 296 kPa. Observed tendency is very similar to those discussed in Fig. 8.32.

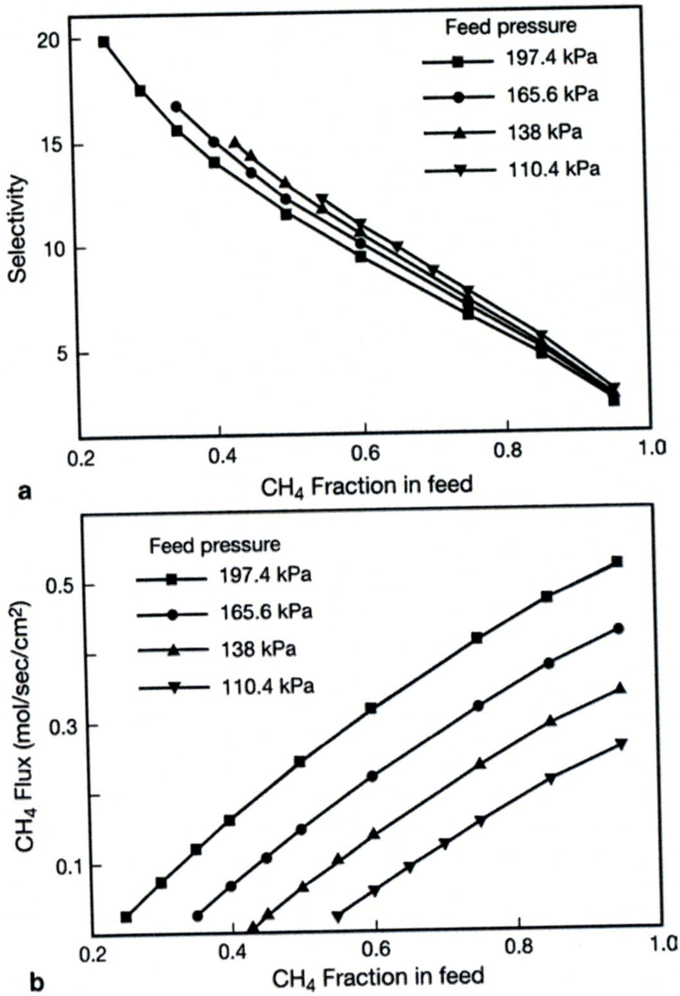

Fig. 8.33 Calculated results for the selectivity and CH_4 flux as functions of CH_4 mole fraction in feed. The feed pressure was fixed to 296 kPa while the tranasmembrane pressure drop was changed. (From [14])

When the flux of CH_4 (1) and H_2 (2) are split into the diagonal contribution and nondiagonal contribution such as

$$J_1 = J_{11} + J_{12} \text{ and } J_2 = J_{22} + J_{21} \tag{8.20}$$

Where J_{11} and J_{22} are diagonal contributions and J_{12} and J_{21} are the nondiagonal contribution, respectively, J_{12}/J_{11} and J_{21}/J_{22} offer, respectively, importance of nondiagonal contribution relative to diagonal contribution in CH_4 and H_2 transport.

Figure 8.34 represents such ratios as functions of CH_4 mole fraction. As CH_4 mole fraction increases, the contribution of nondiagonal contribution increases for H_2 (2) while it decreases for CH_4 (1).

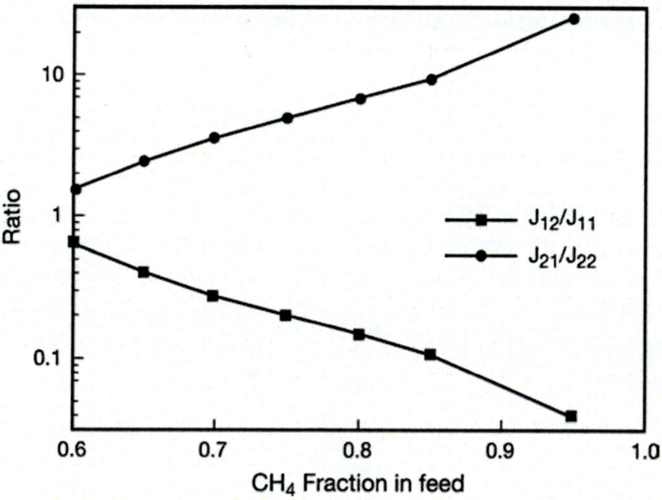

Fig. 8.34 The ratios J_{12}/J_{11} and J_{21}/J_{22} as functions of CH_4 mole fraction in feed. The feed pressure and tansmembrane pressure drop are 296 and 138 kPa, respectively. (From [14])

Table 8.11 Classification of gas and liquid permeation through the membrane pore

Membrane pores	Gas transport	Liquid transport
Micropore	Constrained molecular diffusion	Constrained molecular diffusion, surface transfer
Mesopore	Knudsen diffusion	Viscous flow
Macropore	Viscous flow	Viscous flow

According to Verweij et al. [14], the transport through the membrane pore depends on the pore radius φ_p and the interaction of the permeant and the membrane material. According to the IUPAC terminology and the gas sorption analysis the pores are classified as follows. Micorpore φ_p<2 nm; mesopore 2<φ_p<50 nm; and macropore φ_p>50 nm. Verweij et al. also classified the membrane pore transport, depending on the pore size, as shown in Table 8.11.

8.2.1 *Micropore Transport*

Pore diameters are in the molecular size range, where the distinction between whether the molecules are gas or liquid ceases to exist. The molecules are under the effect of the interaction potential from the pore wall and their transport is according to the hopping diffusion like a solid-state vacancy mechanism. This mechanism can be applied well for amorphous structures and molecules that fit tightly in the pore. Membrane pore system is then considered as a Langmuir lattice of the sites that can be either vacant or occupied by the molecules. In such a case the flux of species l_1 can be written by

$$j_{l_1} = -f_{Ll_1} b^0{}_{l_1} c^{tot} RT[(1-\theta_{l_2})]\nabla\theta_{l_1} + \theta_{l_1}\nabla\theta_{l_2} \tag{8.21}$$

In which b^0 is the mechanical mobility for a molecule on an otherwise empty lattice, c^{tot} the concentration of available sites, R the universal gas constant, T absolute temperature, and $0<\theta<1$ is the average site occupation. Equilibrium between the site occupation and feed concentration on both sides of the membrane can be used as the boundary condition to solve the above equation. Non-equilibrium correction factor $0<f_{L1}<1$ is often very close to unity but can be very small for the molecule of low occupancy (low θ) when the other molecule occupies the sites strongly and hinders the movement of the former molecule. The above generic expression leads to the following two types of microporous and surface diffusion separation:

Type 1: For both molecules θ is very small and nearly equal to zero. This type of behavior is often found for small gas molecules, low pressures and high temperatures.

Type 1 behavior leads to a very simple expression for the diffusion flux for individual gas species l which represents both species 1 and 2.

$$j_l = -b_l{}^0 c^{tot} RT\nabla\theta_l = -\bar{D}\,\nabla c_l \tag{8.22}$$

Which is a simple Fick's first law for a single species with the chemical diffusion coefficient $\bar{D}_l = b_l^0 RT$, which is independent of θ. In type 1 mechanism, the Henry's isotherm $\theta_l = K_l{}^H p_l/p^0$ is applicable so that the diffusion permeance can be written as

$$f_l{}^1 = \frac{b_l{}^0 c^{tot} RT\nabla\theta_l}{p_{lf} - p_{lp}} = \frac{c^{tot}\,\bar{D}_l K_l{}^H}{p^0 X} \tag{8.23}$$

Where p^0 is the standard pressure and X is the apparent membrane thickness. The above equation shows that $f_l{}^1$ that represents the permeance in the micropore is independent of P_l (gas pressure of individual species at either feed or permeate side).

Furthermore, the significant modification of Eq. (8.23) leads to $\alpha_{l_1 l_2} = \alpha_{l_1 l_2}^{s}$, which means the selectivity becomes the ideal selectivity.

Type 2: One of the molecules has a high affinity to the pore wall and low mobility. It may occupy more than 50% of the site and form a percolative network, nearly impermeable for the other gas.

8.2.2 Knudsen Transport

The transport of gases in mesopores generally occurs by the Knudsen diffusion mechanism if the mean free path of the molecule is larger than the pore diameter Φ_p. For this flow resume, the permeance can be written as

$$f_l{}^{Kn} = \frac{\varphi_p \Phi_p}{3\tau X}\sqrt{\frac{8}{\pi RTM_l}} \tag{8.24}$$

In which φ_p is the volume fraction of porosity, τ is the tortuosity of the pore and the M is the molecular weight of the species l (either 1 or 2). The separation factor can be given as $\alpha_{l_2 l_1}^{Kn} = \sqrt{M_{l_2}/M_{l_1}}$. Due to the low selectivity, this mechanism is generally not very interesting for membrane separation. A typical micropore mechanism is much more interesting. Correlation effects can be particularly strong when $\theta \rightarrow 1$ and in single-file diffusion where larger molecules cannot pass each other inside the pore.

8.2.3 *Viscous Flow Transport*

The viscous transport in membrane pores is generally calculated from the Hagen-Poiseuille equation for stationary Newtonian flow in a cylindrical capillary. This leads to the following equations for calculating permeance.

$$f_l^{visc} = \frac{\varphi_p \Phi_p{}^2}{32\tau X} \tag{8.25}$$

Viscous flow is nonselective for molecules. However, the flow field around particles, smaller than Φ_p near the pore entrance may lead to a certain level of size exclusion. Mesoporous membranes that have a charged pore surface in salt solutions may exhibit significant ion retention by a space charge effect if Φ_p is smaller than the Debye length of the solution.

8.2.4 *Discussion on the Gas Transport in the Carbon Nanotubes*

Holt and et al. [3] showed that their DWNT membranes are virtually gas and liquid dense until the caps of the pores are etched open. This means that there are no major processing defects introduced in the early processing stages. They also demonstrated the size exclusion effect by showing the permeation of 1.3 nm $[Ru(bpy)3]^{2+}$ complexes while Au particles with >2 nm diameter were completely blocked by the pore. Therefore, it can be assumed that

1. The pore diameter is 1.6 nm as observed by TEM.
2. Many of the pores are open on both sides.
3. There are no defects with diameters >>2 nm.

The unique gas permeation properties of the DWNT membranes stem from the very weak interaction between the gas molecules and the tube wall when the molecule flows to the axial (z) direction. Skoulidas et al. [4] demonstrated that the potential energy barrier, u^z, for axial translation of CH_4 inside a (10, 10) SWNT is only 6×10^{-4} eV=0.02 $k_B T_{room}$, while in zeolite ZSM-12 micropores, u^z=0.045 eV=1.7 $k_B T_{room}$. This has a profound impact on the dynamics of gas molecules. To the first approximation, CNTs can be thought of presenting a smooth attractive surface (binding energy u^{bind} on the order of 10^{-1} eV) to the gas molecules, whereas zeolite microchannels present a rough, attractive surface.

In pure, non-hydrocarbon gas permeation experiments (H_2, He, Ne, N_2, O_2, Ar, CO_2, Xe), Holt et al. found an $M^{-1/2}$ dependence in permeability. That is, light gases diffuse faster, in proportion to the molecule's thermal velocity $v \propto (k_B T/M)^{1/2}$. The above relationship is usually associated with Knudsen diffusion for the following reasons: The self diffusion coefficient, D^s, of a gas molecules is simply $v\lambda$, where λ is its mean free path. Thus the $M^{-1/2}$ permeability could be explained if one assumes:

1. The true mean free path of the gas molecule is in the Knudsen resume, $\lambda \approx \Phi_p$, where Φ_p is the characteristics of the membrane and independent of the gas species.
2. The permeability is proportional to D^s in a simple fashion.

The problem to use the above Knudsen diffusion for the interpretation of the data is that the magnitude does not work for DWNT membranes; i.e. the measured value of permeability is larger than the permeability calculated by the Knudsen formula by a factor of 10–10^2, and therefore it was called fast transport. The above assumptions (1) and (2) were also shown to be flawed by the molecular dynamics work even before the experiments were performed. The $M^{-1/2}$ dependence may be therefore much more complicated than the Knudsen mechanism.

Figure 8.35a shows molecular dynamics simulation trajectories of CO_2 molecules in a (40, 40) SWNT at 298 K, and Fig. 8.35b shows their density profiles in the radial direction in the pore at different external pressures. Clearly, the gas concentration near the NT wall is very high. Verweij described such a system by using "2D free gas".

Figure 8.35c illustrates the "2D free gas" ideal limit as running a motorcycle gang inside a cylindrical cage. The cage provides centrifugal force support, but can not slow down the motorcycle quickly. The axial mean free path λ of a motorcycle, instead of being limited as $\approx\Phi$ as the Knudsen diffusion model, depends on the adsorbed gas density (the higher the gas density the smaller λ). When two motor cyclists collide, most often both stay near the cage after the collision. Occasionally, the collision sends the motorcycle into a cross-channel flight. Both modes of motion (near-cage spiraling and cross-channel 'ping pong') would have an axial mean free path proportional to $1/C_{ad}\sigma$ where C_{ad} is the density at the peak in Fig. 8.35b, and σ is the collision cross-section as in the kinetic theory of gases. Thus, the self-diffusivity of the '2D free gas' model

$$D^s \approx \left(\frac{k_B T}{M}\right)^{1/2} (C_{ad}\sigma)^{-1} \tag{8.26}$$

From the MD simulations the self diffusivity D^s has been shown to depend sensitively on C_{ad} as the Eq. (8.26) would indicate. However, the simulation of Sholl et al. [4] showed that the chemical diffusion coefficient $\bar{D}$ has been found to be

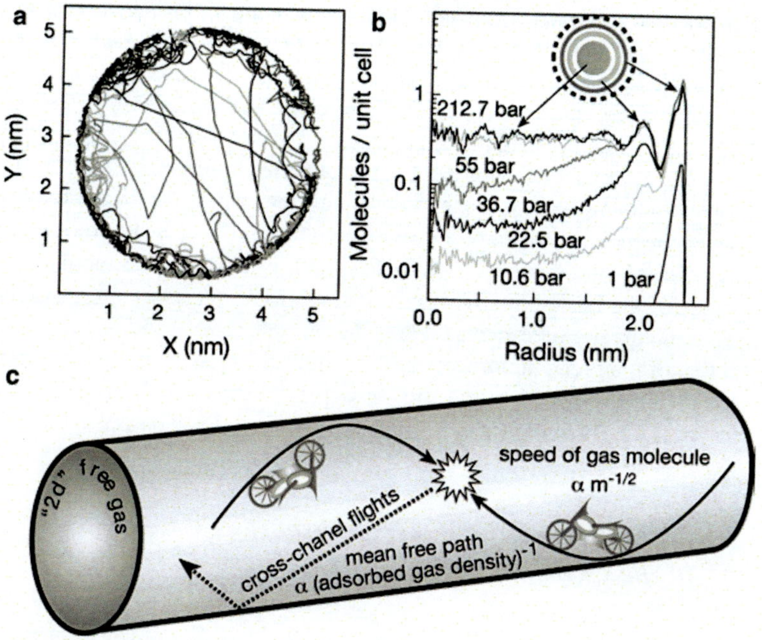

Fig. 8.35 (**a**) Molecular dynamics simulation trajectories of CO_2 molecules in (40, 40) SWNT at 298 K and an experimental partial pressure of 1 bar, (**b**) Simulated density of CO_2 inside a (40, 40) SWNT as a function of the external gas pressure. One unit cell has an axial length of 2.5 Å and consists of 160 carbon atoms, (**c**) Cartoon of the '2D free gas' model, drawing an analogy to running motor cycle gangs (one red and the other green in the original article) inside a cylindrical cage. The motorcycles run on the cage most of the time, collide with other motorcycles frequently near the cage but occasionally send one flying across the channel. The CNT/CO_2 interaction is such that the cage has low friction and high specularity in collision with the motorcycles. (From [14])

practically independent of C_{ad}. This constancy of $\bar{D}$ was found for single component microporous hopping diffusion, as well as in MD simulation of single-file gas diffusion in very narrow pores. Verweij et al. argued that it is the term $\partial \log F/\partial \log C_{ad}$ in the diffusivity that increases with C_{ad} and cancels the C_{ad} in the denominator of Eq. (8.26).

After all the above discussions, it was claimed by Verweij et al. that the ultra-smoothness of the potential energy landscape for axial translation has a profound impact on the overall transport properties of a confined gas molecules.

Fundamentally, this is due to the two features of the CNT: structurally, the atomic planarity of graphene at 0 K; and chemically, the nonpolar nature of the carbon network.

Zeolites and other ceramic membranes, which have charged groups on the pore surface and can polarize the gas molecules, apparently induce a much larger u^z, and therefore belong to a different class. It has been demonstrated that thermal fluctuations of the nanotube (NT) do not significantly alter its gas transport behavior above 1 bar at room temperature.

Although the transport in DWNT membranes is undoubtedly very high, with a diffusivity of $\bar{D}^s \approx 10^{-5}$ m^2/s (very similar to that of the free gas), the industrial usefulness of the membrane depends on the selectivity, in particular, if the selectivity of the membrane can surpass the $M^{-1/2}$ rule of Knudsen diffusion. Verweij et al. argues that it is much related to the adsorption thermodynamics, rather than the kinetics. Going back to the analogy of the motorcycles in Fig. 8.35c, we may assume that both red and green motorcycles have almost the same collisional cross-sections. The mean free path of a red motor cycle would be similar to that of the green motor cycle, i.e. $\lambda_{red} \approx \lambda_{green} \propto 1/(C_{ad,red} + C_{ad,green})$. But if the red is more adsorbed to the surface of the cage, there will be more red motor-cycles running around, marginalizing the green motor-cycles. Thus, the flux of the red motor-cycles will overwhelm that of the green motor-cycles. To achieve the selectivity of > 100, a binding energy difference $\Delta u^{bound} \approx \ln(100) k_B T \approx 0.1$ eV between the two adsorbed species are required. This might be challenging for most of gas pairs.

8.2.5 Discussion on the Water Transport in the Carbon Nanotubes

The small value of ϕ_p/τ, as obtained for the gas flow by the Knudsen flow, gives also a small permeance value when water transport is calculated by Eq. (8.25), applicable for the Poiseiulle flow. However, there is a significant enhancement of permeance over the Hagen-Poiseiulle value by a factor of 560–8400. This effect is ascribed to the significant occurrence of slip and hence, velocity of water at the pore wall is not zero. The occurrence of the slip is again by the very smooth pore wall and the fact that water molecules have little affinity to the hydrophobic nano-tubes. Even though the continuum concept of viscous flow may not be applicable through such a narrow pore of 1.6 nm diameter, yet the above mentioned large enhancement factor demonstrates that water transport is exceptionally fast. The viscous flow stick/non-stick concept may lose its meaning completely to discuss on the transport in the micropore. For example, efficient water transport is well documented for thin zeolite A membranes with pore diameter of $0.4 < \Phi_p < 0.5$ nm, where the water molecules just fit inside the pore and apparently have significant mobility with respect to the pore wall. A representative data for this case is the separation of water from 10% water/ethanol mixture by pervaporation. According to Kondo et al., for a 10 μm thick membrane on a porous multiple support $j_{H_2O} = 0.13$ mol/m^2 s with a separation factor of 47,000 at 120 °C. This shows the significant effect of the slip enhancement for the water molecule. According to Holt et al., for the 2 μm thick membrane DWNT 1, the enhanced pure water flux $j_{H_2O} = 7$–39 mol/m^2 s at $\Delta p = 10^5$ Pa and 120 °C. The DWNT's liquid phase transport has not been demonstrated to be separative when the molecular sizes are smaller than the pore diameter.

Verweij et al. have also quoted the reports of Majumder et al. [15] and Holt et al. [3] on the fast water transport through the CNT membranes at room temperature. These experimental data corroborate the earlier results from the MD simulation done by Hummer et al. [1]. There are however several questions including why does water wet the membrane first of all? A configuration was created in the Hummer et al.'s MD simulation in which water flow through the CNTs is driven by the osmotic pressure difference. Their study showed that the water flow was limited by the particle entry and exit events, and the length of the pore had hardly any effect. For the (6, 6) SWNT studied by Hummer et al., the pore diameter, 0.8 nm, is so narrow that only a single water molecule can be inserted, forming a single file water chain. Verweij also quoted the statement of Truskett et al. [16] that the hydrophobic confining wall reduces the average number of favorable fluid-fluid interactions per molecule, which means it disrupts the hydrogen-bonding pattern in the fluid. Inside the SWNT two hydrogen bonds are lost per molecule compared to bulk liquid water, costing it approximately +0.4 eV in binding energy. The van der Waals interaction between the water and the CNT wall which may partially recuperate the loss is approximately −0.17 eV. But still the binding energy of the water in CNT is ca. 0.2 eV higher than the bulk water as the binding energy distribution depicted in Fig. 8.36a. This makes the insertion of water into the CNT pore improbable. However, Verweij et al. further argues that the contribution of the high energy water molecule is greater in the bulk water due to bulk water's broader distribution. The presence of these high energy water molecules play a disproportionately large role in the statistical mechanics equation for the chemical potential calculation, and eventually the chemical potential in the nanotube becomes less by 0.04 eV than that of the bulk water. Thus, water molecule can enter into the narrow pore.

The water molecules confined in the (6, 6) SWNT may still have considerable entropy at 300 K, since the water molecules can rotate about the nanotube axis

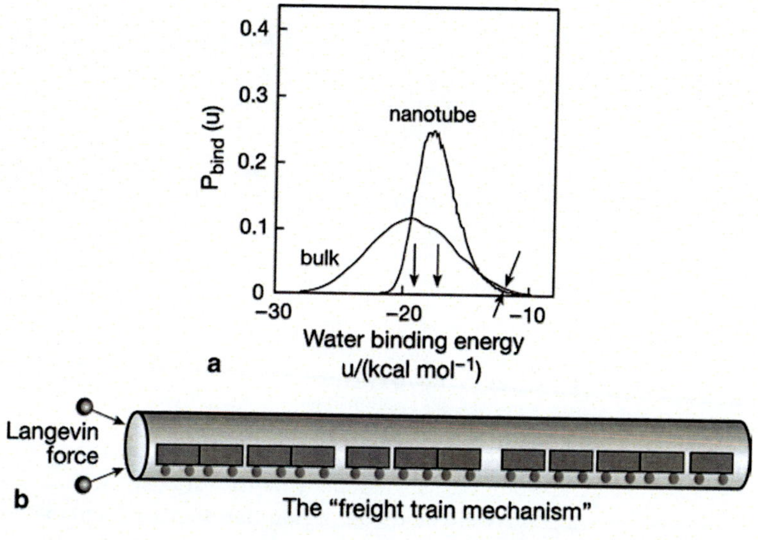

Fig. 8.36 (**a**) Distribution of the water binding energy in bulk liquid water and inside the nanotube at 300 K, (**b**) illustration of the single-file diffusion of water inside a SWNT. (From [14])

to which direction the hydrogen bonds are aligned. The water chains generated in the nanotube can no longer be considered as water of the ordinary sense. In the Poiseiulle flow with a nonslip boundary condition at the pore wall, the dissipation rate comes entirely from the transverse momentum transfer inside the liquid, proportional to the bulk liquid viscosity. However, for the water in the nanotube the dissipation rate originates entirely from water chain/pore wall friction, instead of water–water momentum transfer. Hummer et al. [1] noted that the wall friction seems exceedingly small as in the case of gas diffusion.

Indeed, graphite is an industrial-grade solid-solid lubricant. Verweij et al. describe the movement of water in the nanotube by the singe file diffusion as illustrated in Fig. 8.36b by a chain of freight cars on a rail, even though the water chain inside the nanotube is not truly "solid". The freight cars are pushed to both ends by Langevin thermal forces, which average to pressures. The cars do not necessarily form one long contiguous train from one end to the other but break up and recombine into new trains inside the tube any time.

Abnormality in the structure and dynamics of water inside the sub-2 nm CNTs are often reported. To interpret the fast water movement, there is no physical basis to apply the Poiseuille flow equation which uses the bulk viscosity in solving the equation. One way of modifying the Poiseuille flow model is to incorporate the slip length R_{slip}, the ratio of the translational velocity at the wall, v_z, to the characteristic velocity gradient in the transverse direction, $\partial_y v_z$, i.e. $R_{slip}=v_z/(\partial_y v_z)$. Based on their experimental data, Holt et al. found that they must use a slip length of 100 nm, which is 100 times as large as the pore diameter. This means that the large velocity drop occurs just at the first layer of water at the pore wall. This is just another way of saying that the thin layer of water moves as a solid at the nanotube wall. The key assertion of the Poiseuille flow is that water–water momentum transfer is a main component of the total dissipation. It is likely that this defining characteristic has broken down for the small CNTs. The water-CNT friction has become of paramount importance.

Majumder et al. [15] have shown by MD simulation applied to MWNTs with pore diameter of 7 nm that the Poiseuille law may work well if the friction between the wall and the fluid is so large that no slip occurs. But this was not the case from the experimental data. They have reported the slip lengths that are 10^4 times greater than the pore diameter. Verweij et al. mentioned that water does not flow but translate as shown in Fig 8.36c.

8.3 Carbon Nanofiber Membranes

Electro-spun nanofiber membranes, due to their high surface areas and porosities could be a good candidate for the separation media for various purposes. The high surface area implies that there are numerous adsorption sites for the removal of organic contaminants. In addition, the high porosity means that a smaller driving force is required to let water permeate through the membrane, hence making the process less energy intensive. In a recent work by Singh et al. [17], carbonized electro-spun nanofiber membranes were prepared and tested for their removal of disinfection byproducts (DBPs). Carbon based particles and powders are widely used as adsorption media for organics e.g. DBPs. With this in mind, carbon nanofiber membranes were selected as the membrane adsorption media for the removal of DBPs. In this study, polyacrylonitrile (PAN), a precursor for carbon, was first electro-spun into membranes. These membranes were then heat treated to form carbonized nanofiber membranes (CNMs) and tested for the removal of monochloroacetic acid and chloroform as they represent DBPs. Since use of carbonized electro-spun nanofiber for membrane adsorption is a novel approach for the water treatment, a thorough description of the work is made in this section.

8.3.1 Membrane Preparation

Polyacrylonitrile (PAN) in powder form with an average molecular weight of 1.5×10^5 g/mol and melting temperature of 317 °C was obtained from Sigma-Aldrich Inc. (St. Louis, MO). The PAN powder was heated to 105 °C for 3 h before use, to remove any residual moisture in the powder.

The polymer dope for electro-spinning was prepared by dissolving predetermined amounts of PAN in dimethyl formamide (DMF) and gently stirring the mixture for 24 h at 60 °C to obtain a homogeneous solution. Three polymer solutions of different concentrations; 8, 10 and 12 wt%, were prepared and used in this study. Figure 8.37 shows the schematic of the electro-spinning setup used.

The polymer solutions were placed in a 10 mL syringe. Needles of size 20 G (0.6 mm inner diameter) were first ground to ensure rounded smooth tips and dried before use in each experiment. A high voltage power supply (DW-P503-1C, Beijing Shining Technical & Commercial Centre, Xisanqu, Tiantongyuan, Changping District, Beijing, PR China) with anode connected to the needle and cathode connected to an aluminum plate was used. A syringe pump (74,900 series, Cole-Parmer, Vernon Hills, IL) was employed to supply a constant flow of polymer solution during electro-spinning. The voltage applied ranged from 1 to 18 kV and the polymer solution flow rate used was between 0.6 and 1.2 mL/h. The distance between the needle tip and the aluminum collector plate was kept constant at 19 cm. The temperature and humidity in the fume hood where the electro-spinning was carried out was between 21 ± 3 °C and 25 ± 5% respectively for all experiments. The average collection time was 1 h. Increasing the collection time resulted in thicker membranes. The membrane thickness was measured using an absolute digimatric device (Mitutoyo Corp., Maplewood, NJ).

Collected nanofiber mats were kept overnight in the fume-hood to allow the solvent to evaporate. Subsequently, the mats were placed in an air-circulated oven at 120 °C for 3 h to remove any trace solvent that might be present in the nanofiber mat. After this the membranes were allowed to cool to room temperature and then heated to 250 °C at a heating rate of 5 °C/min. Each membrane was held at 250 °C for at

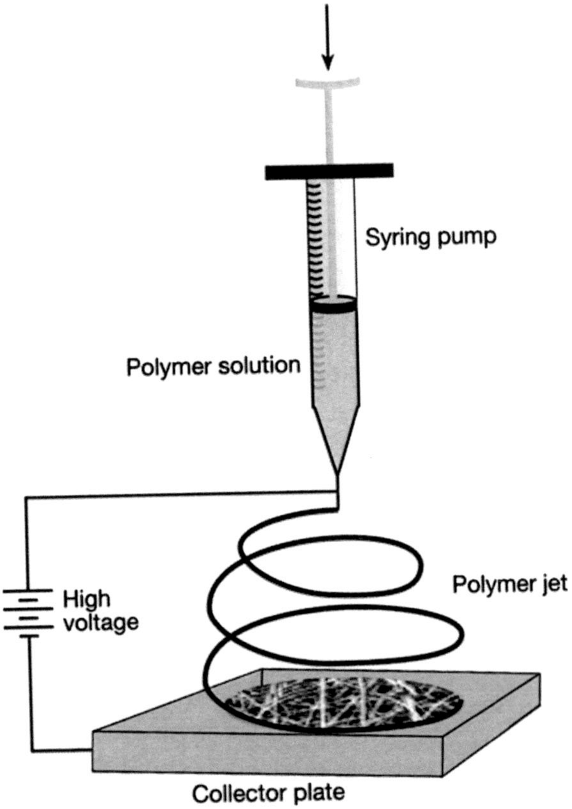

Fig. 8.37 Schematic of self assembled electro-spinning setup used in the experiments. (From [17])

least 4 h. Following this, the mats were carbonized by heating at a rate of 5 °C/min up to a temperature of between 400 and 500 °C and kept for 15 min to 4 h at this temperature, all under a nitrogen atmosphere. Higher heating temperatures of 600–1000 °C resulted in membranes which were crumpled or cracked easily and could not be used for filtration. This is probably due to shrinkage which affects the orientation of the molecular chains.

8.3.2 Membrane Characterization

The static contact angle was measured for each membrane before and after carbonization using the VCA optima surface analysis system (AST Products, Inc., Billerica, MA). For each membrane the contact angle was measured at five different spots and the values were averaged.

Glass transition temperature, T_g, was determined by a differential scanning calorimeter (DSC Q 1000, TA Instruments, New Castle, DE). The T_g value was recorded at the onset of the corresponding heat capacity transition.

Morphology of the membrane was investigated using scanning electron microscopy (SEM) of a model JSM-6400 JEOL (Japan Electron Optics Ltd., Japan). Samples were coated under vacuum with a thin layer (60% gold and 40% palladium) in a Hummer VII sputtering system (Anatech, Springfiled, VA).

Pore size distribution and bubble point measurement of the membranes were carried out using a capillary flow porometer (Porous Materials Inc., USA). The membranes were first wetted with Galwick solution and then placed in the holder. Measurements were made in a wet up, dry up operating mode.

Surface functional groups of the membrane were characterized by Fourier transform infrared (FTIR, Varian 1000, Varian Inc., Palo Alto, CA) spectroscopy in attenuated total reflection (ATR) mode. Membrane surface samples were mounted on face of the crystal surface. Spectra were measured in transmittance mode over a wave length of 600–4000 cm^{-1} at a resolution of 4 cm^{-1}.

8.3.3 Adsorption Tests of Monochloroacetic Acid

Static adsorption experiments were carried out by placing equal weights of carbonized nanofibers into each of 40 mL vials sealed with a septum cap and then adding aqueous solutions of monochloroacetic acid of varied initial concentrations. After the vials were prepared they were placed in a shaker under mild mixing for 2 days at 25 °C. The initial and residual concentrations for each sample were measured after 2 days. The amount of chloroacetic acid adsorbed (mg/g) was calculated from the difference in concentrations between the initial (C_o) and residual or equilibrium (C_c) solutions.

8.3.4 Filtration Conditions and Rejection Measurements

The prepared membranes were cut into circular coupons with an effective filtration area of 3.142 cm^2 (i.e. membrane diameter = 2 cm) and loaded in a self-assembled dead end filtration setup as shown in Fig. 8.38. The dead end cell was connected to a nitrogen cylinder that provided the applied pressure to force the liquid pass through the membrane. Applied pressure was measured and all tests were carried out at a constant pressure of 1 psig, which was carefully controlled. The dead end cell was initially filled with distilled water and the pure water permeation (PWP) flux was determined. The PWP for each membrane was determined by taking the average of at least three measurements. After the membranes were dried, the cell was filled with the feed water consisting of distilled water spiked with known concentrations of chloroform or monochloroacetic acid. The feed water was then filtered through the membrane at the applied pressure of 1 psig, and the permeate concentration at predetermined intervals was measured. Rejection (R) of chloroform and monochloroacetic acid was calculated using the formula in Eq. (8.27).

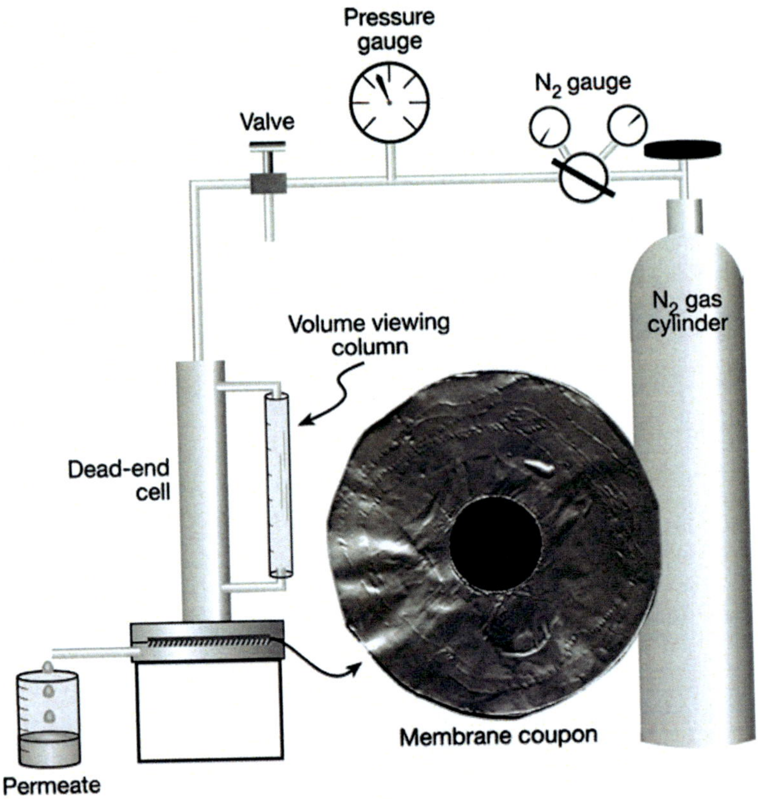

Fig. 8.38 Schematic diagram of dead end filtration cell used in the experiments. (From [17])

$$R = 1 - (c_p - c_w)/(c_f - c_w) \tag{8.27}$$

Where c_p, c_f, and c_w are the concentrations of chloroform or monochloroacetic acid in the permeate, in the feed and the base concentration of the de-ionized (DI) water, respectively. Chloroform concentrations were determined using a total organic carbon (TOC) analyzer using the UV-persulfate method (Phoenix 8000, Rosemount Analytical Inc., Tekmar Dohrmann Division, Santa Clara, CA). Concentration of monochloroacetic acid was determined using a conductivity/TDS meter with RS232C (CON 110 series, Oakton Instruments, Vernon Hills, IL, manufactured by Eutech Instruments, Singapore).

8.3.5 Results and Discussion

The prepared membranes, both before and after carbonization were characterized. A change in the color of the membrane mats during the stabilization and carbonization

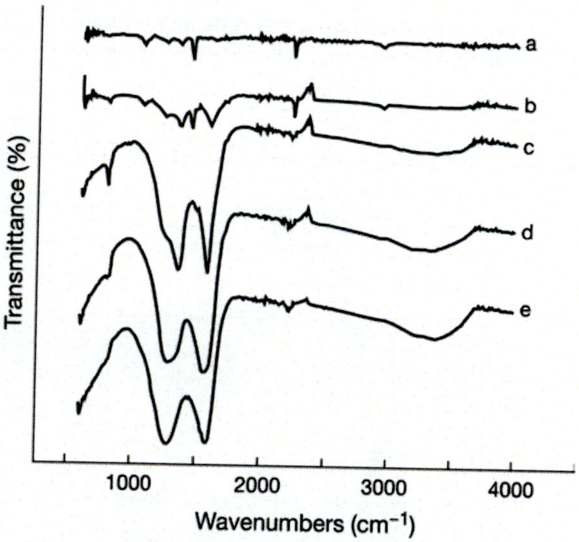

Fig. 8.39 ATR-FTIR spectra of 10% PAN nanofiber membrane under conditions (**a**) air dried at 120 °C, (**b**) stabilized at 250 °C for 6 h, (**c**) carbonized at 400 °C for 4 h, (**d**) carbonize at 500 °C for ¼ h and (**e**) carbonized at 600 °C for ¼ h. (From [17])

process was observed in all experiments. The membranes changed from white after electro-spinning and air drying to reddish brown after stabilization and dark brown/black after carbonization.

ATR-FTIR spectra recorded in the spectral range of 500–4000/cm for PAN dried at 120 °C, PAN stabilized at 250 °C and carbonized between 400 and 600 °C for various intervals are presented in Fig. 8.39. The aliphatic peaks of C-H (bending), C-N (stretching), and C-H (stretching) are observed at 1,454, 2,243 and 2,940/cm, respectively, for PAN nanofiber membranes dried at 120 °C, as shown in Fig. 8.39 for spectra (a). These peaks are still observed when the nanofiber membrane is stabilized at 250 °C as shown in (b). However, for the carbonized membranes, as shown in (c–e), these peaks are not present. Instead, polynuclear peaks at 1,367 and 1,589/cm are observed.

Static contact angle (SCA) measurements carried out on the membranes indicated that all the tested membranes were hydrophobic (SCAs > 90°). The SCAs of PAN membranes air dried at 120 °C were found to be 107 ± 3°. There was a slight increase in SCA when the membranes were subjected to stabilization treatment and carbonization with SCAs of 128 ± 5° and 119 ± 4°, respectively.

Differential scanning calorimetry (DSC) measurements were made in the temperature range of 40–160 °C. Electro-spun PAN nanofiber membranes had a glass transition temperature, T_g, of 90 °C. This is close to 85 °C, the value reported by the supplier. No T_g value was obtainable in the range of temperatures tested for the carbonized naofiber membranes. Rigid aromatic ring structure was found in the carbon nanofiber membrane (CNM) and, as a result, free volume became smaller.

The evidence from the characterization suggests that carbonization of the membranes takes place at temperatures in excess of 400 °C.

Electro-spinning was carried out with PAN solutions of three different concentrations, i.e. 8, 10 and 12 wt%. The nanofiber membranes so obtained were carbonized under identical conditions. In all cases, the membranes were held at 400 °C for 4 h. The SEM images of the membranes before and after carbonization are shown in Fig. 8.40. At least 30 fiber diameters were determined at each condition. Average fiber diameters of the membranes are reported in Table 8.12.

It can be seen from the table that the fiber diameter of the membranes increases with the increasing concentration of PAN before carbonization. At lower PAN concentrations, bead formations were also observed as shown in Fig. 8.40 C and F. However, at higher concentrations, especially at 12 wt% no beads were found.

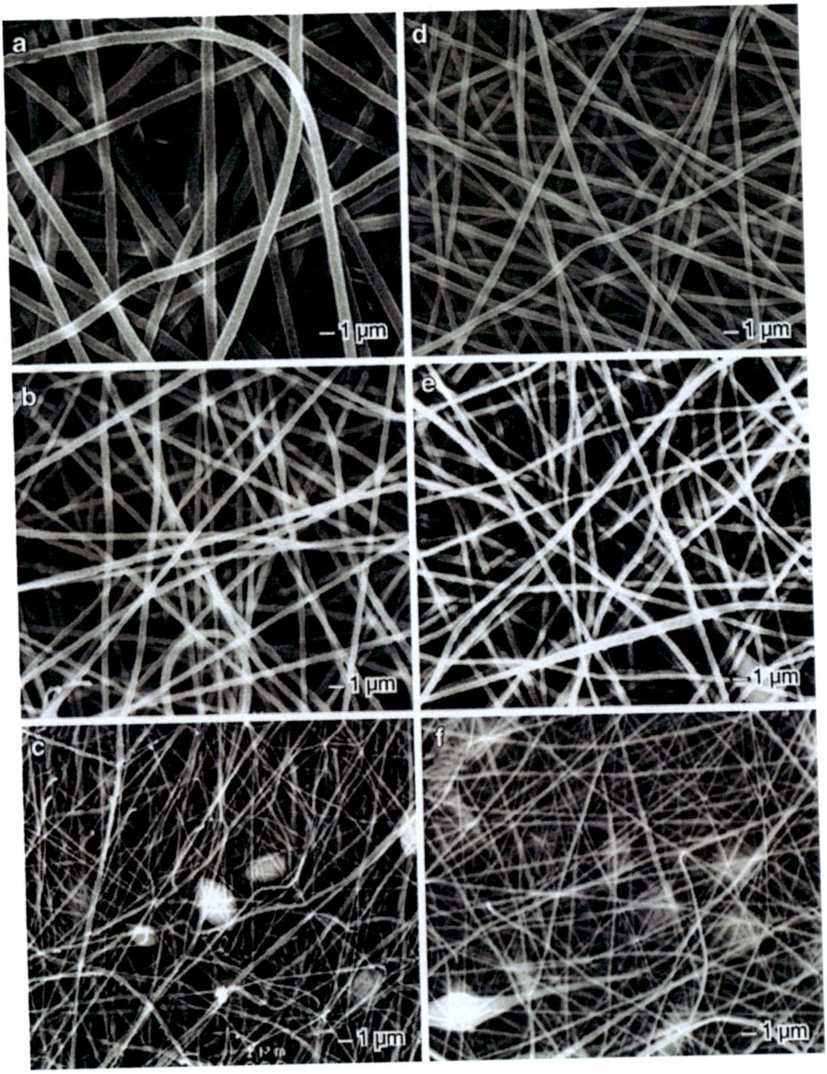

Fig. 8.40 Electro-spun PAN nanofibers before carbonization at (**a**) 12 wt%, (**b**) 10 wt% and (**c**) 8 wt% and after carbonization at (**d**) 12 wt%, (**e**) 10 wt% and (**f**) 8 wt%. (From [17])

Table 8.12 Fiber diameters of PAN electro-spun membranes before and after carbonization

Percentage weight of PAN (wt%)	Fiber diameter, after electro-spinning (nm)	Fiber diameter, after carbonization (nm)
12	640±86	351±83
10	410±34	400±90
8	258±38	245±36

Interestingly, there was a big decrease in fiber diameter after carbonization for the membrane made from 12 wt% PAN, however, no significant change in fiber diameter was observed after carbonization for the membranes made from 10 and 8 wt% PAN. The membranes retain their shape after carbonization; however there is a decrease in weight of the mats due to burn off.

The pore size distribution of the carbonized membranes for polymer concentrations of 12, 10 and 8 wt% are shown in Fig. 8.41a–c, respectively. The mean pore size of the membranes was in the microfiltration range, and the pore size increases with the polymer concentration. These results are in agreement with the fiber diameter of the membranes, which were noted to increase with the polymer concentration. A monodispersed pore size distribution was observed for 10 and 12 wt% polymer concentration, whereas a bidispersed pore size distribution was observed for the 8 wt% polymer concentration. It is hypothesized that this is due to the presence of beads in the 8 wt% polymer concentration, hence there are pores that are blocked by the beads, which results in a bidispersion of the pore size distribution.

Three coupons of carbonized membranes of final thickness 0.18±0.02 mm and carbonized at 400 °C for 4 h, were tested for the removal of chloroacetic acid. The PWP of the membranes at 1 psig were measured to be 5002±220 L/m^2 h. All membranes were filtered with 100 mg/L of monochloroacetic acid (MCAA) and the rejection with volume of permeate collected was determined as shown in Fig. 8.42.

From the graph it is seen that the rejection of monochloroacetic acid averages approximately between 5 and 10% for the three coupons tested. More significantly, although the coupons were prepared and tested separately, the rejection of monochloroacetic acid does not differ considerably. Hence, the results are fairly reproducible.

Removal mechanism of the membranes can be explained as follows. The feed concentration in the dead end cell at the end of each experiment was measured and the concentration of monochloroacetic acid was determined. No change in concentration of the final feed compared with the initial feed concentration was noted in all the experiments. Accumulation of monochloroacetic acid on the feed side due to rejection or back diffusion would result in an increase in concentration of the feed at the end of the experiment. Given the large pore size of the membranes, approximately 0.3–0.44 μm, it was inferred that the removal of monochloroacetic acid was due to its adsorption on the surface of the nanofibers rather than its rejection due to the size exclusion.

The removal efficiency of the membranes for various concentrations of monochloroacetic acid was tested. Membranes of the same thickness and carbonization conditions i.e. 0.18 mm and heated to 400 °C for 4 h were tested. All membranes

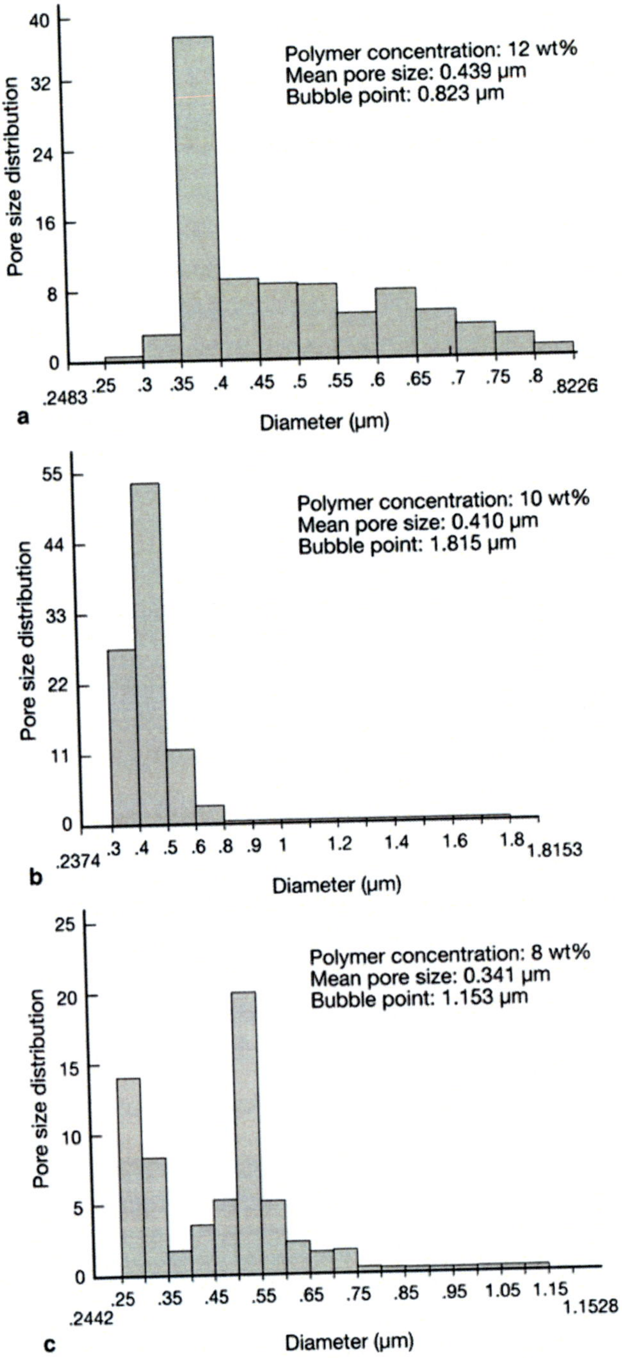

Fig. 8.41 Pore size distribution profile of carbonized membranes made from different PAN concentrations; (**a**) 12 wt%, (**b**) 10 wt%, (**c**) 8 wt%. (From [17])

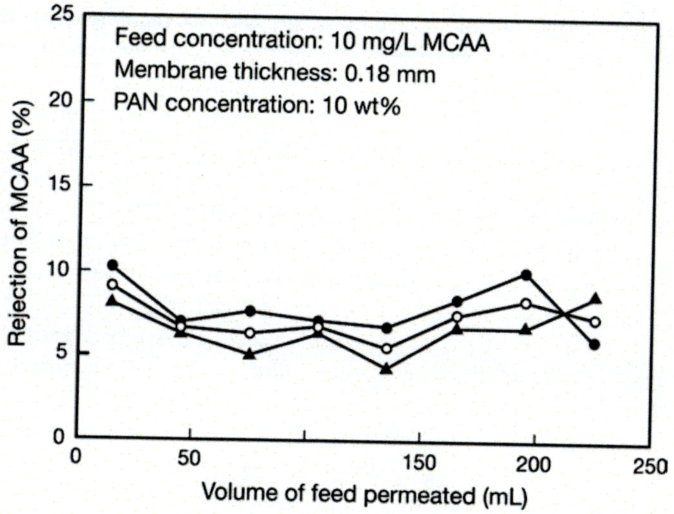

Fig. 8.42 Rejection of monochloroacetic acid (MCAA) using three coupons of carbonized membranes prepared and tested under the same conditions. (From [17])

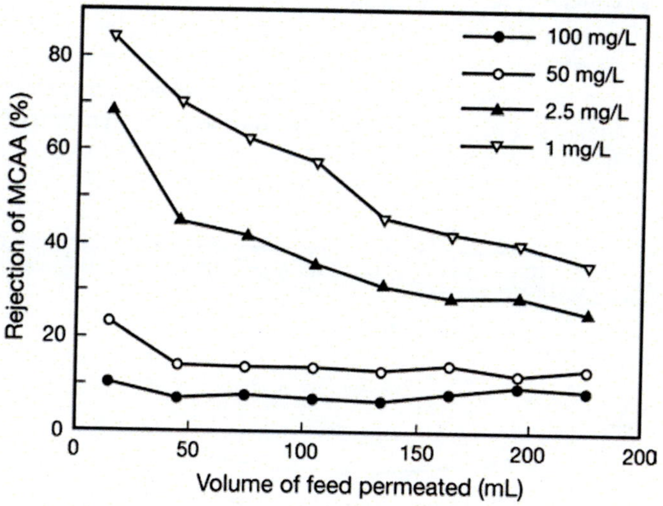

Fig. 8.43 Removal efficiency of carbonized membranes for 100, 50, 2.5, and 1 mg/L concentration of monochloroacetic acid (MCAA) in the feed water. (From [17])

had a PWP of approximately 4500 ± 200 L/m^2 per h. Four different feed waters of monochloroacetic acid concentrations 100, 50, 2.5 and 1 mg/L were separately tested. The removal efficiency of the membranes is shown in Fig. 8.43.

The removal efficiency of monochloroacetic acid by the membrane decreases with its concentration in the feed water. At low concentrations of 1 mg/L high removal efficiency of approximately 85% is observed, which decreases to about

40% after passing through 225 mL of feed through it. However, at higher concentrations e.g. 100 mg/L, the amount rejected is approximately 9%. In addition, at higher concentrations of monochloroacetic acid the rejection reaches a stable value quickly as compared to when the monochloroacetic acid concentration is smaller. This can be understood in terms of the adsorption kinetics of the membrane. The lower concentration of monochloroacetic acid implies that more can be adsorbed on the membrane as compared to when a higher concentration is used, which quickly saturates the adsorption sites on the membrane surface. Hence, through optimization of the membrane it might be possible to improve the rejection efficiency of the membranes.

PAN solutions of three different concentrations, i.e. 12, 10, and 8 wt%, were electro-spun and carbonized. Carbonization conditions were kept constant as per the previous sections. The thickness of the membranes was carefully controlled by adjusting the electro-spinning time such that the final thickness of the carbonized membranes was 0.20 ± 0.02 mm. The PWP of the membranes measured are 3,800, 4,500 and 4,600 L/m^2 h for the 8, 10 and 12 wt% PAN membranes, respectively. The measured filtration fluxes are in agreement with the mean pore size of the membranes, which increases with the polymer concentration. Feed water containing 80 mg/L monochloroacetic acid was then filtered through each membrane and its rejection with volume of feed permeated is plotted in Fig. 8.44.

The general rejection trend of monochloroacetic acid is similar for all membranes. Initially, a high rejection of monochloroacetic acid is observed, which declines reaching a plateau subsequently. From the graph it is noted that carbonized 12 wt% PAN membranes have the highest monochloroacetic acid removal efficiency. This is likely due to the absence of beads as found in the 8 wt% PAN membranes and the smaller carbonized fiber diameter compared to the 10 wt% membranes.

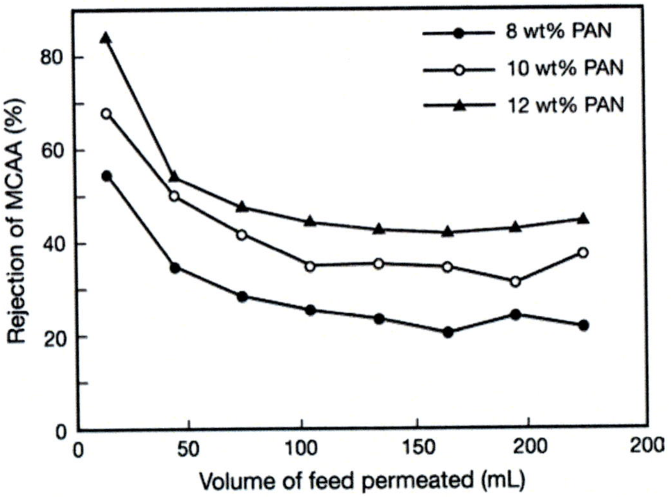

Fig. 8.44 Rejection of monochloroacetic acid (MCAA) using carbonized membranes of initial PAN concentration of 8, 10 and 12 wt%. (From [17])

Further experiments were then carried out with the 12 wt% electro-spun nanofiber membranes.

Since adsorption was postulated as the main removal mechanism of monochloroacetic acid from the water, increasing membrane thickness would result in greater removal capacity. Thicker membranes were produced by increasing the electro-spinning time and folding several layers of electro-spun membranes together followed by heat treatment to form a thick mat for membrane filtration. Two carbonization conditions i.e. (CC-1) heated from 250 to 400 °C at 5 °C/min and held at 400 °C for 4 h or (CC-II) heated from 250 to 500 °C at 5 °C/min and held at 500 °C for 15 min were carried out on separation membranes. Using these processes, carbonized membranes of thickness 0.48 mm under two conditions described above were prepared. Prepared membranes were tested with monochloroacetic acid of concentration 100 mg/L in water.

Membranes prepared under carbonization conditions (CC-I) and (CC-II), have PWP of 4251 ± 94 and 6006 ± 500 L/m^2 h, respectively. The higher PWP rate in membranes of carbonization condition (CC-II) is a result of the smaller reduction in fiber diameter observed for these membranes as compared to those of carbonization condition (CC-I). Although in carbonization condition (CC-II) as the carbonization temperature increased, the carbonization time was reduced significantly. It is hypothesized that this prevented more burn-off of the fiber resulting in a smaller reduction of the fiber diameter and hence a higher porosity. Correspondingly, the membranes prepared using carbonization condition (CC-I) resulted in a higher removal efficiency of monochloroacetic acid as compared to membranes prepared using carbonization condition (CC-II) as shown in Fig. 8.45.

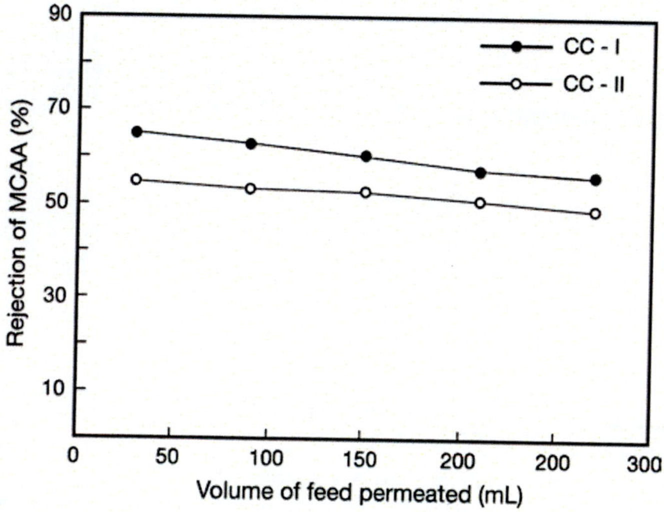

Fig. 8.45 Rejection of monochloroacetic acid (MCAA) using a thicker membrane prepared using two different carbonization conditions i.e. (CC-I) heated from 250 to 400 °C at 5 °C/min and held at 400 °C for 4 h and (CC-II) heated from 250 to 500 °C at 5 °C/min and held at 500 °C for 15 min. (From [17])

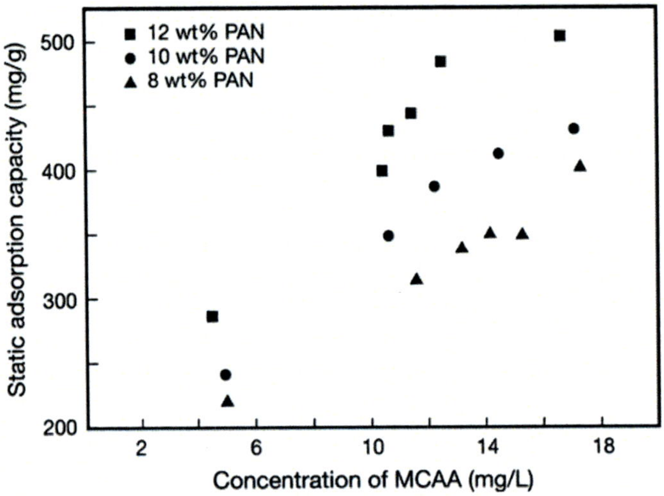

Fig. 8.46 Adsorption capacity of monochloroacetic acid (MCAA) on carbonized nanofiber membranes made from PAN solutions of different concentrations. The membranes were prepared by heating to 250 °C for 6 h and then carbonizing at 600 °C for 4 h under nitrogen. (From [17])

As noted, the rejection of monochloroacetic acid was around 65–56% and 55–49% for the membrane carbonized using condition (CC-I) and condition (CC-II), respectively. The rejection obtained with this thicker membrane is approximately 3 times that of the thinner membrane (0.18 mm), hence the rejection increases almost proportionately with the thickness of the membrane.

The adsorptions of monochloroacetic acid on carbonized membranes of different polymer concentrations are shown in Fig. 8.46. From the graph it is observed that the adsorption capacity increases with the polymer concentration within the range tested. The adsorption capacity for this 12 wt% carbonized PAN in the range 4–18 mg/L was 287–504 mg/g. The adsorption capacity for 10 wt% carbonized PAN in the range 5–17 mg/L was 242–431 mg/g. For the carbonized 8 wt% PAN the adsorption capacity in the range of 5–17 mg/L was 221–402 mg/g. This was due to its high concentration of beads for 8 wt% PAN, which resulted in fewer adsorption sites for the removal of monochloroacetic acid. The data was found to fit the Freundlich adsorption isotherm shown below,

$$(x/m) = kC^{1/n} \tag{8.28}$$

Where x and m are the masses of the adsorbate and adsorbent, respectively, C the equilibrium concentration of adsorbate and k and n constants. The regression coefficients R^2 and constants k and n are shown in Table 8.13.

Attempts were made to regenerate the filters after use in the membrane adsorption process. Several chemical treatment methods were carried out on the used membranes e.g. immersing them in warm water, sodium hydroxide (NaOH) and hydrochloric acid (HCl). Three membranes were tested for each chemical treatment

Table 8.13 The values of regression coefficient (R^2) and the constants k and n for the Freundlich equation curve fitting

PAN concentration (wt%)	R^2	k	N
12	0.966	145.6	2.21
10	0.992	111.7	2.06
8	0.981	105.4	2.21

method used. Therefore, altogether 9 membrane coupons were used in the experiments. The membranes were all prepared using the same conditions, i.e. 10 wt% PAN and carbonized at 400 °C for 4 h. The thickness of the membranes was approximately 0.24 ± 0.02 mm. All membranes were first tested for rejection of monochloroacetic acid of concentration 20 mg/L. 50 mL of solution was passed through each membrane and the rejection was measured. After the membranes were used for the filtration of monochloroacetic acid solution, the membranes were removed and rinsed with deionized water before being placed in a petri dish containing either (1) deionized (DI) water at 80 °C, (2) 0.05 M NaOH at room temperature, (3) 0.05 M HCl at room temperature for 1 day each. After this the membranes were removed from the petri dish and rinsed with DI water before being dried at 105 °C for 1 day. Subsequently, 50 mL of monochloroacetic acid solution of the same concentration, i.e. 20 mg/L was filtered through the membrane and the rejection determined. The rejection before and after each treatment are shown in Fig. 8.47.

The regeneration potential of the membrane is calculated using the ratio of the rejections (rejection after treatment/rejection before treatment). A higher regeneration potential indicates a more effective chemical treatment method. For soaking in

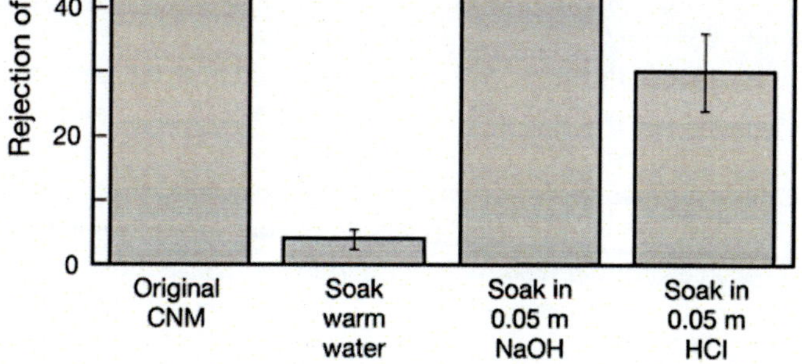

Fig. 8.47 Rejection of monochloroacetic acid (MCAA) before regeneration treatment and after regeneration treatment by soaking in warm water (80 °C), 0.05 M NaOH, 0.05 M HCl. All membranes were prepared from 10 wt% PAN solution and stabilized at 250 °C for 6 h and then carbonized at 400 °C for 4 h. (From [17])

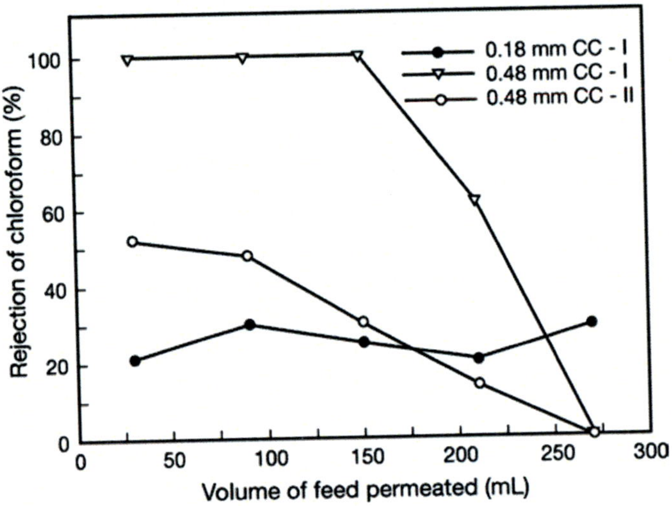

Fig. 8.48 Removal efficiency of chloroform of concentration 100 mg/L using membranes of thickness (0.18 mm, CC-I) 0.18 mm carbonized at 400 °C for 4 h, (0.48 mm, CC-I) 0.48 mm carbonized at 400 °C for 4 h, (0.48 mm, CC-II) 0.48 mm carbonized at 500 °C for 15 min. (From [17])

warm water (80 °C) the regeneration potential was approximately 0.1. Regeneration potential when treated with base and acid are 1.2 and 0.6, respectively. Membranes when soaked in NaOH were able to achieve the same if not higher rejection of chloroacetic acid. However, soaking in warm water was not a good way of regeneration. The data indicates that the regeneration of the membranes can be achieved by soaking them in a basic solution followed by rinsing.

The removal of chloroform using carbonized membranes was then tested. Dilute solutions of chloroform dissolved in DI water were prepared and used as feed water. Membranes of thickness 0.18 and 0.48 mm produced using carbonization conditions (CC-I) and (CC-II) as described above were tested for the remove al of 100 mg/L of chloroform dissolved in DI water. The results are shown in Fig. 8.48.

The removal efficiency of chloroform using a membrane of thickness 0.18 mm was approximately 25% and remained relatively constant throughout the experiment. When the membrane thickness was increased to 0.48 mm the initial rejection was higher, but decreased with an increase in the permeate. Membranes produced using carbonization condition (CC-I) also resulted in better chloroform rejection than carbonization condition (CC-II). For the thicker membranes produced using carbonization condition (CC-I), the initial rejection of chloroform was 100%. However, this value decreased after 150 mL was permeated. Based on this data, the value of chloroform adsorbed per unit weight of the membrane with the carbonization condition of 400 °C for 4 h can be evaluated. The measured membrane weight/area is 8.612×10^{-3} g/cm^2. From the data, it is known that the membrane becomes saturated after passing 150 mL of chloroform solution with a concentration of 100 ppm through it. Therefore, 0.015 g chloroform is adsorbed by

2.71×10^{-2} g (3.142 cm$^2 \times 8.612 \times 10^{-3}$ g/cm^2) of the membrane. This corresponds to (0.015/0.027055)=0.554 g (554 mg) of chloroform adsorbed per unit g of CM. It should be noted that this is a conservative estimate of the adsorption capacity of the carbonized membrane, as the membrane is not completely saturated after 150 mL is permeated as seen in Fig. 8.48. This is higher than the adsorption capacity of commercial activated carbon and activated carbon filter, which ranges between 74 and 128 mg/g for an initial chloroform concentration of 90 mg/L at 30 °C.

To further test the removal efficiency of the membrane for chloroform and monochloroacetic acid, denser membranes of thickness 0.6 mm were prepared. The concentration of chloroform and monochloroacetic acid was also reduced to 20 mg/L. The rejection of the species using membranes produced by carbonization conditions (CC-I) and (CC-II) are shown in Fig. 8.49.

As noted from above, chloroform is effectively removed from the water regardless of the carbonization condition. On the other hand, the membranes are still not able to achieve complete rejection of the monochloroacetic acid. The rejection of monochloroacetic acid for the membranes prepared using condition (CC-I) and condition (CC-II) are between 86–66% and 69–43%. The reasons for the better adsorption of chloroform compared to the monochloroacetic acid are due to its smaller molecular size and lower solubility in water (higher hydrophobicity). Furthermore, the adsorption of monochloroacetic acid is likely to be affected by the pH of the water, which affects the molecules' dissociation degree. This was not addressed as all experiments were carried out at a constant pH of 6. Similar observations have been noted for the lower haloacetic acid adsorption on granular activated carbon as compared with trihalomethanes [18].

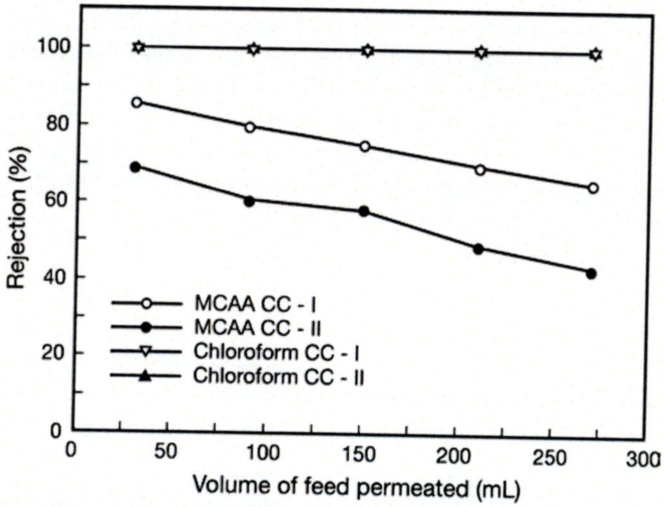

Fig. 8.49 Rejection of chloroacetic acid and chloroform using membranes of thickness 0.60 mm of 12 wt% PAN produced using carbonization conditions (CC-1) 400 °C for 4 h, (CC-II) 500 °C for 15 min. (From [17])

8.3.6 *Conclusions*

The removal of disinfection by-products (DBPs) was attempted using carbonized nanofilter membranes. Chloroform and monochloroacetic acid were used as the model for DBP compounds. The main mechanism of removal of the monochloroacetic acid and chloroform was observed to be due to adsorption. The removal efficiency of the species increased when the feed concentration decreased or the surface area of the carbonized nanofiber membrane increased.

Chloroform was more effectively removed from the water as compared to monochloroacetic acid using the carbonized membranes. This is likely due to its smaller size and much lower solubility in water. The carbonized membranes were able to achieve at least 554 mg/g removal of chloroform through filtration at very low pressure. The preliminary results suggest that carbonized nanofilter membranes could provide a useful means for the removal of trace concentrations of DBPs. Further optimization of the carbonized nanofiber, by controlling the shrinkage so that higher temperature can be achieved and by oxidation of the surface such as that activated carbon nanofilter membranes can be produced, would result in better performance.

8.4 Mixed Matrix Membranes (MMMs)

Mixed matrix material has been developed with the ultimate goal of encapsulating the molecular sieve entities within a high-performance continuous polymer matrix. In gas separation technology, this hybrid material stimulated a great research interest due to its potential to overcome unresolved limitations in conventional membrane materials. It is believed that this emerging approach is viable to synergistically combine the good features of both phases and overcome limitation in each medium. Porous nature of molecular sieve material provides the justification of high productivity and selectivity in mixed matrix membrane (MMM). The polymer phase determines the minimum separation performance of the membrane and provides the desirable mechanical properties and economical processibility of MMM. However, currently, poor polymer-sieve adhesion becomes one of the critical issues in mixed matrix membrane development causing the realization of an ideal MMM becoming more challenging.

8.4.1 *Membranes Filled with Activated Carbons or CMSs*

Anson et al. fabricated a MMM that consists of acrylonitrile-butadiene styrene (ABS) copolymer as the polymer matrix and activated carbon (AC) as the inorganic filler for CO_2/CH_4 separation [19]. As the fillers, two different kinds of activated carbon AC1 and AC2 were used. As Fig. 8.50 shows AC1 has a micropore structure

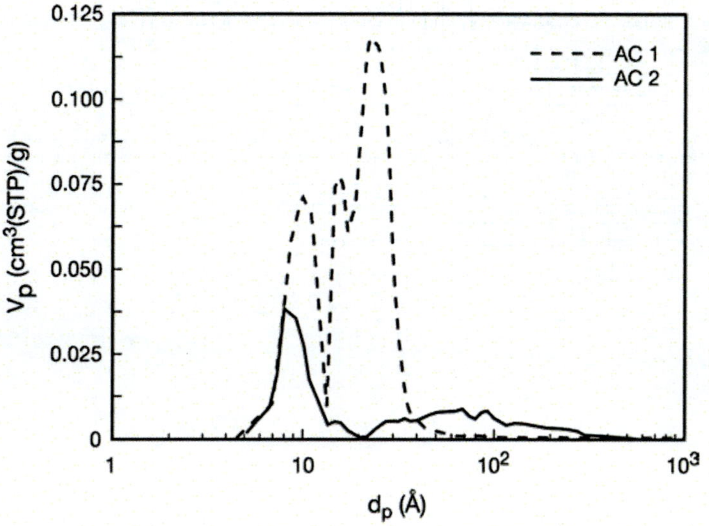

Fig. 8.50 Pore size distribution of AC1 and AC2 activated carbons obtained by the DFT method. (From [19])

($7<d_p<30$ Å) while AC2 includes both micro-and meso-pores with a wider distribution ($7<d_p<400$ Å). The particles sizes are 0.90 and 4.47 μm for AC1 and AC2, respectively.

Figure 8.51 depicts the SEM pictures of AC1 incorporated MMMs. Tight interfacial contact between the particle and the polymer seems to exist. Figure 8.52 shows the SEM pictures of AC2 incorporated MMMs. When the AC2 content is as high as 33%, contacts between the carbon clusters and between void spaces surrounding

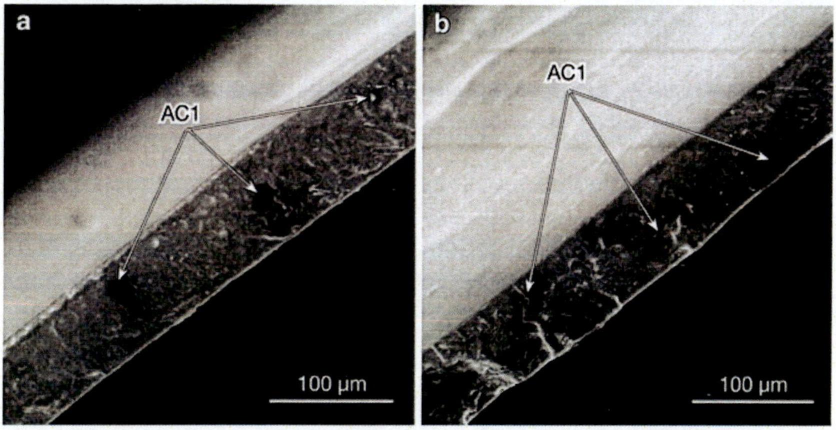

Fig. 8.51 SEM micrographs of the cross-section of MMMs; (**a**) MMM containing 2% AC1, (**b**) MMM containing 7% AC1. (From [19])

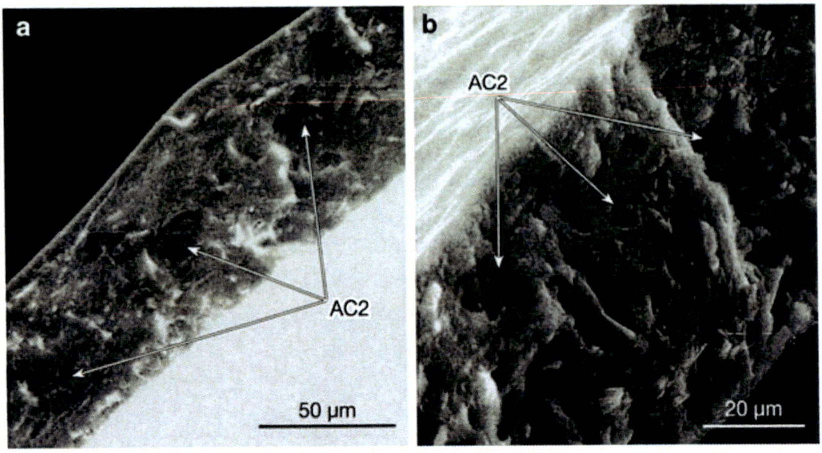

Fig. 8.52 SEM micrographs of the cross-section of MMMs; (**a**) MMM containing 25% AC2, (**b**) MMM containing 33% AC2. (From [19])

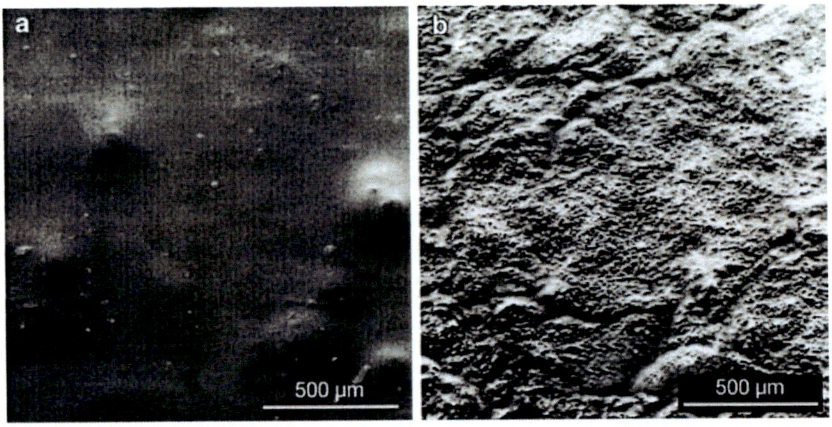

Fig. 8.53 SEM micrographs of the top surface of MMMs; (**a**) MMM containing 10% of AC1, (**b**) MMM containing 33% AC2. (From [19])

the clusters take place. Figure 8.53 displays the SEM pictures of the top surface of the membrane. When 10% of AC1 is incorporated the surface is smooth and the particles are embedded into the polymer. However, as high as 33% carbon (AC2) concentration makes the surface very rough.

From Table 8.14 it is clear that both permeability and selectivity increase with an increase in the amount of AC in the membrane. Furthermore, the incorporation of AC2 produces more effective results than the incorporation of AC1.

Although carbon molecular sieves (CMSs) and zeolites offer very attractive permeation properties and the selectivites are significantly higher than that of polymeric materials, processing challenges and high costs are still hindering their industrial applications.

Table 8.14 CO_2 permeability and CO_2/CH_4 selectivity of activated carbon incorporated ABS membranes

	ABS	AC1 conc. in ABS-AC1 MMM (wt%)				AC2 conc. in ABS-AC2 MMM (wt%)			
		2	5	7	10	20	25	33	40
CO_2 permeability, (Barrer)	2.87	4.31	5.43	7.96	10.81	7.49	9.82	13.16	20.50
CO_2/CH_4 selectivity	24.11	26.6	29.7	31.8	34.3	28.4	32.5	41.0	50.5

For example, it is estimated that a zeolite membrane module would cost around US\$ 3,000/m^2 of active area compared to US\$ 20/m^2 for existing gas separation polymeric hollow fibre membrane modules. In addition, zeolite and CMSMs require high temperature treatment. Even though the problems involved in the manufacturing process are solved, there remain still issues such as handling, which will greatly limit their usefulness. MMMs were proposed as an alternative approach to combine the benefits of high selectivity and productivity of CMSM and the desirable mechanical properties and economical processing capability of polymeric membranes by incorporating the molecular sieves into the base polymeric membrane [20].

In Vu et al.'s work, the CMSMs were laboratory made by the high temperature pyrolysis of the PI, Matrimid® 5218. The polymer solution (Matrimid® 5218 5–10 wt% in dichloromethane) was cast into a film on a glass surface either by pouring the solution into a circular metal ring or with a uniform thickness film applicator. The film thickness was 2 mils. To remove the residual solvent the film was heated at 100 °C in a vacuum oven. The polymer film was then pyrolysed in a quartz tube furnace. The detailed description of the pyrolysis procedure is given elsewhere [21, 22]. After the pyrolysis the membranes were subjected to permeation experiments.

The CMS films were further crushed into fine particles by a ball mill/mixer after being dried at 250 °C. After milling, the CMS particles were heated at 250 °C at least overnight prior to any characterization experiments. The purpose of the heat treatment is to remove moisture that may have been sorbed during storage. The size distribution of the particle ranged from submicron to 10 μm.

MMMs were prepared by casting the slurry of fine CMS particles (15–20 wt%) dispersed in the polymer solution. The fine CMS powder (0.2–0.4 g) was dispersed in dichloromethane (3 mL) in a vial (40 mL) via sonication. After sonication, the vial was allowed to stand for 6 h to let larger particles settle to the bottom. The finer particles still remaining in dichloromethane were decanted into a new 40 mL vial with a pipette. Generally, 80–90% of the original CMS particles were captured in the new vial. After sonication, the particles were primed by adding about 10 wt% of the total amount of polymer into the CMS slurry. The slurry was then well mixed by rolling the vial on rotating parallel rollers for at least 12 h. The remaining amount of the polymer was finally added to the slurry, which was again mixed on the rotating rollers for 12 h. Normally, the solid content (CMS + polymer) was 15–20 wt%.

A stainless steal film applicator was used to cast the polymer solution into a film under a controlled atmosphere of near saturated solvent vapor. The solvent was removed slowly over a 12 h time period. The resulting CMS/polymer film had the

Table 8.15 Permeation properties of the CMS films pyrolyzed from flat polymer films of Matrimid® 5218 [data taken from 20]

CMS	Pyrolysis condition	Permeability (Barrer)				Selectivity	
		CO_2	CH_4	O_2	N_2	CO_2/CH_4	O_2/N_2
CMS 550-2	550 °C (2 h) vacuum	1250	20	435	50	63	8.7
CMS 550-8	550 °C (8 h) vacuum	375	4.2	166	18.2	89	9.1
CMS 800-2	800 °C (2 h) vacuum	43.5	0.21	24	1.8	200	13.3

Pure gas permeation at 50 psia for CO_2 and CH_4, 21 psia for O_2, and 79 psia for N_2, temperature 35 °C

thickness of 30–60 μm. The film could be peeled off easily. Finally, the film was further dried after initial evaporation at a temperature of about 100 °C for at least 12 h in a vacuum oven to remove the solvent.

Table 8.15 summarizes the permeation properties of the CMS membranes.

Among these CMSs, CMS 800-2 is particularly interesting because of its high selectivities and still reasonable permeabilities. CMS-800-2 had the skeletal density of 1.69 g/cm^3, micropore volume of 0.186 cm^3/g and the bulk density of 1.29 g/cm^3.

An attempt was made to confirm the similarity in the permeability of the CMS powder and the CMS film. For example, Fig. 8.54 shows the WAX diffraction spectra of CMS in both film and powder from. As can been seen from the figure, the

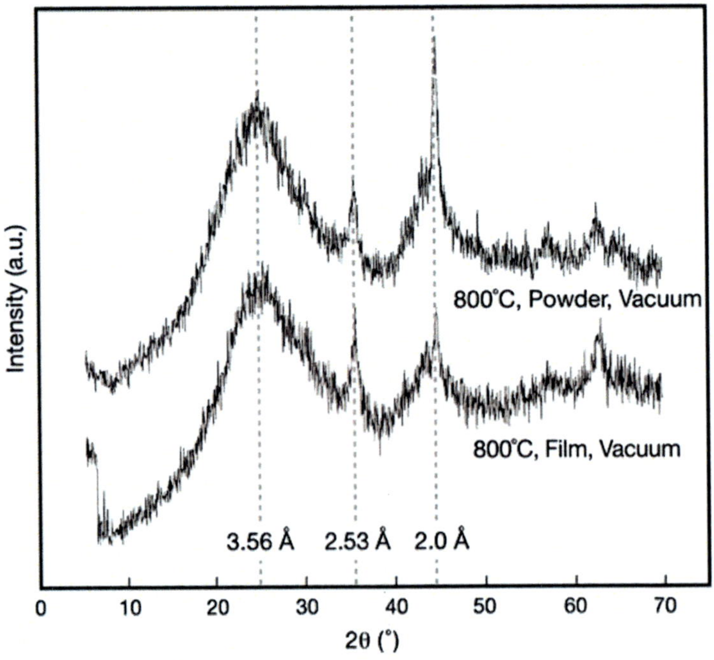

Fig. 8.54 WAXD spectra of CMS materials formed via pyrolysis with different polymer precursor forms (film vs. powder). Precursor polymer (Matrimid® 5218) and pylorysis temperature protocol are the same for both forms. (From [20])

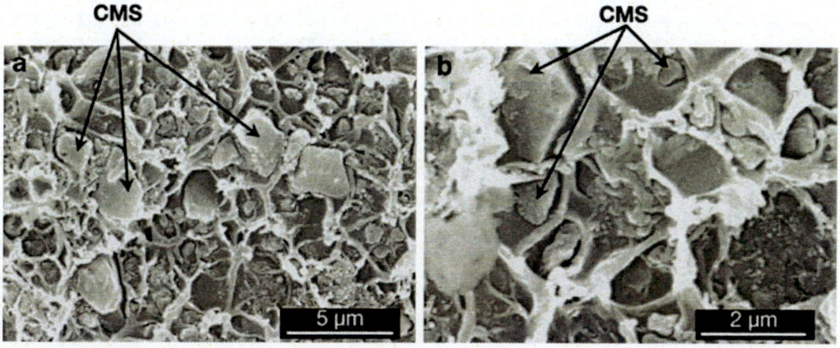

Fig. 8.55 SEM cross-sectional image of early MMMs containing carbon molecular sieve insert (CMS 800-2). Continuous polymer matrix is Ultem® 1000 with CMS loading of 35 vol%. Poor polymer-sieve contact is observed. (From [20])

characteristic peaks of the diffraction spectra were the same for both forms of CMS, confirming that the structures of CMSs in both forms are similar.

Three problems arose in the initial stage of the membrane fabrication; surface pattern effects, sedimentation aggregation and poor polymer sieve contact. The first problem arose by the rapid evaporation of solvent that causes the temperature gradient. The surface tension driven force forms convective flow cells that result in uneven clustering of the particles and appearance of irregular pattern on the membrane surface. Sedimentation of CMS particles led to aggregation with nearby sieve particles. Lastly, the incompatibility between the CMS particles and polymer was also observed. Figure 8.55 shows the SEM images of an MMM fabricated in the earlier stage of the research, revealing submicron (0.1–0.2 μm) gap between the particle and the surrounding polymer matrix due to poor adhesion.

Sonication and decantation of the smaller CMS particles and forming more viscous mixed matrix slurries all contributed to the significant reduction of sedimentation and aggregation problem. Sizing of the CMS particles with a small amount of the base polymer also promoted the particle-polymer adhesion. Slow solvent evaporation also helped solve all three problems. Figure 8.56 shows the membrane after improvement. It shows smaller CMS particles (mostly submicron range) and better distribution of particles with better adhesion between the polymer and the particle.

The results of DSC experiments are given in Fig. 8.57. With CMS loading T_g increases both for Ultem® 1000 and Matrimid® 5218. These results indicate that the incorporation of the particles can affect the T_g of the continuous polymer phase. This is due to the fact that steric constraint was imposed by the CMS particles on neighboring, otherwise mobile polymer segments and thus increases the T_g.

In Fig. 8.57, Ultem® 1000 showed 2–5 °C increase from the pure Ultem® 1000 by the loading of CMS. In contrast all Matrimid® 5218 loaded membranes show an increase in T_g of ca. 15 °C. What caused this difference is currently not clear. Table 8.16 summarizes the pure gas permeation data of the mixed matrix membranes.

For both polymers, the permeabilities of the MMM are generally intermediate between the constituents. For the fast gases, the permeability of O_2 and CO_2

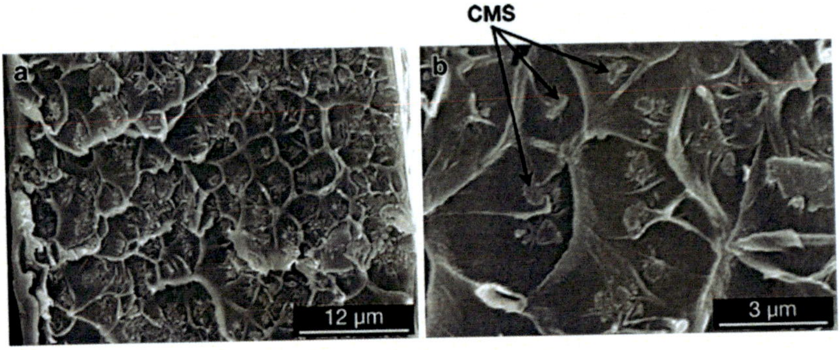

Fig. 8.56 SEM cross-sectional image of early MMMs containing caron molecular sieve insert (CMS 800-2). Continuous polymer matrix is Ultem® 1000 with CMS loading of 19 vol%. Improved polymer-sieve contact is observed. (From [20])

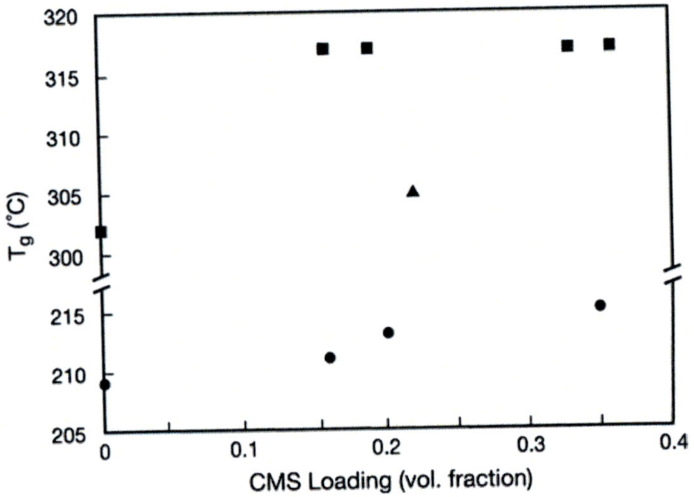

Fig. 8.57 Increases in glass transition temperature, T_g, over a range of CMS loadings in the mixed matrix film; (•) Ultem® mixed matrix film, (■) Marimid® mixed matrix film, (▲) Marimid® mixed matrix film (primed with Ultem®, 3.8 wt%). (From [20])

increases by up to 42% (O_2) and 26% (CO_2), respectively, for MMM of Matrimid® 5218 over pure Matrimid® 5218, and by up to 187% (O_2) and 209% (CO_2), respectively, for MMM of Ultem® 1000 over the pure Ultem® 1000. CO_2/CH_4 selectivity was enhanced from 35.3 (pure Matrimid® 5218) and 38.8 (pure Ultem® 1000) to 51.7 (for 36 vol% CMS in Matrimid® 5218) and 53.7 (for 35 vol% CMS in Ultem® 1000). Similar results were obtained for O_2/N_2 selectivity data. These results were further plotted in Figs. 8.58 and 8.59. It can be noticed that the data are moving towards the desired direction (towards upper right corner) as the CMS loading is increased.

Table 8.16 Permeation properties of the mixed matrix films using Ultem® 1000 and Matrimid® 5218 as matrices for various CMS 800-2 loadings (data taken from [20])

Mixed matrix film	Permeability (Barrer)				Selectivity	
	CO_2	CH_4	O_2	N_2	CO_2/CH_4	O_2/N_2
Matrimid® 5218	10.0	0.28	2.12	0.32	35.3	6.6
CMS 800-2	44.0	0.22	22.0	1.65	200	13.3
CMS 17 vol%	10.3	0.23	2.08	0.29	44.4	7.1
CMS 19 vol%	10.6	0.23	2.41	0.35	46.7	7.0
CMS 33 vol%	11.5	0.24	2.70	0.38	47.5	7.1
CMS 36 vol%	12.6	0.24	3.00	0.38	51.7	7.9
Ultem® 1000	1.45	0.037	0.38	0.052	38.8	7.3
CMS 800-2	44.0	0.22	22.0	1.65	200	13.3
CMS 16 vol%	2.51	0.058	0.56	0.071	43.0	7.9
CMS 20 vol%	2.90	0.060	0.71	0.090	48.1	7.9
CMS 35 vol%	4.48	0.083	1.09	0.136	53.7	8.0

Upstream pressure, 50 psia: temperature, 35.0 °C

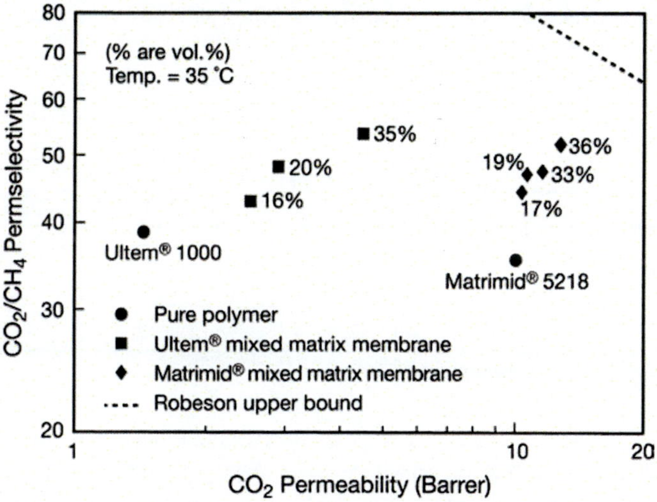

Fig. 8.58 CO_2/CH_4 permeation properties of MMMs using Ultem® 1000 and Matrimid® 5218 as matrices for various loadings of CMS particles. Pure gas permeation measurements were done at 50 psia upstream pressure and at 35 °C

MMM was also fabricated by priming the CMS particles in a small quantity of Ultem® 1000 and then the primed particles were dispersed into the Matrimid® 5218 solution. The use of Ultem® 1000 for the priming polymer and Matrimid® 5218 for the polymer matrix was possible since these two polymers are miscible; i.e., the DSC measurement gives only a single value of T_g for the mixture of these two polymers. As well, the MMM also exhibits only one T_g intermediate between those of Ultem® 1000 and Matrimid® 5218. The performance data of this MMM from the single gas permeation test is shown in Table 8.17.

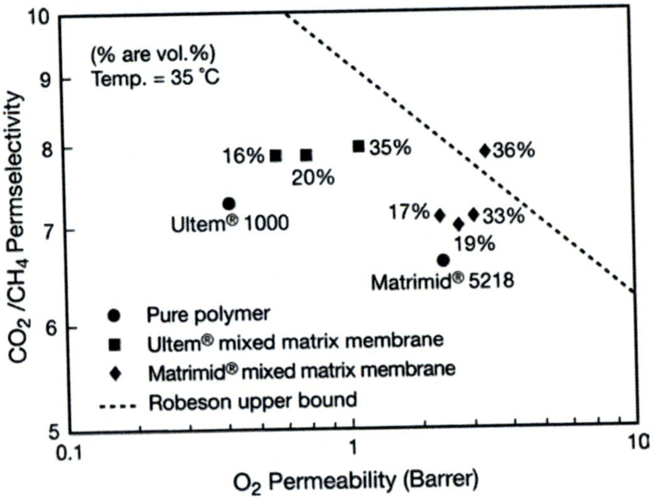

Fig. 8.59 O_2/N_2 permeation properties of MMMs using Ultem® 1000 and Matrimid® 5218 as matrices for various loadings of CMS particles. Pure gas permeation measurements were done at 50 psia upstream pressure and at 35 °C

Table 8.17 Permeation properties of MMM using a small amount of Ultem® 1000 as a priming polymer and Matrimid® 5218 as the polymer matrix with 22 vol% loading of CMS 800-2

Constituents	Permeability (Barrer)				Selectivity	
	CO_2	CH_4	O_2	N_2	CO_2/CH_4	O_2/N_2
Ultem® 1000	1.45	0.037	0.38	0.052	38.8	7.3
Matrimid® 5218	10.0	0.28	2.12	0.32	35.3	6.6
CMS 800-2	44.0	0.22	22.0	1.65	200	13.3
MMM	9.40	0.21	2.14	0.293	45.2	7.3

Feed gas pressure, 50 psia; temperature 35 °C

The data for MMM is intermediate between the two component polymers. Effect of CMS is found in the enhanced selectivity. Mixed gas permeation was examined at the pressures up to 900 psia and with a gas mixture 10% CO_2/90% CH_4, using a Ultem® 1000-CMS MMM (35 vol% CMS). The data are summarized in Fig. 8.60. Pure gas permeation data are also included in Fig. 8.60. Both CO_2 permeability and CO_2/CH_4 permselectivity are greater for MMM than for pure Ultem® 1000. Both decrease gradually as the feed pressure is increased.

Table 8.18 further shows the effect of heat treatment at a temperature above T_g made on a defective SMM membrane with 35 vol% of CMS loading.

Heat treatment resulted in remarkable increase in the selectivity. However, the permeability decreased even to a level lower than the Ultem® 1000 membrane without incorporating CMS.

Vu et al. further attempted to analyze their data based on the Maxwell and the Bruggeman models [23]. The mathematics of modeling the gas transport through MMMs is rather complex. Several models have been proposed. A model useful to

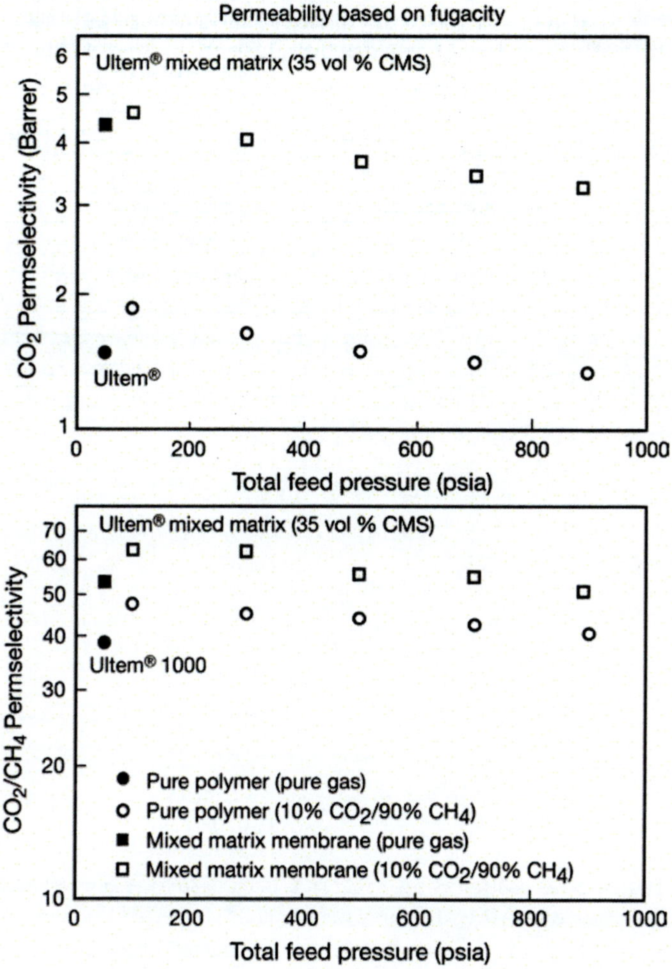

Fig. 8.60 Mixed gas permeation results for Ultem® 1000 based MMM containing 35 vol% CMS. Experimental conditions are: feed gas composition, $CO_2/CH_4 = 10/90$ and temperature 35 °C. (From [20])

Table 8.18 Effect of above T_g thermal annealing (at 250 °C) of a defective Ultem® 1000 MMM fabricated in the early stage of the work

Mixed matrix film	Permeability (Barrer)				Selectivity	
	CO_2	CH_4	O_2	N_2	CO_2/CH_4	O_2/N_2
Ultem® 1000	1.45	0.037	0.38	0.052	38.8	7.3
CMS 800-2	44.0	0.22	22.0	1.65	200	13.3
MMM defective	2.74	0.097	0.60	0.095	28.2	6.3
MMM after heat treatment	0.88	0.018	0.27	0.035	48.3	7.7

Feed gas pressure 50 psia; temperature 35 °C

predict the permittivity of a dielectric was presented by Maxwell [24] in 1873 and this model is by far the most well known among models when applied to the prediction of gas transport through MMMs. According to the model the permeability of MMMs can be calculated for a dilute dispersion of ellipsoids by

$$P_{eff} = P_c \left[\frac{nP_d + (1-n)P_c - (1-n)\phi_d(P_c - P_d)}{nP_d + (1-n)P_c + n\phi_d(P_c - P_d)} \right] \tag{8.29}$$

Where P_{eff} is the effective permeability of a gas penetrant in a MMM with a volume fraction of ϕ_d of the dispersed phase d in a continuous matrix phase c, P_c and P_d represent the permeability in the continuous and dispersed phase, respectively, and n is the shape factor of the dispersed particles. The limit of $n=0$ corresponds to the parallel transport of gas through side by side layers of continuous and dispersed phase. In this case Eq. (8.29) is simplified by

$$P_{eff} = (1-\phi_d)P_c + \phi_d P_d \tag{8.30}$$

The limit of $n=1$ corresponds to the transport of gas through the two phases connected in series:

$$P_{eff} = \frac{P_c P_d}{(1-\phi_d)P_d + \phi_d P_c} \tag{8.31}$$

The result for dilute suspension of spherical particles is known as the Maxwell equation:

$$P_{eff} = P_c \left[\frac{P_d + 2P_c - 2\phi_d(P_c - P_d)}{P_d + 2P_c + \phi_d(P_c - P_d)} \right] \tag{8.32}$$

The Maxwell equation as written in Eq. (8.32) has been used by many researchers to describe the behavior of MMMs. It was however later suggested that the Maxwell equation is applicable for the low loading level of dispersed phase since it was assumed that the streamline around the dispersed articles were not affected. Instead the Bruggeman model was recommended in which the effect of the particles added to the dilute suspension is considered. The Bruggeman equation is given by

$$\left[\frac{(P_{eff}/P_c) - (P_d/P_c)}{1 - (P_d/P_c)} \right] \left(\frac{P_{eff}}{P_c} \right)^{-1/3} = 1 - \phi_d \tag{8.33}$$

Equations (8.32) and (8.33) show fairly good agreement up to the particle loading of ca. 20%. Beyond the above loading level, the particle addition to the dilute dispersion takes effect.

After calculating the effective permeability for the individual gases, the ideal separation factor can simply obtained by

$$\alpha_{A/B} = \frac{(P_{eff})_A}{(P_{eff})_B} \tag{8.34}$$

Table 8.19 Permeation properties of pure polymer matrices and the CMS 800-2 used for the insert (This is an excerpt of data given in Table 8.17.)

MMM components	Permeability (Barrer)				Ideal separation factor	
	CO_2	CH_4	O_2	N_2	CO_2/CH_4	O_2/N_2
Matrix phase						
Matrimid® 5218	10.0	0.28	2.12	0.32	35.3	6.6
Ultem® 1000	1.45	0.037	0.38	0.052	38.8	7.3
Sieve phase						
CMS 800-2	44.0	0.22	22.0	1.65	200	13.3

In Vu et al.'s work the results from these two models are compared with the permeability and the ideal separation factor oftained experimentally from MMMs.

The CMS material used (called CMS 800-2) was formed from the vacuum pyrolysis of Matrimid® 5218. The glassy polymer matrices used were Ultem® 1000 and Matrimid® 5218. Performances of the synthesized MMMs were evaluated for the permeation tests of pure gases (O_2, N_2, CO_2 and CH_4) and the mixed gas (10% CO_2/90% CH_4). Permeability measurement was at 50 psia upstream pressure and 35 °C.

Table 8.19 shows the permeation properties of pure polymer matrices and the CMS used as the insert.

Using the data given in Table 8.19 for the MMM components, the performances of the MMMs were predicted by the Maxwell equation (Eq. 8.32) and the Bruggeman equation (Eq. 8.33) and the results compared with the experimental data in Table 8.20. Table 8.20 shows clearly the tendency observed in the theoretical prediction agrees with that of the experimental data. Similar data were obtained for the O_2/N_2 gas pair.

The Bruggeman model always predicts higher gas permeability and gas selectivity than the Maxwell model. According to the authors, these results suggest that the Bruggeman model predicts that flow patterns around the CMS particles can become distorted by nearby CMS particles. This model envisions that some pen-

Table 8.20 Comparison of model predictions with the experimental data with respect to CO_2/CH_4 gas pair

MMMs	CO_2 permeability (Barrer)			CO_2/CH_4 ideal separation factor		
	Experimental	Maxwell model	Bruggeman model	Experimental	Maxwell model	Bruggeman model
Matrimid® 5218						
17 vol% CMS	10.3	13.0	13.1	44.4	47.7	48.1
19 vol% CMS	10.6	13.3	13.5	46.7	49.3	50.0
33 vol% CMS	11.5	16.5	17.0	47.5	63.1	65.0
36 vol% CMS	12.6	17.2	17.8	51.7	66.2	68.6
Ultem® 1000						
16 vol% CMS	2.51	2.21	2.33	43.0	44.2	45.8
20 vol% CMS	2.90	2.40	2.59	48.1	45.3	47.7
35 vol% CMS	4.48	3.44	4.23	53.7	50.6	59.1

etrants travel alternate streamlines (when other nearby CMS particles are not present) through the more permeable nearby sieves instead of all gas penetrants surrounding the sieve continuing on normal streamlines through the matrix polymer around the sieve particle.

As the CMS loading increases, more alternate faster streamlines become available and affect a larger percentage of bypassing gas penetrants, which would have ordinarily traveled through the polymer matrix around the sieve particle. This effect will become more obvious when the permeability of the CMS is much greater than that of the polymer matrix. In the above two examples of polymer matrices, Ultem® 1000 has much smaller permeabilities, and hence should show greater effect of the CMS loading. In fact, the Bruggeman model deviates most from the Maxwell model at the highest CMS loading (35%), which amounts to 20%.

The selectivity of Ultem® 1000 for CO_2/CH_4 gas pair were depicted by Fig. 8.61. The predicted values are close to experimental values at lower CMS loading of 20%. At 35% loading the deviation of predicted values from the experimental is quite significant. In particular, the Bruggeman model predicts a much higher selectivity value. The agreement of model prediction and the experimental data is poor for Matrimid® 5218 (Fig. 8.62). Even though the experimental data show enhancement of both permeability and ideal separation factor by the incorporation of CMS, the enhancement is not as much as the models predict. This may be ascribed to the local rigidification in the vicinity of the sieve particle, causing the reduced permeability. This is further evidenced by the much larger increase in T_g observed for Matrimid® 5218 (ca. 15 °C) than for Ultem® 1000 (2–5 °C) by the incorporation of CMS. Furthermore, T_g of Ultem® 1000 is much lower than Matrimid® 5218, thus Ultem® 1000 is much more flexible and has less chance of rigidification.

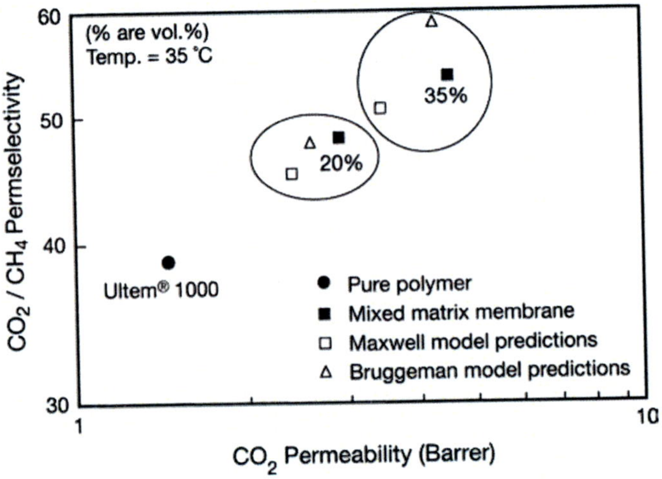

Fig. 8.61 Comparison of experimental data for permeability and ideal separation factor for CO_2/CH_4 gas pair with the predicted values by the Maxwell and the Bruggeman model (Ultem® 1000/CMS MMMs). (From [23])

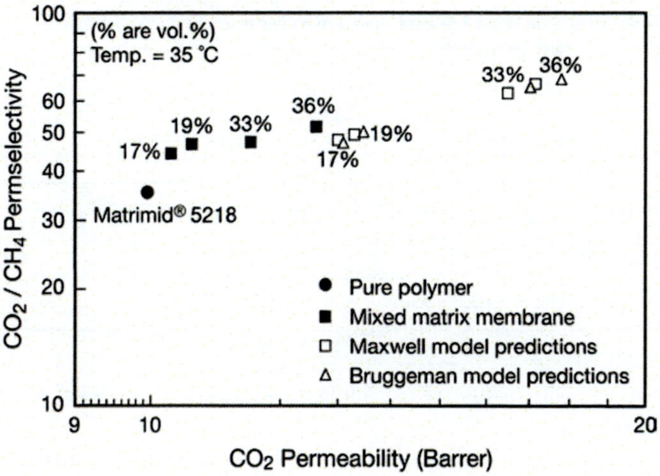

Fig. 8.62 Comparison of experimental data for permeability and ideal separation factor for CO_2/CH_4 gas pair with the predicted values by the Maxwell and the Bruggeman model (Matrimid® 5218/CMS MMMs). (From [23])

Based on the rigidification concept, Mahajan [25] has developed an approach in which the Maxwell model is used twice. The polymer region in the vicinity of the CMS particle is assumed to have reduced permeability due to the immobilization effect. This is an extension of the concept given for the semi-crystalline polymer. When crystallites are present within an amorphous phase the chain mobility of the amorphous phase appears to be reduced leading to high activation energy of diffusion. Michaels et al. introduced chain immobilization factor β by which the overall diffusion coefficient D is given by

$$D = \frac{D^*}{\tau\beta} \tag{8.35}$$

where D^* is the diffusion coefficient for the amorphous phase and τ is the tortuosity.

In applying the Maxwell model twice, the following two cases are considered; i.e., (1) the CMS dispersed phase and the surrounding region of the rigidified polymer matrix; (2) the pseudo single phase "insert" comprising the CMS particle and the rigidified region and the bulk polymer matrix region. In this approach, there are three adjustable parameters; i.e. CMS particle diameter, the chain immobilization factor and the thickness of the domain of rigidification. Among these parameters CMS particle diameter and chain immobilization factor are assumed to be 1 μm and 3, respectively. Then, the thickness of the domain of rigidification becomes the only adjustable parameter to fit the calculated data to the experimental ones. The results of the model fitting with the thickness of the rigidification domain equal to 0.075 μm are given in Table 8.21.

The better fitting with the two steps Maxwell model than the single step Maxwell model together with the T_g elevation in the presence of CMS in Matrimid®

Table 8.21 Comparison of the experimental data for Matrimid® 5218 MMM for the CO_2/CH_4 separation with prediction from a modified three phase model of the Maxwell model

MMM	CO_2 permeability (Barrer)			CO_2/CH_4 ideal separation factor		
	Experi-mental	Maxwell model	Two step Maxwell model	Experi-mental	Maxwell model	Two step Maxwell model
Matrimid® 5218						
17% CMS loading	10.3	13.0	11.0	44.4	47.7	44.2
19% CMS loading	10.6	13.3	11.2	46.7	49.3	45.4
33% CMS loading	11.5	16.5	12.1	47.5	63.1	55.3
36% CMS loading	12.6	17.2	12.3	51.7	66.2	57.5

5218 polymer matrix may justify the assumption of the rigidified region around the inserted CMS particles.

Vu et al. further investigated the modeling for the binary CO_2/CH_4 mixtures. The performance was first examined using the gas feed containing 10% CO_2 and 90% CH_4 with feed gas pressure up to 900 psig. Then the Maxwell model was applied to calculate the permeability and ideal separation factor based on the data from the single gas experiments.

For glassy polymers, the Dual Mode (DM)/Partial Immobilization Model (PIM) has been developed to predict the sorption and diffusion process of gas permeation though membranes [26]. The model proposes that the sorption occurs in the dual sorption mode, consisting of the Henry's law and the Langmuir mode. The diffusion through these modes is also different and the gas molecules in the Langmuir mode are partially immobilized. According to the DM/PIM, the permeability of component gases A and B is given by

$$P_A = k_{DA} D_{DA} \left(1 + \frac{F_A K_A}{1 + b_A f_A + b_B f_B}\right) \tag{8.36}$$

$$P_B = k_{DB} D_{DB} \left(1 + \frac{F_B K_B}{1 + b_A f_A + b_B f_B}\right) \tag{8.37}$$

Where $F_A = D_{HA}/D_{DA}$, $F_B = D_{HB}/D_{DB}$, $K_A = (C'_{HA} b_A/k_{DA})$, $K_B = (C'_{HB} b_B/k_{DB})$. f corresponds to the upstream fugacity of the gas. D_D and D_H are the diffusion coefficient of the gas in the Henry and Langmuir site, respectively. k_D is the Henry's law constant and C'_H and b are the Langmuir capacity constant and affinity constant, respectively. Since all the parameters involved are known for pure CO_2 and CH_4 from the respective pure gas experiment, the permeabililty for the binary mixture can be calculated from Eqs. (8.36) and (8.37). These parameters were previously obtained for Ultem® 1000 for both CO_2 and CH_4 as summarized in Table 8.22.

For the CMS 800-2 membrane there is no theoretical tools and associated parameters to predict the permeability for the mixed gases. The only available data are the permeability data for the pure gas and mixed gas system as shown in Table 8.23.

Table 8.22 Dual mode sorption and dual mode transport parameters for CO_2 and CH_4 for Ultem® 1000 at 35 C

Dual mode parameters	Ultem® 1000	
	CO_2	CH_4
Dual mode sorption		
k_D (cm³(STP)/cm³ atm)	0.916	0.223
C_H'(cm³(STP)/cm³)	23.38	7.08
b (atm^{-1})	0.399	0.139
b (psia^{-1})	0.0272	0.00946
Dual mode transport		
D_D (10^{-9} cm²/s)	10.4	1.10
D_H (10^{-9} cm²/s)	0.582	0.0715
$F=(D_H/D_D)$	0.056	0.065

According to Table 8.23 permeabilities for CO_2 and CH_4 were assumed to be 36.0 and 0.20 Barrer, regardless of the partial pressure of the constituent gases.

Figure 8.63 shows the prediction for the mixed gas system using the peremabilities evaluated by Eqs. (8.36) and (8.37) for Ultem® 1000 and the permeabilites mentioned above for CMS 800-2. CO_2 permeability as well as CO_2/CH_4 ideal separation factor are plotted versus the total gas pressure. The data calculated by the transport models agree with the experimental data reasonably well. As expected the values calculated by the Bruggeman's model are generally higher than those calculated by the Maxwell model.

Formation of "sieve-in-cage" morphology is the most common indicator of poor polymer-sieve contact and MMMs with the morphology exhibit higher permeability but poor selectivities approaching those of the native polymer or even worse. Since enhancement of slectivities by molecular sieves can only be achieved in the absence of these defects, concerned efforts have been directed to enhance the adhesion between these two phases [27].

Polyvinylpyrrolidone (PVP) is a common chemical used as additive in casting solution for the preparation of polysulfone (PS) membranes by the phase inversion technique. PVP is also known to be an established thermoplastic sizing in composite technology. The effectiveness to use PVP as a sizing agent to promote adhesion between inorganic substrate with polymer matrix has been extensively reported in fiber reinforced polymer matrix composite development.

Rafizah and Ismail explored the modification of CMS particles using sizing technique [28]; MMMs comprising polysulfone (PS) Udel® P-1700 and 30 wt%

Table 8.23 Comparison of pure gas and mixed gas (10% CO_2/90% CH_4) permeation results for CMS 800-2 membrane at 35°C

CMS 800-2	Permeability (Barrer)		CO_2/CH_4 ideal separation factor
	CO_2	CH_4	
Pure gas permeation	43.5	0.21	200
Mixed gas permeation	36.0	0.20	180

Pure gas experiments were performed at 50 psig while the mix gas performance was performed at 90 psig.

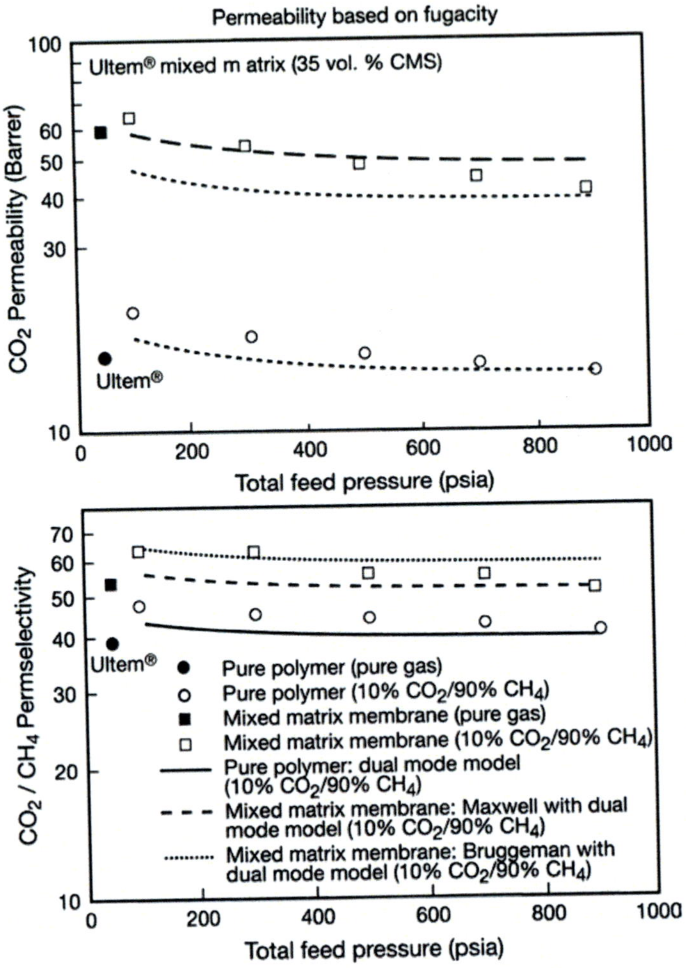

Fig. 8.63 Comparison of experimental mixed gas (10% CO_2/90% CH_4) permeation and separation results with the results calculated by Maxwell and Bruggerman transport model. (From [23])

CMS particles (<25 μm) have been fabricated and characterized. CMS particles were treated in PVP kollidone 15 (PVP K-15) sizing bath solution (1–10 wt% in isopropanol) prior to embedment into the matrix solution to enhance matrix-sieve interfacial adhesion; i.e., PVP K-15 powder was dissolved in isopropanol to produce dilute solution with the concentration of 1–10 wt% PVP. A predetermined amount of CMS particles was added to the PVP solution and stirred at 30 rpm for 1 h. The sized CMS cake was rinsed with isopropanol to remove unadsorbed PVP before further drying in a vacuum oven at 50 °C for 24 h.

First of all, 25 wt% PS in NMP was prepared and in a separate flask, a predetermined amount of PVP-sized CMS was wetted with a small amount of NMP. Then,

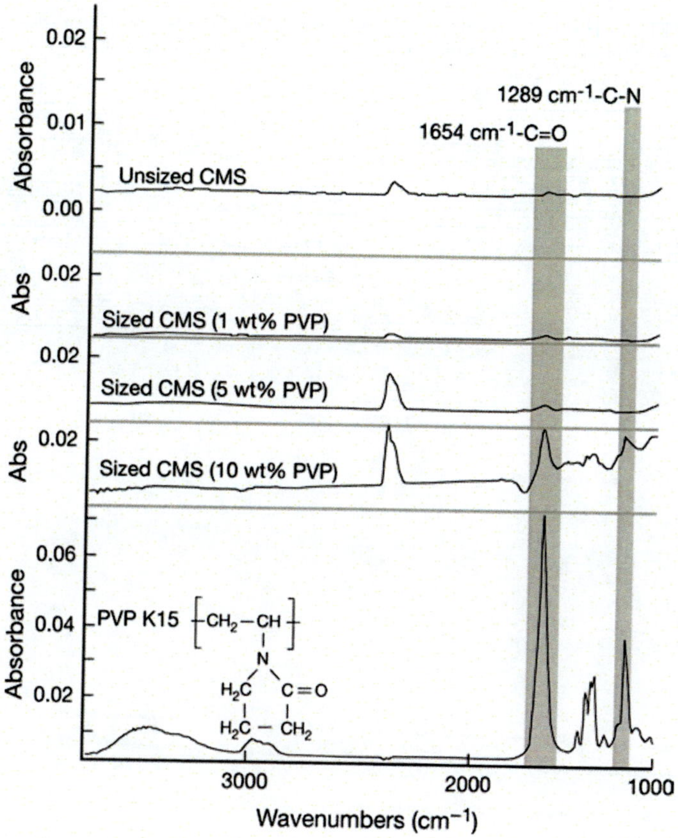

Fig. 8.64 IR spectra for PVP K-15, unsized CMS and PVP K-15 sized CMS. (From [28])

the PS solution was added to the second flask containing the wet CMS under vigorous stirring. The film casting was done at elevated temperatures using a specifically designed heating protocol. The thickness of workable membranes was 60–70 μm. The cast film was kept in a vacuum oven at 60 °C prior to characterization.

The FTIR spectra are given in Fig. 8.64. The infrared absorption bands at 1,654 and 1,289 cm^{-1}, respectively, correspond to the amide I carbonyl (–C=O) band and amide III (–C–N) stretching band, respectively, of PVP. The surface of unsized CMS, on the other hand, was almost inert with no apparent appearance of the characteristic peaks. The presence of the characteristic peaks of PVP was observed in the sized CMS spectra. This finding supports that PVP was successfully deposited on top of carbon surface.

A qualitative assessment was conducted by using FESEM images in order to compare the morphology of the fabricated MMM containing PVP-sized CMS and MMM containing untreated CMS, as shown in Fig. 8.65. Both of these membranes were loaded with 30% of CMS. In all images, CMS particles (<25 μm) were

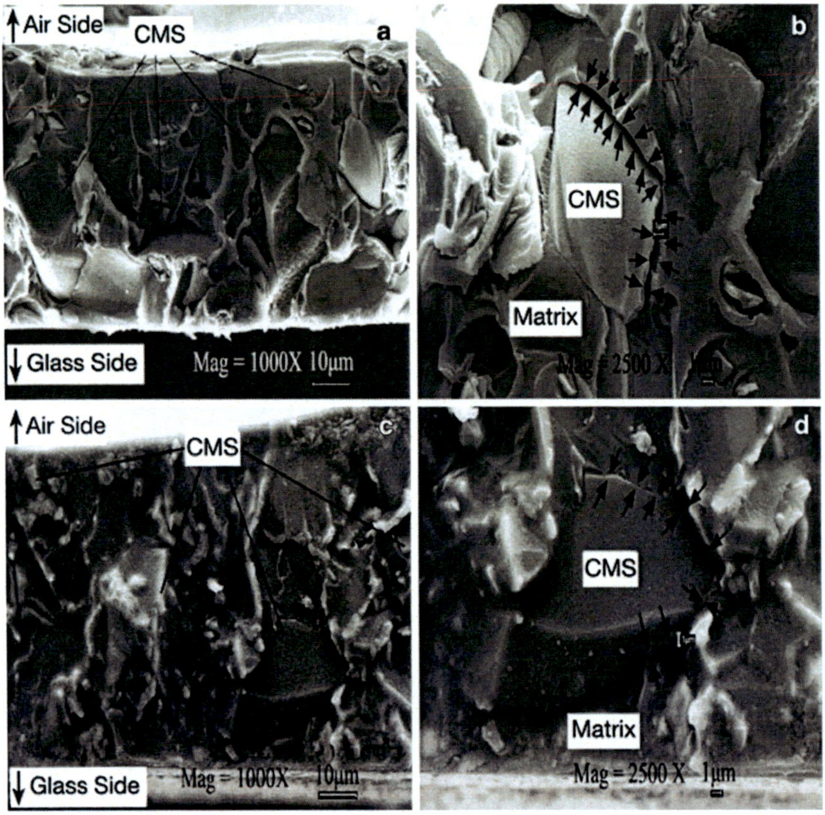

Fig. 8.65 Comparison of FESEM micrographs for the cross-section of PS-CMS MMM with 30 wt% CMS loading; (**a**) containing unmodified CMS under 1000× magnification, the white bar indicates 10 μm, (**b**) containing unmodified CMS under 2500× magnification, the white bar indicates 1 μm, (**c**) containing PVP-sized CMS under 1000× magnification, the white bar indicates 10 μm, (**d**) containing PVP-sized CMS under 2500× magnification, the white bar indicates 1 μm. (From [28])

randomly distributed and securely entrapped within 60–70 μm thick PS layer. It is revealed that the sizing with PVP had a dramatic effect on the morphology of PS-CMS-MMM. The existence of sub-micron scaled voids could be easily distinguished in the cross-sectional images of unmodified MMM (Fig. 8.65a, b). The shaded arrows pointing at the interphase region highlight the occurrence of void surrounding the CMS particle in MMM that contains unsized CMS. On the other hand, the incorporation of PVP-sized CMS into the PS matrix resulted in an almost void-free morphology (Fig. 8.65c, d). The PS matrix adhered intimately on the carbon surface suggesting that the compatibility of CMS with the PS matrix improved significantly by PVP-sizing. It is speculated that the weak interaction between PS matrix and inert CMS surface caused the void formation. The stress generation during the membrane vitrification process could easily detach the weakly bound

Table 8.24 Comparison of PS/unsized-CMS and PS/PVP-sized CMS MMMs for O_2 and N_2 permeation

Membrane	CMS loading (wt%)	Permeability (Barrer)		α_{O_2/N_2}
		O_2	N_2	
PS	0	1.58	0.29	5.50
PS-CMS without sizing	30	6.77	1.82	3.69
PS-CMS with sizing	30	6.52	1.08	6.05

polymer chains from the CMS surface. In contrast, PVP K-15 sizing layer that was formed on the outer surface of the CMS particles promoted the interaction with the surrounding PS matrix by introducing more reactive groups such as $-C=O$ that can form a specific interaction with the surrounding sulfonate group of PS.

The gas permeation experimental results are shown in Table 8.24.

It is clear from Table 8.24 that the incorporation of CMS particles increased the gas permeability significantly. However, the selectivity decreased when the CMS was not sized. On the other hand, when the CMS was sized the selectivity increased, indicating the removal of the void spaces between the CMS particle and PS matrix.

8.4.2 Membranes Filled with Carbon Nanotubes (CNTs)

Recent theoretical work of Skoulidas et al. [4] has reported the results from the atomic simulation for both self- and transport- diffusivities of light gases such as H_2 and CH_4 in CNTs and zeolites. They reported that the transport rates in CNTs will become orders of magnitude faster than in zeolites, which is the result of the inherent smoothness of the nanotubes. They have also suggested that if the carbon nanotubes are used as membranes the flux-selectivity properties will far exceed those of any conventional membrane materials. To verify this prediction, nano-composite membranes consisting of SWNTs embedded in a poly(imide siloxane) copolymer were fabricated and their transport characteristics investigated by Kim et al. [29]. While the siloxane segment enhances the interfacial contact, the polyimide component is expected to impart the mechanical integrity of the membrane.

6FDA-6FpDA-PDMS copolymer was chosen to be poly(imide siloxane) polymer matrix that was synthesized in the laboratory. The chemical formula of the polymeric material is shown in Fig. 8.66. According to GPC experiments, M_n of the co-polymer was 62,330 g/mol and M_w was 1,31,300 g/mol. M_n of PDMS monomer

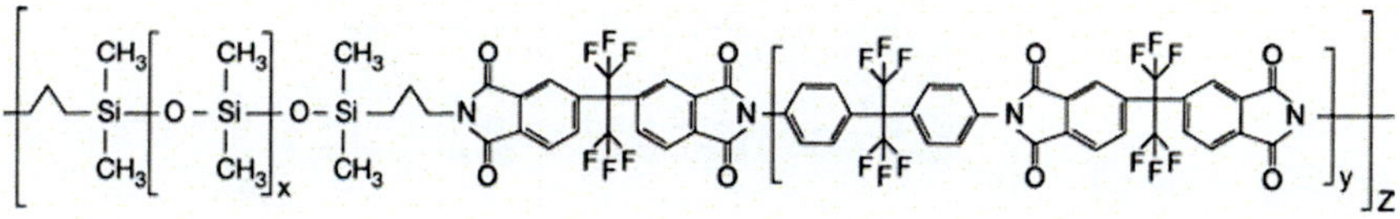

Fig. 8.66 The chemical formula of 6FDA-6FpDA-PDMS. (From [29])

was determined by 1H-NMR to be 885.5 g/mol. The weight per cent of PDMS segment was 41%.

Both closed-ended and open-ended SWNTs were purchased from Carbon Solutions, Inc. The average diameter of nanotubes was 1.4±0.2 nm and purified as described in the original paper. CNT was cut into shorter length by transferring the purified open-ended CNTs to 100 mL of 3:1 mixture of concentrated H_2SO_4 (98 vol%) and HNO_3 (70 vol%) solution. The mixture was sonicated in a water bath for 4 h at 40 °C and then diluted with deionized water. The solution was centrifuged and brown solution was decanted. After repeating the process several times the CNTs were filtered and dried. Approximately 1.3 g of the copolymer was dissolved in 20 mL of THF and mixed for 24 h. Afterward, predetermined mass of CNTs (2–10 wt%) was added to the copolymer solution. During the mixing process the volume of the solution decreased to ca. 10 mL and became viscous. The solution was poured into 6.35 cm diameter Teflon coated pan. After 2 days of slow solvent evaporation, the solution film was placed in a vacuum oven and dried under vacuum at 100 °C for 12 h.

Figure 8.67 shows the FESEM images of the CNT samples on a copper support. CNTs are closely entangled and impurities of metal clusters originated from the catalyst stick on the surface of the ropes (a). These metal impurities can block gas flow through the CNTs if they are on the tips of the ropes without being removed. Samples in (b) were purified and most of the impurities were removed. However these samples are highly entangled and long. To be used in MMM these long nano-

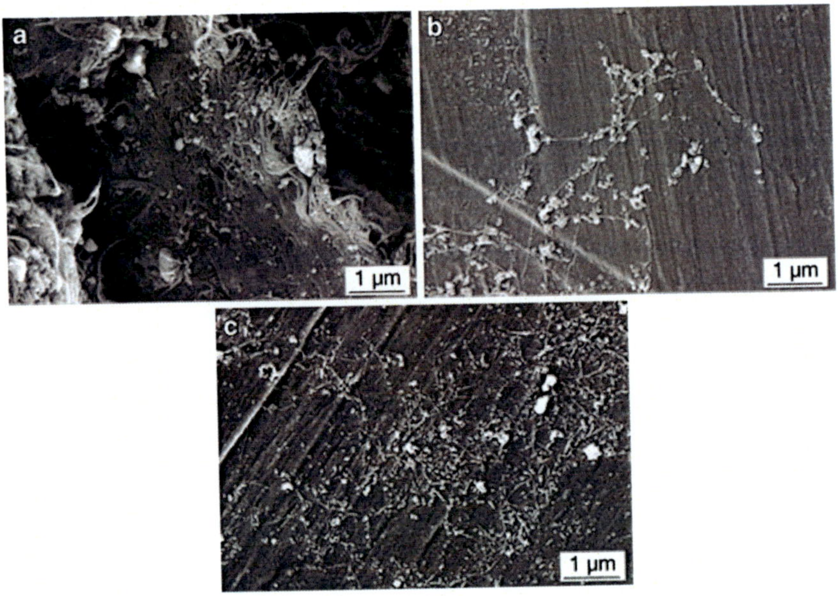

Fig. 8.67 FESEM images of CNTs; (**a**) before purification, (**b**) after purification, (**c**) after shortening. (From [29])

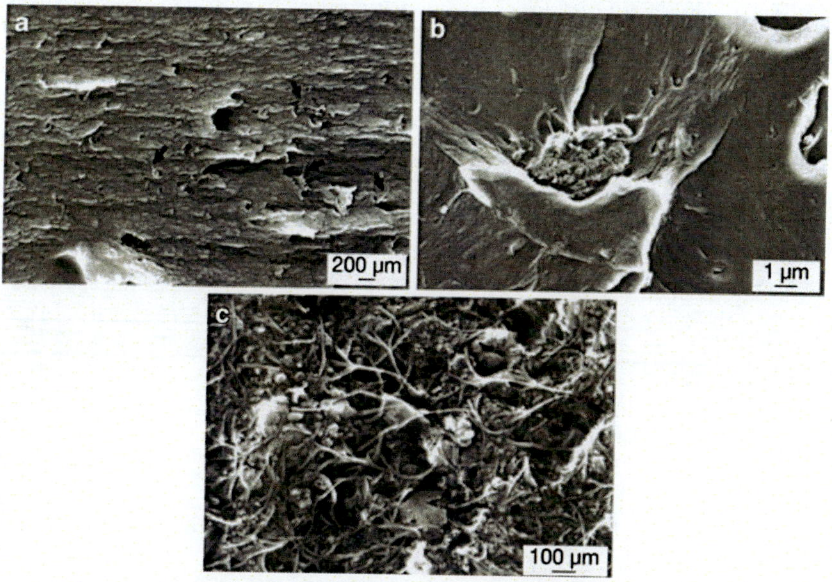

Fig. 8.68 FESEM cross-sectional images of MMMs; (**a**) MMM with 2 wt% CNTs loading, (**b**) and (**c**) MMM with 10 wt% CNTs loading. (From [29])

tubes with high entanglement should be cut into shorter pieces with opened ends. The MMMs were observed by FESEM. Cross-sectional images in Fig. 8.68 show that CNTs are well dispersed in the polymer matrix when the CNT loading was as low as 2 wt%. However at 10 wt%, the tubes were agglomerated and formed domains in the polymer (Figs 8.68b, c).

In Table 8.25, gas permeabilities of poly(imide-siloxane) membranes with and without CNT loading are compared. Thus, incorporation of CNT decreased the He flow rate. On the other hand, N_2 permeability seems to be statistically unchanged by CNT loading.

Solubility and diffusivity were also compared for membranes without and with CNT loading. The results shown in Table 8.26 indicate that diffusivity decreased by incorporating CNTs. This is consistent with the incorporation of closed-ended CNTs that behave as impermeable filler. The solubility for N_2 increased and as a result there was no change in permeability statistically. Thus, incorporation of CNT lowered the He permeability (Table 8.25) and hindered the diffusion path of N_2, lowering its diffusivity (Table 8.26).

Table 8.25 Gas permeabilities of pure poly(imide-siloxane) membrane and poly(imide-siloxane)-carbon nanotube MMM

CNT loading (wt%)[a]	Permeability (Barrer)	
	He	N_2
0	85.85	18.48 ± 1.56
1	63.80	19.51 ± 0.24

[a] Closed-ended carbon nanotube was used

Table 8.26 Diffusivity and solubility of N_2 in the pure poly(imide-siloxane) membrane and poly(imide-siloxane)-carbon nanotube MMM

CNT loading (wt%)[a]	Diffusivity (cm^2/s)	Solubility (cm^3(STP)/cm^3polymer atm)
0	175.92	0.08
1	127.04	0.12

[a] Closed-ended CNT was used

Table 8.27 Permeability of various gases for the poly(imide-siloxane) membranes without CNT loading and with CNT loading to different degrees

CNT loading (wt%)	Permeability (Barrer)[a] He	H_2	CO_2	O_2	N_2	CH_4
0	59.54	70.57	166.02	32.24	11.99	28.19
2	66.95 (12.45%)	79.62 (12.82%)	190.67 (14.85%)	36.57 (13.43%)	14.42 (20.27%)	34.16 (21.18%)
10	67.92 (14.08%)	79.15 (12.16%)	191.30 (15.23%)	39.81 (23.48%)	17.83 (48.71%)	36.71 (30.22%)

[a] Figures inside the bracket are % increase in permeability by CNT loading

Table 8.27 shows the permeability data for different gases. Interestingly, the gases are classified in two groups. One, the fast permeating gasses including He, H_2 and CO_2. The other, the slowly permeating gases including O_2, N_2 and CH_4. For fast gases, the permeability increased by 12–15% when for 2% of CNT was loaded but somehow the further increase in CNT loading did not increase the permeability. On the other hand, for slower gases, permeability increased by 13–21% at 2% CNT loading and at 10% CNT loading the permeability increased further to 23–49%.

For small molecules such as He and H_2, the low loading of CNT increased the permeability of these gases by letting the gases pass through the CNT particles. The high permeability through the CNT particles is thus felt by the gas molecules. However, The permeability of polymer matrix for these gases are already very high, and the increase in the permeability by further increase in CNT loading is offset by the increased tortuosity around the entangled CNT's domain. In the case of CO_2, plasticization effect increases the free volume and the permeability in the polymer matrix. Hence the effect of CNT loading on CO_2 permeability looks similar to that of He and H_2 gas.

The diffusivity data summarized in Table 8.28 shows tendencies similar to the permeability; for CO_2 gas the diffusivity increase stops at 2% CNT loading while for the gases such as O_2, N_2 and CH_4, the diffusivity keeps increasing up to the CNT loading of 10%. Hence it can be concluded that the increase in permeability of latter gases are due to the increase in diffusivity when the gases are allowed to pass through the CNT particles.

Solubility data given in Table 8.29, however, show hardly any change by CNT loading.

Another work on CNT based MMM was performed by Cong et al. [30]. The mechanical strength of polmeric membranes is one of the limitations on their ap-

Table 8.28 Diffusivity of various gases for the poly(imide-siloxane) membranes without CNT loading and with CNT loading to different degrees

CNT loading (wt%)	Diffusivity ($\times 10^8$ cm^2/s)[a]					
	He	H_2	CO_2	O_2	N_2	CH_4
0	–	–	78.04	147.74	83.29	64.37
2	–	–	88.21 (13.03%)	164.30 (11.21%)	107.36 (28.90%)	76.16 (18.32%)
10	–	–	88.60 (13.53%)	204.47 (38.39%)	149.37 (79.34%)	90.52 (40.63%)

[a] Figures inside the bracket are % increase in diffusivity by CNT loading

Table 8.29 Solubility of various gases for the poly(imide-siloxane) membranes without CNT loading and with CNT loading to different degrees

CNT loading (wt%)	Solubility (cm^3(STP)/cm^3 polymer atm)					
	He	H_2	CO_2	O_2	N_2	CH_4
0	–	–	1.62	0.17	0.11	0.33
2	–	–	1.64	0.17	0.10	0.34
10	–	–	1.64	0.15	0.09	0.34

plications. CNTs are very effective in reinforcing polymeric materials but it is unknown whether they degrade the membrane's gas separation performance. In their work, a novel CNT composite membrane for CO_2/N_2 separation has been reported. Brominated poly(2,6-diphenyl-1,4-phenylene oxide)(BPPO$_{dp}$), which has better intrinsic permeation properties (P_{CO_2} ca. 80 Barrer, $\alpha CO_2/CH_4$ ca. 30 at 25 °C) and mechanical properties than poly(2,6-dimethyl-1,4-phenylene oxide)(PPO), formed more uniform and flexible membranes with CNTs than PPO.

Three types of MWNTs (o.d.: 3–10 nm, length: 0.1–10 µm, >90%; o.d.: 40–60 nm, length: 0.5–500 µm, >90%; o.d.: 40–70 nm, length: 0.5–2 µm, >90%) were purchased from Aldrich and used as received. High purity SWNTs (prepared by high pressure carbon monoxide flow method (HighP$_{CO}$)>88%, od: 0.8–1.2 nm, length: 0.1–1 µm) were obtained from Carbon Nanotechnologies Inc., USA. Bromination of PPO$_{dp}$ was done by adding bromine solution in chloroform into the polymer solution in chloroform. Bromination was 100% from NMR spectrum. Carboxyl modified SWNT was prepared by the following method.

The SWNTs were suspended in a 3 M HNO_3 solution and kept there for about 10 h in reflux until the evolution of NO_2 vapors subsided. The solid product was separated by centrifuge and then washed successively using deionized water, 3 M HCl, and again deionized water. After being dried under vacuum at 50 °C, a black powder was obtained. Vibration peaks of carboxyl group were observed at 1,730 and 3,425 cm^{-1} in the IR spectrum.

BPPO$_{dp}$ (0.6 g) was dissolved in 8 mL of chloroform with stirring in a beaker (polymer concentration is 4.8 wt%). CNTs (SWNT or MWNT) were added slowly into the polymer solution at 2.0, 5.3, 9.9 and 20.5 wt% of the polymer, respectively. The mixture was vigorously stirred at a rotation speed of 1,150 ppm for 15 min to

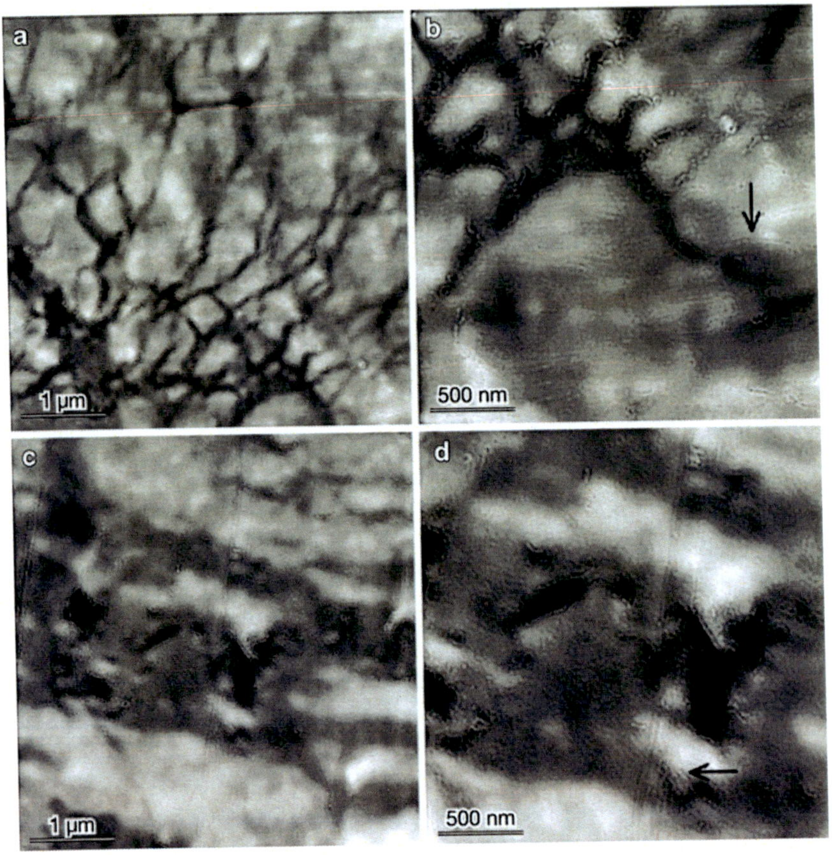

Fig. 8.69 Morphology of $BPPO_{dp}$/CNTs with 5 wt% SWNTs (**a**, **b**), and 5 wt% MWNTs (**c**, **d**). (From [30])

disperse CNTs. The mixture was then cast onto a clean and dry glass plate at 25 °C. After the solvent was evaporated, the resulting membrane was peeled off and stored in a desiccator for characterization and performance testing. The thickness of the membrane was from 50 to 90 µm. The characterization was done by transmission electron microscope (TEM) and mechanical property test.

The TEM images of the $BPPO_{dp}$/SWNTs and $BPPO_{dp}$/MWNTs are shown in Fig. 8.69. The figure shows some CNT aggregated (black parts) and some dispersed (white parts). Mechanical mixing therefore could not completely eliminate the nanoparticle aggregation. The membranes were flexible when CNT content was less than 9 wt%. As shown in Table 8.30, the tensile modulus of $BPPO_{bd}$ increased by 67% and 44%, respectively, for the 5% addition of SWNT or MWNT.

The permeability and selectivity of the membranes are shown in Fig. 8.70.

The $BPPO_{dp}$ membrane had CO_2 permeability of 78 Barrer with a selectivity of 30. The addition of CNTs initially increased the CO_2 permeability; the permeability became 155 Barrer at 9 wt% addition of SWNT and 148 Barrer at 5 wt% addition of

Table 8.30 Mechanical properties of $BPPO_{dp}$/CNT nanocomposite membranes

Membrane[a]	Elongation (%)	Tensile modulus (GPs)
$BPPO_{dp}$	4	0.9
$BPPO_{dp}$/SWNT	2	1.5
$BPPO_{dp}$/MWNT	3	1.3

[a] 5 wt% of CNT was added

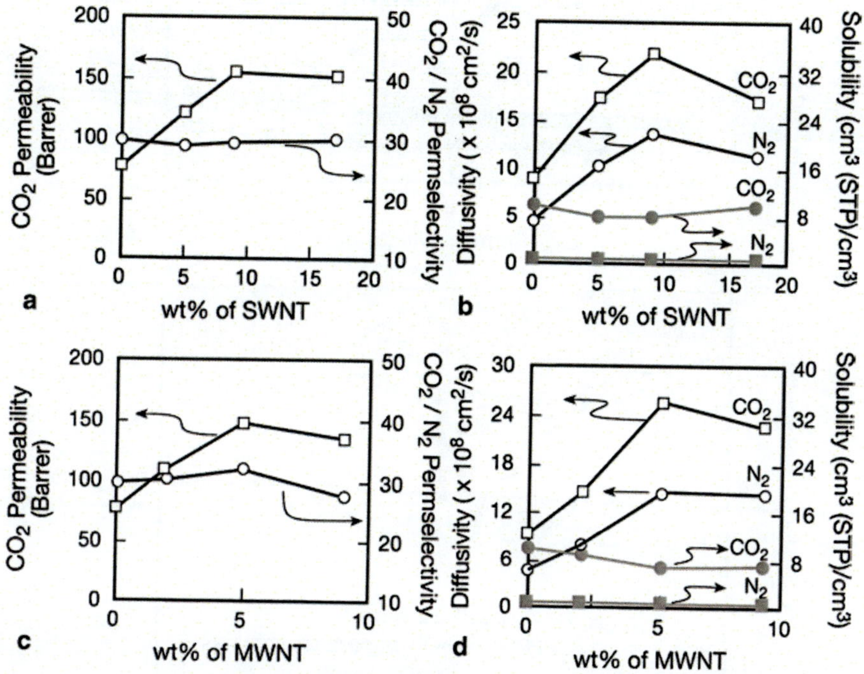

Fig. 8.70 CO_2/N_2 separation performance of $BPPO_{dp}$/SWNTs (**a**, **b**), and $BPPO_{dp}$/MWNTs (**c**, **d**) membranes (test conditions: 10 psig and 25 °C, from [30])

MWNT, respectively. Hence, the addition of both CNTs doubled CO_2 permeability. On the other hand, the addition of CNTs had little impact on the selectivity of the membrane.

Thus, the addition of CNTs had favorably influenced the permeability and the mechanical strength of the membrane. Further investigation (Fig. 8.70c, d) revealed that the addition of CNTs did not influence significantly the solubility of the membranes for both CO_2 and N_2 gases. However, the diffusivity was substantially increased. At 10 wt% addition of SWNT and 5 wt% addition of MWNT the permeability as well as diffusivity started to decrease.

Figure 8.71 shows the CO_2 peremability versus CO_2/N_2 selectivity of the $BPPO_{dp}$/SWNTs and $BPPO_{dp}$/MWNTs nanocomposite membranes. The figure shows that the data were above the Robeson's reference line. (From [30]).

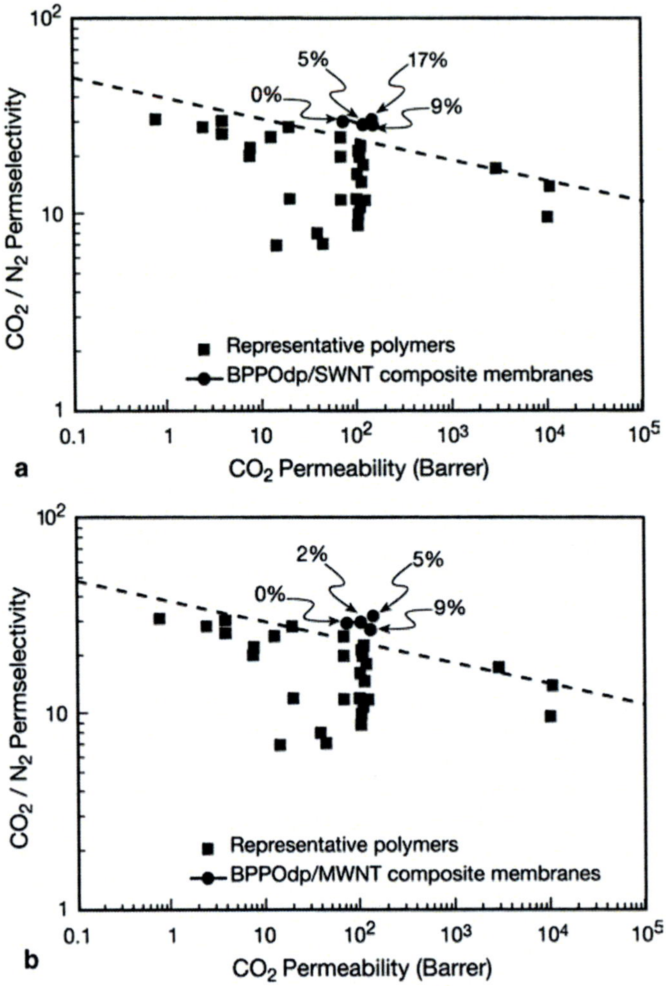

Fig. 8.71 The separation properties of (**a**) $BPPO_{dp}$/SWNTs and (**b**) $BPPO_{dp}$/MWNTs membranes. (From [30])

As Table 8.31 shows the size of CNTs had hardly any effects on the membrane gas separation performance.

The treatment of CNTs by strong acids could successfully introduce carboxyl group onto their surface. As well the compatibility between the CNTs and the polymer matrix was improved considerably as Fig. 8.72 shows.

However, as Table 8.32 shows, there was hardly any improvement of membrane separation performance.

The results shown in Table 8.32 indicate that the increase in gas permeability is not because of the disturbed polymer chain packing in the presence of the CNT particles in this case, as was suggested by other researchers (for example [31]).

Table 8.31 Effect of MWNT size on the permeability and selectivity of $BPPO_{dp}$/MWNTs nanocomposite membranes

Size of MWNT				Gas permeation performance		
o.d. (nm)	i.d. (nm)	Length (μm)	Purity (%)	CO_2 permeabilty (Barrer)	N_2 permeability (Barrer)	α_{O_2/N_2}
3–10	1–3	0.1–10	90+	153	5.4	28
40–60	5–10	0.5–500	90+	148	4.7	31
40–70	5–40	0.5–2	95+	134	4.0	34

Fig. 8.72 Images of $BPPO_{dp}$/5 wt% surface modified SWNTs composite membrane. (From [30])

Table 8.32 Gas separation performance of $BPPO_{dp}$/SWNT nanocomposite membranes with and without surface modification

Membranes[a]	CO_2 permeability (Barrer)	N_2 permeability (Barrer)	α_{CO_2/N_2}
$BPPO_{dp}$	78	2.6	30
$BPPO_{dp}$/SWNT without modification	123	4.3	29
$BPPO_{dp}$/SWNT with modification	79	2.6	30

[a] SWNT loading 5 wt% of polymer; Gas permeation test at 10 psig and 25°C

Meanwhile, the results shown in Fig. 8.70 and Table 8.32 suggest that the increased gas permeability is not due to the gas transport inside the nanotubes. The inner diameters of MWNTs and SWNTs are 5–10 nm and 0.4–1 nm, respectively, far above the kinetic diameters of CO_2 (0.33 nm) and N_2 (0.364 nm). The maximum length of MWNTs is 500 μm, while the membrane thickness was 50–90 μm. If the CNTs are oriented perpendicular to the membrane surface, the pores in the CNTs would act as pinholes. The permeability would increase enormously and selectivity would go down. Besides, the increase in CNT loading would result in the permeability increase. None of them was observed experimentally for the studied $BPPO_{dp}$/SWNTs and $BPPO_{dp}$/MWNTs nanocomposite membranes.

$BPPO_{dp}$ is not compatible with pristine CNTs, as evidenced by the CNT aggregation in the polymer matrix. Thus the $BPPO_{dp}$ chains could not fall onto the CNT outer surfaces tightly to form narrow gaps surrounding the CNT particles. Gas molecules may pass through these gaps. Thus, the apparent diffusivity may increase. The gap should be sufficiently narrow to maintain the selectivity practically intact. The surface modified CNTs are more compatible to the polymer matrix and the chance of forming the narrow gap between the CNT particles and the polymer matrix is less. Thus the permeability did not increase with the CNT surface treatment.

Kim et al. reported another example of MMM development using CNTs as the dispersed inorganic phase [32]. In the investigation of Kim et al., novel nanocomposite membranes containing SWNTs inside a PS matrix were prepared and characterized. The main purpose of the study is to construct highly permeable and selective membranes containing CNTs inside a polymer matrix that could easily be scaled up to large area membranes. The novel nanocomposite membranes consist of well-dispersed SWNTs inside a commercial PS matrix. The SWNTs were functionalized with long chain alkyl amines to facilitate the dispersion and were added to the matrix in different amounts. The structure, the absence of defects, and the properties of the SWNT/PS MMMs were characterized by FESEM, dynamic mechanical testing, sorption studies and gas permeation measurements. Atomic simulation studies of gas sorption in SWNT bundles are presented to compare with experimental sorption measurements. Comparison of experimental results and simulations will help characterize the SWNT samples used in the experiments and provide an estimate for the fraction of nanotubes that are open and accessible to sorption of gases.

Electric arc-discharged and HiP_{CO} SWNTs were purchased from Carbon Solutions Inc. and Carbon Nanotechnologies Inc, respectively. Raw CNT materials produced by electric-arc method were treated by multistage purification (a combination of wet oxidation and acid treatments) to purify SWNTs from impurities and cut into small length with 3:1 mixture of concentrated H_2SO_4 (98 vol%):HNO_3 (70 vol%). HiP_{CO} SWNTs were purified by a two step purification procedure, the first one is to rebundle the nano-tube ropes and the second one is to remove metals and amorphous carbon. Table 8.33 summarizes the properties of SWNTs so prepared. After purification of the SWNTs, the surface area, micropore volume, and total pore volume increased without changing the pore size distribution.

To produce soluble SWNTs, an octadecylammonium (ODA) and SWNT-carboxylate zwitter ion was produced by the following procedure. Shortened SWNTs

Table 8.33 Pore structures of SWNTs

SWNT	Surface area[a] (m^2/g)	Micropore[b] volume (cm^3/g)	Total pore[c] volume (cm^3/g)	Pore diameter[d] (nm)
Raw electric arc	174	0.006	0.23	1.2
Raw HiP_{CO}	281	0.003	0.77	1.5
Purified electric arc	464	0.051	0.63	1.2
Purified HiP_{CO}	1032	0.165	0.83	1.5

[a] BET surface area; [b] t-plot; [c] BJH pore volume; [d] H-K method

were heated with ODA at 393 K for 96 h. After cooling to room temperature, the black colored ODA and SWNTs mixture was washed with THF and filtered through a membrane filter (0.2 μm). Because the unreacted ODA was expected to block the entrance to the channel of SWNTs, ODA was further removed by Soxhlet extraction in ethanol at 393 K for 10 days. The zwitter ion is soluble in chloroform.

PS (Udel P-3500, Solvey) was degassed at 413 K for 3 h under vacuum. Then, 0.6 g of PS was dissolved in 3 mL of chloroform and stirred for one day leading to a viscous solution. The membranes were cast onto a glass substrate using a Doctor Blade. The cast film was allowed to be dried with uniform thickness without curling by slowing the solvent evaporation using a glass cover. The solution was dried for one day. The membrane was further vacuum dried at 458 K (glass transition temperature of the polymer) for 1 h and cooled down to the room temperature.

To prepare a MMM which includes 10 wt% of functionalized SWNTs in PS, approximately 0.45 g of the pure PS was dissolved in 3 mL of chloroform and mixed for 24 h. A predetermined mass of the purified SWNTs (0.05 g) was dissolved in 4.5 g of chloroform. This SWNT solution was added to the polymer solution and the mixture was allowed to mix for 6 h at room temperature. Following this time period, the mixture was sonicated for 10 min to allow for good mixing. This process was repeated several times. The membrane casting procedure is the same as described above.

The grand canonical Monte Carlo simulation technique was used to model the gas absorption process in SWNT bundles. The paramters used in the simulation are as follows:

For H_2, Lennard-Jones parameters are: $\varepsilon=34.2$ K and $\sigma=0.296$ nm
For methane, $\varepsilon=161.35$ K and $\sigma=0.372$ nm
For O_2, a two site LJ model with O–O bond length of 0.1208 nm, $\varepsilon=52$ K and $\sigma=0.299$ nm
For CO_2, EPM2 potential model developed by Harris and Yung [33]
The solid-fluid potentials: $\varepsilon=28.0$ K, $\sigma=0.34$ nm

Potentials were truncated at >15 Å, the simulation cell was greater than 30 Å with imposed periodic boundary conditions. The number of molecules adsorbed in the system was calculated as a function of the chemical potential (bulk pressure) of the gas. Total (absolute) adsorption, N_{tot}, was calculated and the excess amount adsorbed, N_{ex}, was computed by

$$N_{ex} = N_{tot} - V_{free}\rho_{bulk}$$

Where V_{free} is the volume not occupied by the nanotubes and ρ_{bulk} is the number density of the gas in the bulk. The excess adsorption isotherm is directly compared with the experimentally measured isotherm.

In Fig. 8.73 the cross-sectional image of the PES MMM with the untreated SWNT shows a non-uniform dispersion of the SWNT with formation of agglomeration and cluster, while Fig. 8.74a shows that SWNTs were well distributed throughout the polymer matrix, No agglomeration was found at the higher magnification (Fig. 8.74b). When the SWNT loading was increased from 10 to 15%, even the

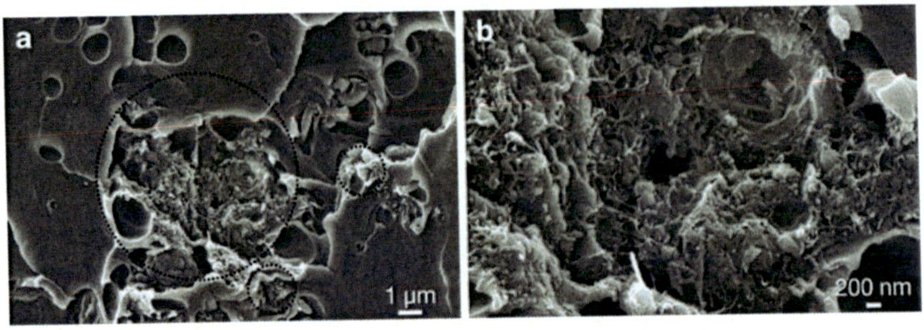

Fig. 8.73 Cross-sectional FESEM images of PES/unmodified SWNT MMM with 10% SWNT loading. (**a**) lower magnification. Dotted circle indicates formation of SWNT cluster, (**b**) higher magnification. Arrow indicates SWNTs in the polymer matrix. (From [32])

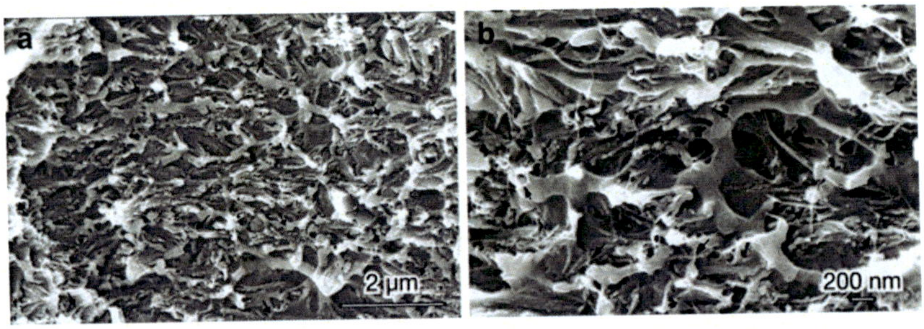

Fig. 8.74 Cross-sectional FESEM images of PES/modified SWNT with 10% SWNT loading (**a**) lower magnification, (**b**) higher magnification. The arrow indicates the SWNTs in the polymer matrix. (From [32])

treated SWNT was not uniformly dispersed and dense layers of SWNTs were found (Fig.8.75a, b). However, there are still regions where SWNTs were well dispersed (Fig. 8.75c).

Figure 8.76a displays dynamic mechanical thermal analysis (DMTA). The storage modulus of PS increased with an increase in the SWNT loading. Figure 8.76b shows the δ peaks. There is a slight shift in T_g as well as the increase in broadening in the δ peak as the SWNT loading increases. The broadening can be attributed to the local heterogeneity in nanocomposites with well dispersed nanotubes. Hence, it is possible that favorable interaction between the SWNTs and the polymer matrix were formed and they created an interfacial zone of polymer segments with distributions in mobility.

Figure 8.77 shows the experimental adsorption isotherm data for the Electric-arc discharged SWNTs purified by the wet oxidation method (Fig. 8.77a) and HiP_{CO} SWNTs purified by the two-stage purification step (Fig. 8.77b) and the MD simulation results (Fig. 8.77c). The simulations were performed assuming heterogeneous bundle of nanotubes that included large interstitial sites due to imperfect packing of the nanotubes. The bundle consisted of 45 nanotubes of various diameters. The mean diameter of the tubes in the bundle is about 13.6 Å, which is the diameter

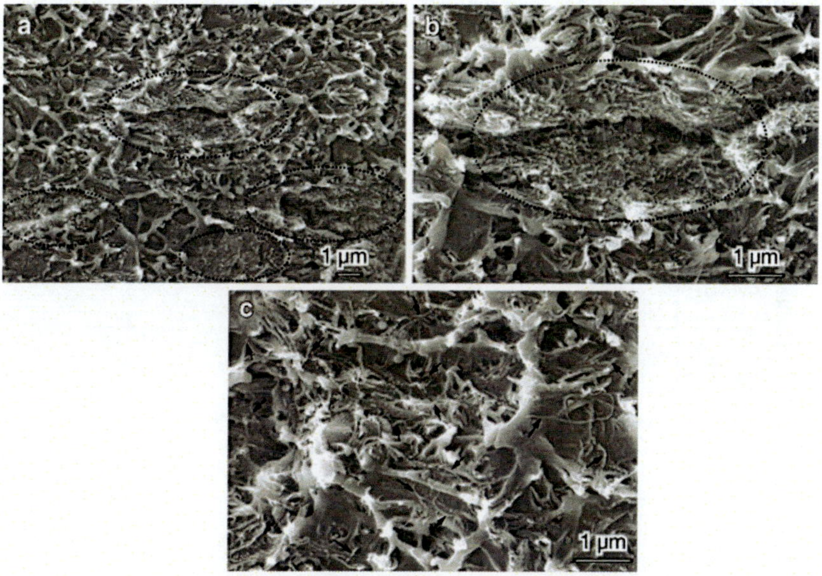

Fig. 8.75 Cross-sectional FESEM images of PES/modified SWNT with 15% SWNT loading (**a**) lower magnification, (**b**) higher magnification showing the SWNT layers, (**c**) region without SWNT layers. *Dotted circles* and *arrows* indicate dense SWNT region and well dispersed SWNTs in the polymer matrix, respectively. (From [32])

of a (10, 10) SWNT. The experimental adsorption data for the HiP_{CO} SWNTs are much higher than those for the electric-arc discharged SWNTs and the data are close to the results from the MD simulation. The order in the amount adsorbed was $H_2<O_2<CH_4<CO_2$ for all cases, indicating the MD simulation can predict the adsorption isotherm very well. The lower amount of adsorption of electric-arc discharged SWNTs than the HiP_{CO} SWNTs is due to the following two reasons. (1) The purified electric-arc nanotubes are partially closed or blocked by impurities or internal adsorption, which would be consistent with the lower surface area and micro-porous volume seen in the electric-arc nano-tubes when compared to the purified HiP_{CO}. (2) The carboxylic acid group created by the acid cutting procedure blocks the entrance to the interior of the CNTs.

The CO_2 adsorption at 308 K given in Fig. 8.78 includes the simulation based on the assumption that the adsorption inside the CNTs are not allowed. Thus, the adsorption takes place only at the external surface of the CNTs and in the large interstitial defect sites. The simulation shows adsorption isotherm higher than experimental data for the Electric-arc discharged SWNTs. This indicates that in the real system some external and interstitial defect sites are also partially blocked due most likely to impurities, amorphous carbon, or residual functional groups.

Figure 8.78 also shows excellent agreement between the experimental data from the HiP_{CO} SWNTs and the simulation data based on the 47% opening of the CNTs.

Figure 8.79 displays the adsorption isotherms for the gases N_2, O_2 and CH_4 for the PS MMM membranes containing 0, 2.5, 5 and 15 wt% of functionalized SWNTs

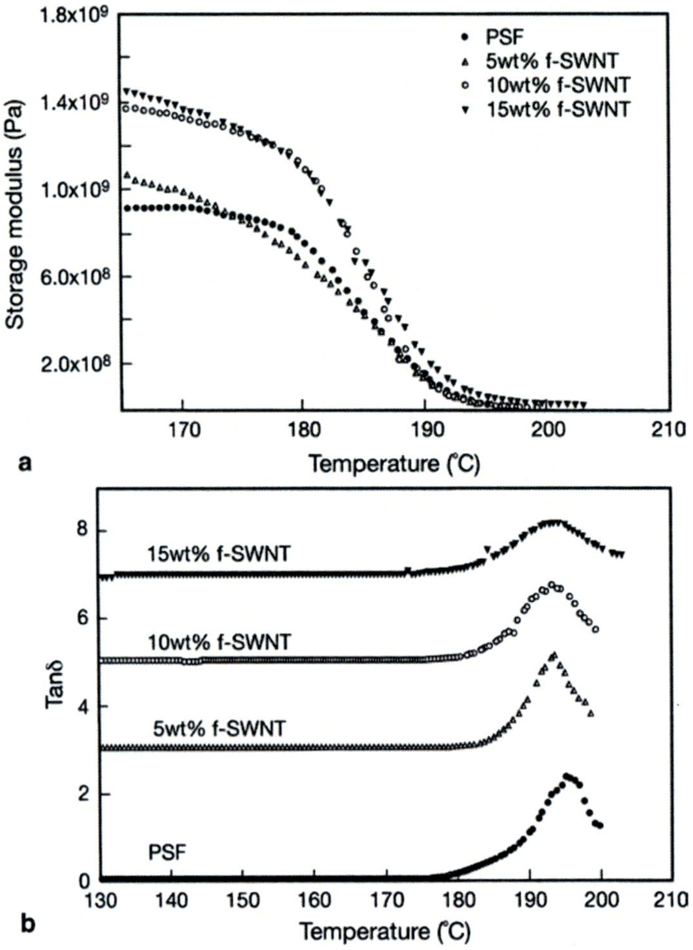

Fig. 8.76 DMTA results for pure PS and functionalized SWNT/PS MMMs (**a**) Storage modules, (**b**) shifted tan δ. (From [32])

at 308 K. The dashed lines on Fig. 8.79 are the data predicted by atomic simulation for the 15 wt% loading of the functionalized SWNTs based on the pure PS and SWNTs materials and the additivity rule. The adsorption of the gas increases with the increase of SWNTs loading. The measured adsorption isotherm for the gases N_2 and O_2 are lower than the values predicted by atomic simulation. This may be due to the change in the PS structure in the zone surrounding the SWNT particles because of the interaction between the particle and the macromolecules. Thus, the presence of SWNTs increases the gas adsorption but the increase is counteracted by the constrained polymer chain packing at the SWNT/polymer interface.

The data from the gas permeatin experiments are given in Table 8.34. The data demonstrates the permeability increases significantly by the 5 wt% loading of functionalized SWNT but there is only slight increase or no increase in the permeability

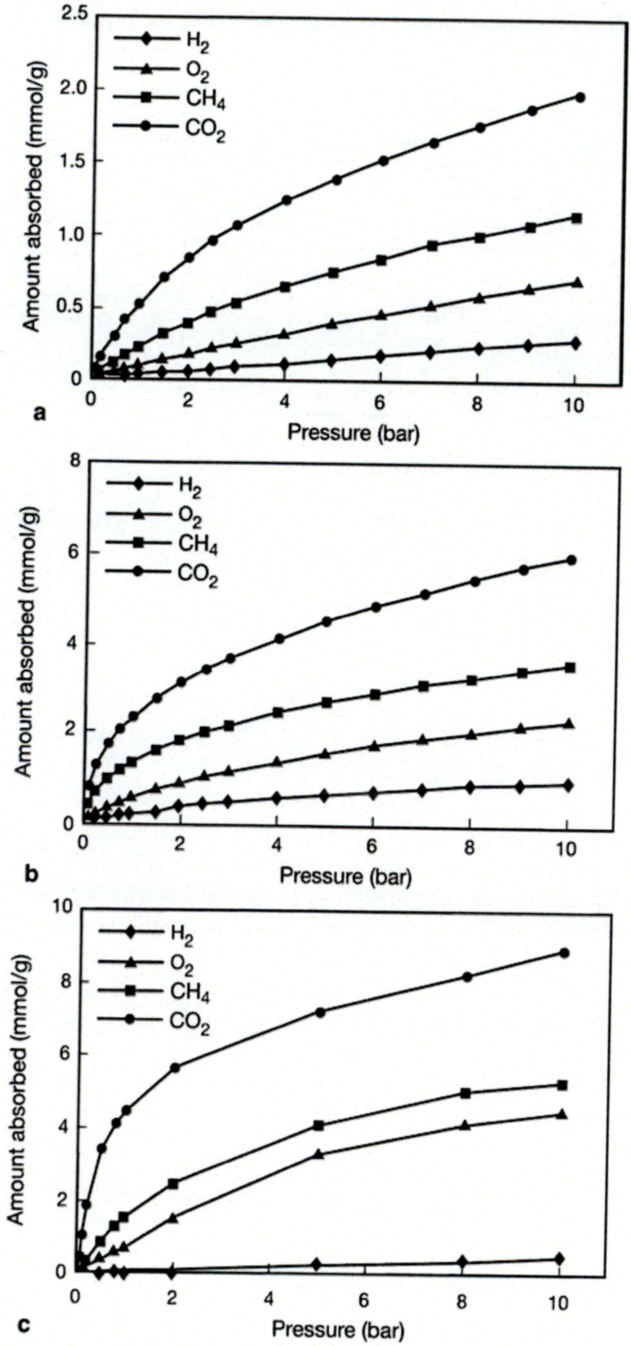

Fig. 8.77 Adsorption isotherm for various gases from experiments (**a** and **b**) and from molecular dynamic simulation. (**a**) Electric arc-discharged SWNTs purified by the wet oxidation method, (**b**) HiP_{CO} SWNTs purified by the two-stage purification step, and (**c**) atomic simulations. (From [32])

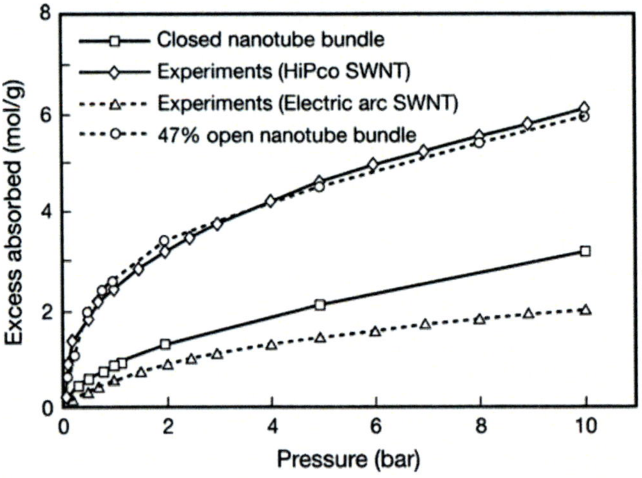

Fig. 8.78 Comparison of closed and partially open nanotubes for the adsorption of CO_2 at 308 K. (From [32])

when the loading is increased from 5 to 10 wt%. Further increase in the loading causes a decrease in the permeability in most of the gases.

The diffusivity data were obtained by two different methods; i.e. from the time-lag in the gas permeation and from the sorption kinetics. The results from both methods are plotted versus SWNT loading in Fig. 8.80. Even though the absolute values obtained from the two different methods are different, the trends observed are the same; i.e., there is a significant increase in diffusion coefficient from 0 to 5 wt% loading while the diffusion coefficient either leveled off or decreased slightly by the further increase in SWNT loading.

The sorption data obtained by time-lag method and by the gravimetric sorption method are presented in Fig. 8.81. The solubility showed little change with the change in the SWNT loading.

Table 8.35 shows the ideal separation factors for various gas pairs. Interestingly, the ideal separation factor tends to decrease for He/CH_4 and CO_2/CH_4 gas pairs while it tends to increase for the gas pair CH_4/N_2. This is due to the enhanced CH_4 transport through the MMM when at the higher functionalized SWNT loadings. Chen and Sholl [5] argued in their simulation study that SWNT membranes would become highly selective for CH_4 over H_2 due to the significant adsorption selectivity of CH_4 over H_2. This argument is also valid for other gas pairs in which CH_4 is involved.

8.5 Other Inorganic Materials Blended in Precursors

This section deals with MMM which is comprised of an inorganic component that is blended in CMS. Barsema et al. prepared a flat sheet Ag-functionalized CMS membranes from blends of P84 co-polyimide and a sulfonated poly(ether ether ketone)

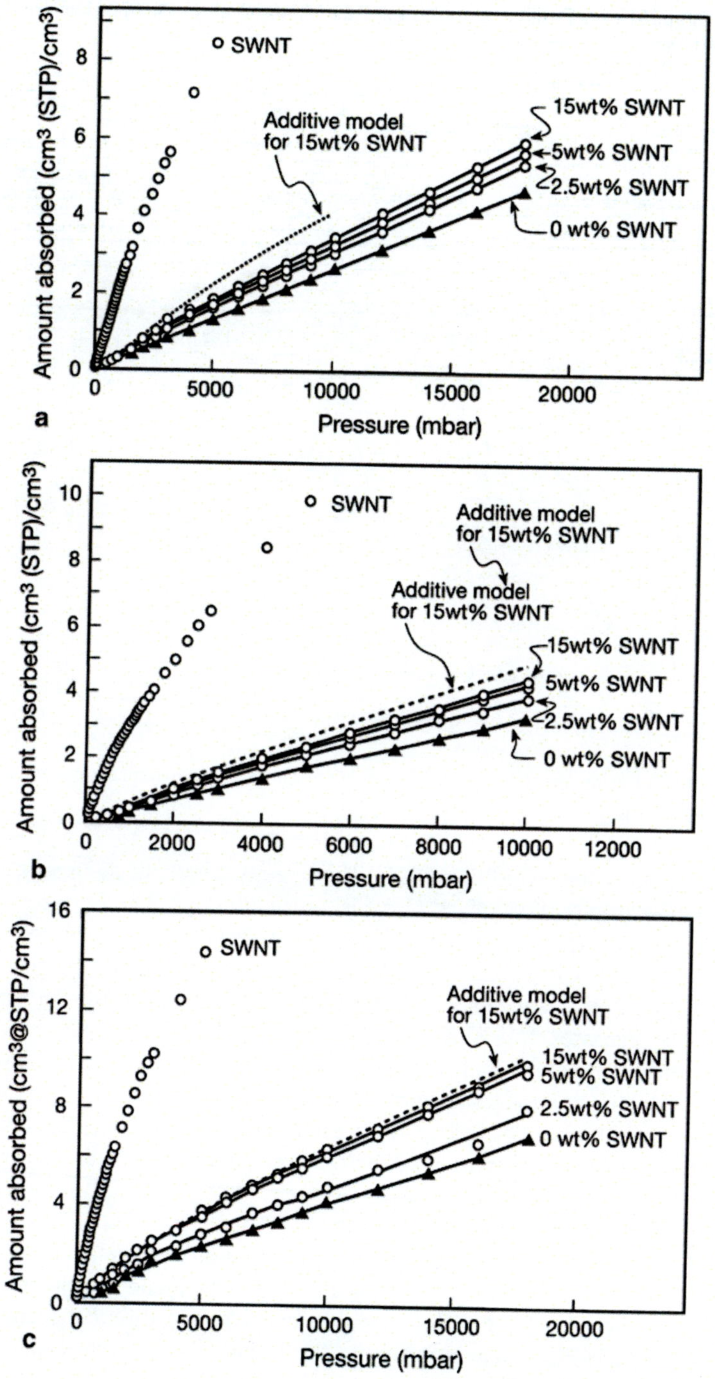

Fig. 8.79 Adsorption isotherms of PS MMMs containing 0, 2.5, 5 and 15 wt% of functionalized SWNTs; (**a**) N_2, (**b**) O_2 and (**c**) CH_4. (From [32])

Table 8.34 Permeability of various gases for the PS membrane and the PS MMM loaded with functionalized SWNTs

Membrane	Permeability (Barrer)				
	He	CO_2	O_2	N_2	CH_4
PS	7.88	3.90	0.84	0.17	0.17
5% SWNT loading	10.20	5.12	1.16	0.23	0.27
10% SWNT loading	10.27	5.19	1.23	0.23	0.28
15% SWNT loading	8.88	4.52	1.11	0.22	0.28

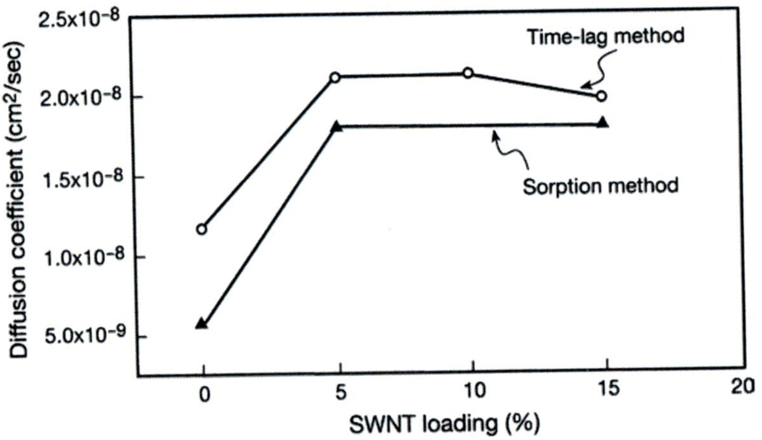

Fig. 8.80 CO_2 diffusion coefficients in PS MMMs with different loadings of functionalized SWNTs. (From [32])

with a Ag^+ counter ion (AgSPEEK) [34]. There are two parts of the report. In one part, Ag nano-particles were mixed to the PI precursor dope. The Ag incorporated CMS membranes outperformed the CMS membranes without Ag incorporation in O_2/N_2 separation. It was speculated that the superior performance of Ag incorporated CMS was due to the creation of selective bypasses around the Ag nanoparticles. In these bypasses, which arise from the imperfect cohesion of Ag and carbon matrix, separation is enhanced through the O_2 surface flow occurring at the surface of Ag particles due to stronger adsorption of O_2. As well, these bypasses seem to offer a significant increase of volume available for gas diffusion. The authors claim that this speculation was supported by their experimental sorption and gas permeation data.

In the second part, P84 co-polyimide (Fig. 8.82b) was blended with AgSPEEK (Fig. 8.82a). The polymer blend was then pyrolyzed at different temperatures. Figure 8.83 shows the thermogram of P84, AgSPEEK and the blends in which the wt% of AgSPEEK is changed. AgSPEEK shows a considerably different thermogram from P84. As AgSPEEK content decreases the blend's thermogram comes closer to the P84's thermogram. The gas permeation performance of these membranes is shown in Figs. 8.84 and 8.85. According to Fig. 8.84 a maximum in permeability

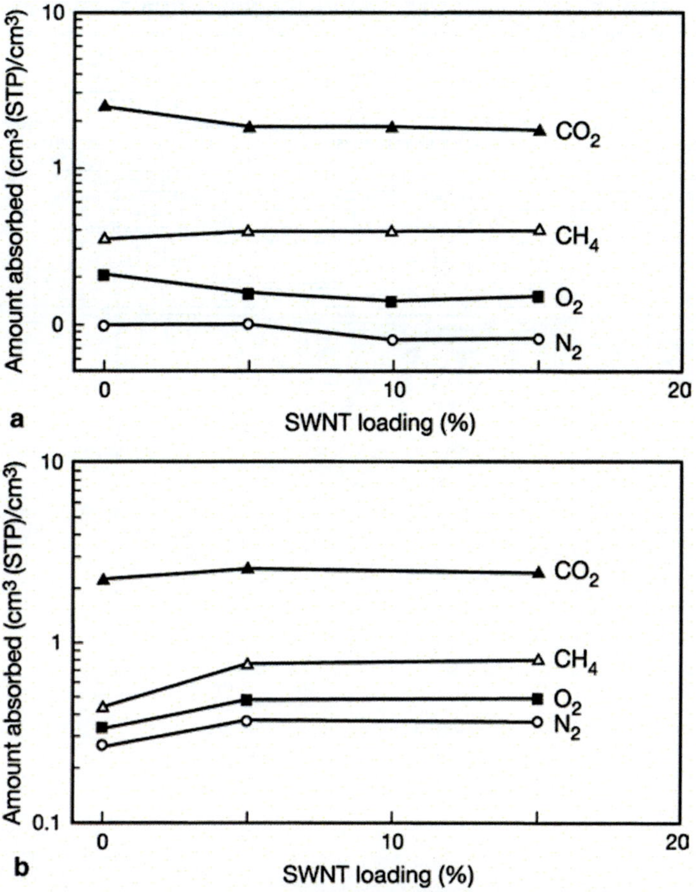

Fig. 8.81 Solubility coefficients of PS MMMs versus functionalized SWNT loading at 308 K and 4 bar; (**a**) time-lag method, (**b**) gravimetric sorption method. (From [32])

Table 8.35 Ideal separation factors of PS MMMs with different loadings of functionalized SWNTs for different gas pairs

Membrane	He/CH_4	CO_2/CH_4	O_2/N_2	CH_4/N_2
PS	47.5	23.55	5.07	1.00
5 wt% SWNT loading	37.51	18.82	5.04	1.17
10 wt% SWNT loading	36.43	18.1	5.35	1.21
15 wt% SWNT loading	31.66	16.09	5.10	1.27

was reached at the silver content of 2.46 wt%. Higher concentration of Ag leads to the formation of Ag layer on the membrane surface reducing the permeability. On the other hand, permselectivity kept increasing as the silver content increased. The selectivity of CO_2 over O_2 remained almost constant as silver content changed since they (CO_2 and O_2) behaved similarly in permeability vs. silver content curve. To

Fig. 8.82 Chemical formula of (**a**) SPEEK and (**b**) P84 co-polyimide. (From [34])

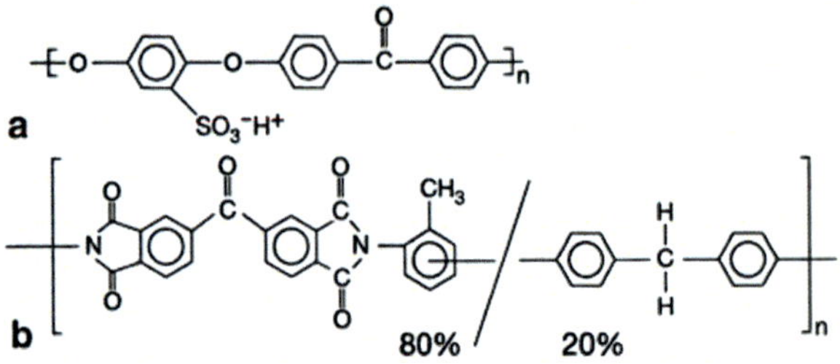

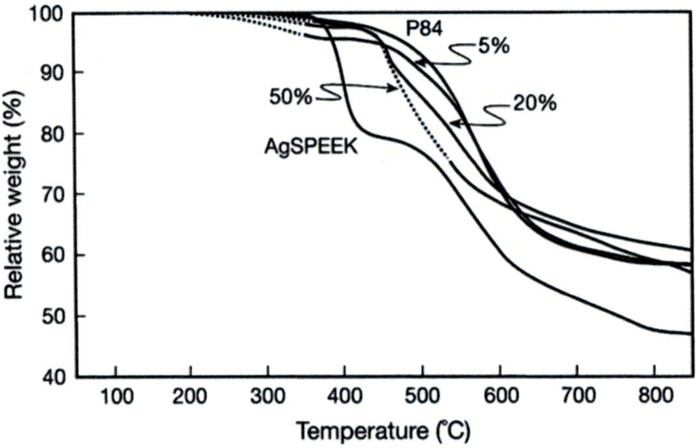

Fig. 8.83 Relative weight decrease as function of temperature for various compositions of AgSPEEK/P84 blends measured by TGA. (From [34])

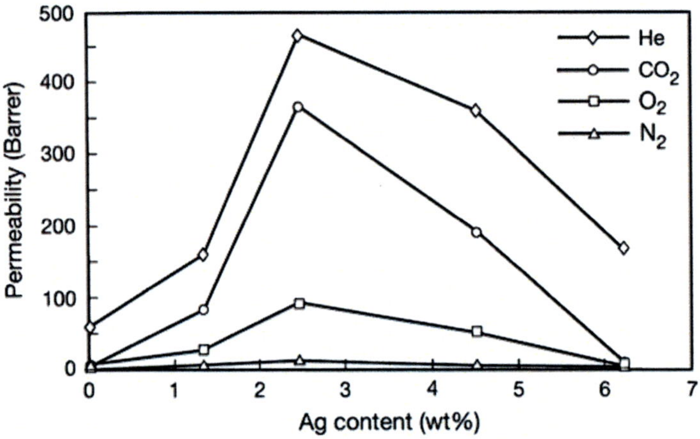

Fig. 8.84 Permeability of (◊) He, (○) CO_2, (□) O_2, (Δ) N_2 versus Ag content at 2 bar and 25 °C. (From [34])

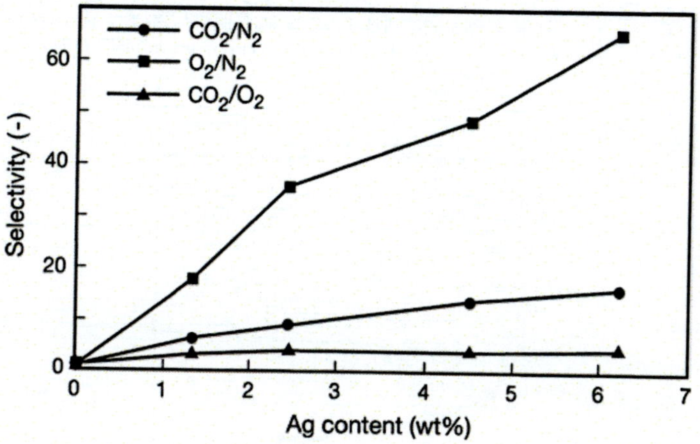

Fig. 8.85 Selectivity of (●) CO_2/N_2, (■) O_2/N_2 and (▲) CO_2/O_2 versus Ag content at 2 bar and 25 °C. (From [34])

obtain a good distribution of Ag throughout the carbon matrix, it is essential to have a homogeneous blend of P84 and SPEEK. Temperature-modulated DSC assured the homogeneous blend of two polymer components.

A novel carbon/ZSM-5 nanocomposite membrane was successfully prepared by incorporating nano-sized ZSM-5 into PI precursor. The rational behind the development of the novel membranes is as follows. ZSM-5 possesses uniform pores of ca. 0.55 nm and highly temperature resistant. Hence the structure will not be destroyed during the high temperature pyrolysis. As well, the space created between the nano-particle and the carbon matrix may help increase the permeability. The carbon deposited at the ZSM-5 pore entrance will also help increase the permselectivity. Thus, the dry zeolite ZSM-5 was dispersed in DMSO solvent by ultrasonic treatment for 2 h and the suspension was mixed with the polyamic acid (PAA) solution derived from pyromellitic dianhydride (PMDA) and 4, 4′-oxydianiline (ODA). Following the prepolymerization at 298 K for 10 h, the solution was cast and dried. The membrane was then heated to 373, 473 and 623 K in Ar stream. The membrane was further pyrolysed at 873 K. XRD patterns shown in Fig. 8.86 obviously show that the zeolite ZSM-5 was incorporated in the carbon membrane. Figure 8.87 shows the nitrogen adsorption curved at 77 K for membrane PI Z2 (ZSM-5 content in PI 9.1%) and PI Z3 (ZSM-5 content in PI 16.7%). They are the intermediate type between Type 1 and Type 2 according to the IUPAC classification, suggesting ultra-micropores are present in the membranes. The figure also shows the pore size distribution for both membranes.

Finally, Table 8.36 shows the permeation properties of the PI membrane, carbonized PI membrane and three PI-ZSM-5 composite membranes.

Table 8.36 shows that the permeability increases significantly at higher zeolite loading, indicating that a well-established ultramicroporous structure is present in the nanocomposite membrane. In the meantime, the selectivity of oxygen to nitro-

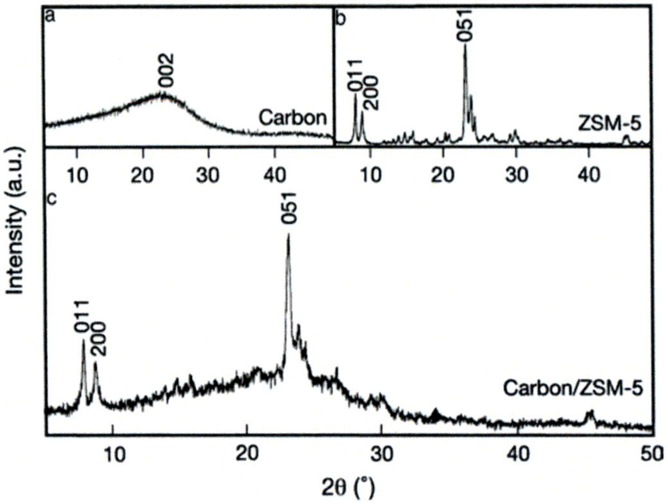

Fig. 8.86 XRD patterns of (**a**) CMS membrane, (**b**) ZSM-5, and (**c**) ZSM-5 blended CMS membrane (PI Z2). (From [35])

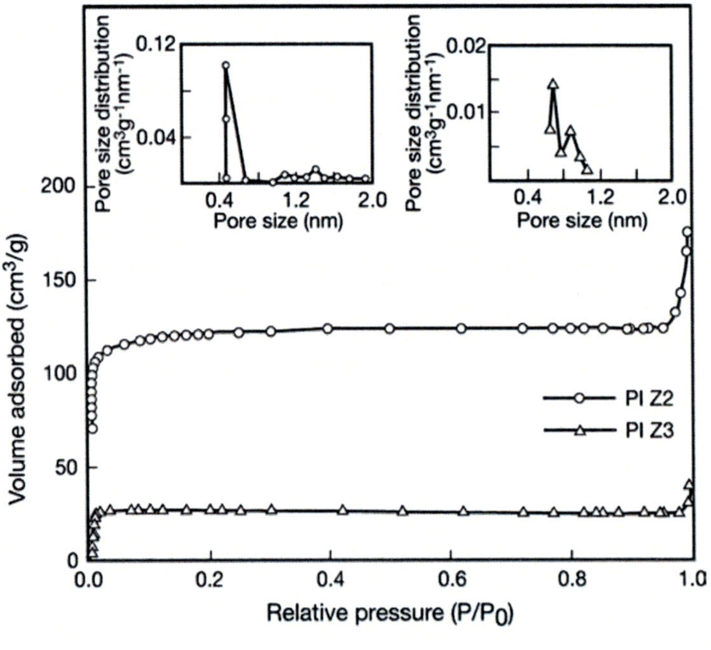

Fig. 8.87 Nitrogen adsorption isotherms of nanocomposite carbon membranes at 77 K, insert showing pore size distribution calculated by the HK method. (From [35])

Table 8.36 Permeation properties of carbon/ZSM-5 composite membranes at 298 K

Membranes	Permeability (Barrer)		Selectivity
	O_2	N_2	O_2/N_2
PI membrane	0.16	0.04	4.2
PI membrane, carbonized	2.21	0.16	13.8
Composite membrane ZSM-5, 4.8%	70.0	5.85	12
Composite membrane ZSM-5, 9.1%	431.0	30.0	14.4
Composite membrane ZSM-5, 16.7%	499.0	39.6	12.6

gen remained in the range 12–14.4. It is well known that zeolite ZSM-5 is nonselective porous zeolite. Most likely, the zeolite pore window was narrowed by the deposited carbon during the pyrolysis process.

References

1. Hummer G, Rasaiah JC, Noworyta JP (2001) Water conduction through the hydrophobic channel of a carbon nanotube. Nature 414 (6860): 188-190
2. Karla A, Garde S, Hummer G (2003) Osmotic water transport through carbon nanotube membranes. Proc Natl Acad Sci USA 100 (18): 10175-10180
3. Holt JK, Park HG, Wang Y, Stadermann M, Artyukhin AB, Grigoropoulos CP, Noy A, Bakajin O (2006) Fast mass transport through sub-2-nanometer carbon nanotubes. Science 312 (5776): 1034-1037
4. Skoulidas AI, Ackerman DM, Johnson JK, Sholl DS (2002) Rapid transport of gases in carbon nanotubes. Phys Rev Lett 89 (18): 185901-1 – 185901-4
5. Chen HB, Sholl DS (2006) Predictions of selectivity and flux for CH_4/H_2 separations using single walled carbon nanotubes as membranes. J Membr Sci 269 (1-2): 152-160
6. Ackerman DM, Skoulidas AI, Sholl DS, Johnson JK (2003) Diffusivities of Ar and Ne in carbon nanotubes. Mol Sim 29 (10-11): 677-684
7. Chen HB, Johnson JK, Sholl DS (2006) Transport diffusion of gases is rapid in flexible carbon nanotubes. J Phys Chem B 110 (5): 1971-1975
8. Kim S, Jinschek JR, Chen H, Sholl DS, Marand E (2007) Scalable fabrication of carbon nanotube/polymer nanocomposite membranes for high flux gas transport. Nano Lett 7 (9): 2806-2811
9. Mi W, Lin YS, Li Y (2007) Vertically aligned carbon nanotube membranes on macroporous alumina supports. J Membr Sci 304 (1-2): 1-7
10. Hinds BJ, Chopra N, Rantell R, Andrews R, Gavalas V, Bachas LG (2004) Aligned multi-walled carbon nanotube membranes. Science 303 (5654): 62-65
11. Furukawa S, Nitta T (2000) Non-equilibrium molecular dynamics simulation studies on gas permeation across carbon membranes with different pore shape composed of micro-graphite crystallites. J Membr Sci 178 (1-2): 107-119
12. Wu Z, Liu Z, Wang W, Fan Y, Xu N (2008) Non-equilibrium molecular dynamics simulation on permeation and separation of H_2/CO in nanoporous carbon membranes. Sep Purif Technol 64 (1): 71-77
13. Chen H, Sholl DS (2006) Prediction of selectivity and flux for CH_4/H_2 separations using single walled carbon nanotubes as membranes. J Membr Sci 269 (1-2): 152-160
14. Verweij H, Schillo MC, Li J (2007) Fast mass transport through carbon nanotube membranes. Small 3 (12): 1996-2004

15. Majumder M, Chopra N, Andrews R, Hinds BJ (2005) Nanoscale hydrodynamics: Enhanced flow in carbon nanotubes. Nature 438 (7064): 44
16. Truskett TM, Debenedetti PG, Torquato S (2001) Thermodynamic implications of confinement for a waterlike fluid. J Chem Phys 114 (5): 2401-2418
17. Singh G, Rana D, Matsuura T, Ramakrishna S, Narbaitz RM, Tabe S (2010) Removal of disinfection by-products in water by using carbonized nanofiber membranes. Sep Purif Technol 74 (2): 202-212
18. Tung HH, Unz RF, Xie YF (2006) HAA removal by GAC adsorption. J. AWWA 98 (6): 107-112
19. Anson M, Marchese J, Garis E, Ochoa N, Pagliero C (2004) ABS copolymer-activated carbon mixed matrix membranes for CO_2/CH_4 separation. J Membr Sci 243 (1-2): 19-28
20. Vu DQ, Koros WJ, Miller SJ (2003) Mixed matrix membranes using carbon molecular sieves I. Preparation and experimental results. J Membr Sci 211 (2): 311-334
21. Vu DQ, Koros WJ, Miller SJ (2002) High pressure CO_2/CH_4 separation using carbon molecular sieve hollow fiber membranes. Ind Eng Chem Res 41 (3): 367-380
22. Steel K (2000) Carbon membranes for challenging gas separations. PhD Dissertation, The University of Texas at Austin, TX, USA
23. Vu DQ, Koros WJ, Miller SJ (2003) Mixed matrix membranes using carbon molecular sieves II. Modeling permeation behaviour. J Membr Sci 211 (2): 335-348
24. Maxwell JC (1873) Treatise on Electricity and Magnetism. Oxford University Press, London, UK
25. Mahajan R (2000) Formation, characterization and modeling of mixed matrix membranes materials. Ph.D. Dissertation, The University of Texas at Austin, TX, USA
26. Chern RT, Koros WJ, Yui B, Hopfenberg HB, Stannet VT (1984) Selective permeation of CO_2 and CH_4 through Kapton polyimide: Effects of penetrant competition and gas-phase nonideality. J Polym Sci Polym Phys Ed. 22 (6): 1061-1084
27. Mahajan R, Burns R, Schaeffer M, Koros WJ (2002) Challenges in forming successful mixed matrix membrane with rigid polymeric materials. J Appl Polym Sci 86 (4): 881-890
28. Rafizah WAW, Ismail AF (2008) Effect of carbon molecular sieve sizing with poly(vinyl pyrrolidone) K-15 on carbon molecular sieve-polysulfone mixed matrix membrane. J Membr Sci 307 (1): 53-61
29. Kim S, Pechar TW, Marand E (2006) Poly(imide siloxane) and carbon nanotube mixed matrix membanes for gas separation. Desalination 192 (1-3): 330-339
30. Cong H, Zhang J, Radosz M, Shen Y (2007) Carbon nanotube composite membranes of brominated poly(2,6-diphenyl-1,4-phenylene oxide) for gas separation. J Membr Sci 294 (1-2): 178-185
31. Merkel TC, Freeman BD, Spontak RJ, He Z, Pinnau I, Meakin P, Hill AJ (2003) Sorption, transport, and structural evidences for enhanced free volume in poly(4-methyl-2-pentyne)/fumed silica nanocomposite membranes. Chem Mater 15 (1): 109-123
32. Kim S, Chen L, Johnson JK, Marand E (2007) Polysulfone and functionalized carbon nanotube mixed matrix membranes for gas separation: Theory and experiments. J Membr Sci 294 (1-2): 147-158
33. Harris JG, Yung KH (1995) Carbon dioxide's liquid-vapor coexistence curve and critical properties as predicted by a simple molecular model. J Phys Chem 99 (31): 12021-12024
34. Barsema JN, Van Der Vegt NFA, Koops GH, Wessling M (2005) Ag-functionalized carbon molecular-sieve membranes based on polyelectrolyte/polyimide blend precursors. Adv Funct Mater 15 (1): 69-75
35. Liu Q, Wang T, Qiu J, Cao Y (2006) A Novel carbon/ZSM-5 nanocomposite membrane with high performance for oxygen/nitrogen separation. J Chem Soc Chem Commun: 1230-1232.

Chapter 9
Applications of Carbon-based Membranes for Separation Purposes

9.1 Application in Gas Separation and RO/NF/UF/MF

As stated in Chap. 6, most of separation experiments using CMSMs have been performed for gas separation. Many examples have already given throughout the foregoing chapters. In contrast, only few papers have been published for ultrafiltration (UF) [1, 2].

The most important large application of carbon membrane seems in the production of low cost and high purity nitrogen from air. Other examples are separation of hydrogen from gasification gas, purification of methane [3]. In addition, it is used to recover a valuable chemical (H_2) from a waste gas stream without further compression of the feed gas while rejecting a substantial portion of the hydrocarbons [4].

Carbon membranes are also promising candidate for the separation of light alkenes/alkanes especially for propene/propane separation, since it has been reported that the carbon membrane showed an excellent propene/propane permselectivity. It is expected to be superior to other methods such as distillation, adsorption and absorption owing to its low energy consumption. It also showed a good performance for the 1,3-butadiene/n-butane separation [5]. Considering that separation of light alkenes/alkanes has been recognized to be a key technology in the petrochemical industry [6], the carbon membrane can give a great contribution to petrochemical industry. Since a large number of papers on CMSMs applications in gas separation and the few available papers on UF applications have already been shown, this chapter is dedicated to the applications of CMSMs in other separation processes.

9.2 Vapor Separation

There is a report on the Carbon Membrane Separator for elimination of SF_6 from gas insulated electrical utilities [7]. In the electrical power industry 80% of 6,500–7,500 metric t SF_6 produced worldwide is used as a gaseous dielectric in circuit breakers, gas-insulated substations and switchgear. Besides its high price, SF_6 is a very potent greenhouse gas. A possible way to capture and reuse the SF_6 in the

A. F. Ismail et al., *Carbon-based Membranes for Separation Processes*,
DOI 10.1007/978-0-387-78991-0_9,

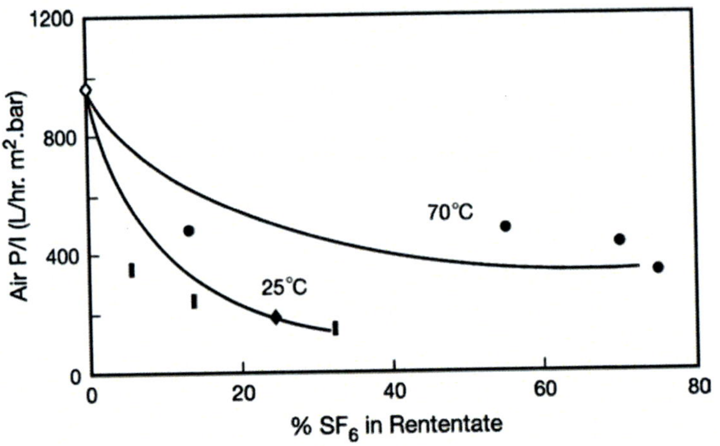

Fig. 9.1 Permeance of air versus % SF_6 in the retentate at different temperatures. (From [7])

vented gas is to use gas/vapor separation by membrane. Attempts were made to fabricate a CMSM to separate PFC gases (SF_6, C_2F_6, etc.) from various "fast" gases such as air, CO_2 etc. Carbonized hollow fiber membranes from Carbon Membrane Inc., Israel, was used for this purpose. Pores of the CMSMs are too large and the partial blocking of the pores is necessary. This is achieved by coating the inner side of hollow fibers with a thin dense layer of carbon by CVD process. The CVD coating process is followed by the process of gradually opening the passages throughout the membrane cross-section (the process not disclosed). Pore size is so adjusted that small O_2 (3.28 Å) and N_2 (3.60 Å) are effectively separated from the large molecule SF_6 (5.02 Å) by sieve mechanism. Feed air contaminated by SF_6 was supplied on the shell side of the hollow fiber module and the clean air was collected on the bore side as the permeate. Figure 9.1 indicates the permeance of the air as the function of SF_6 concentration in the (feed/retentate). The permeance decreases with SF_6 concentration since the driving force for the air permeation decreases.

9.3 Pervaporation

The separation of acetic acid and water was studied by pervaporation using CMS filled-polydimethyl siloxane (PDMS) membrane by Li et al. [8]. CMS powder (particle size 1.5–2.0 μm) was incorporated in the PDMS membrane by dispersing CMS in the solution before annealing. The membrane thickness was 120 and 160 μm, respectively, for PDMS membrane without and with CMS.

The permeation properties of the CMS filled and unfilled membranes are given in Fig. 9.2. There is a maximum in pervaporation flux at the CMS content of 20 wt%. The separation factor also showed the maximum at the 20% loading. Figure 9.3 indicates an increase in pervaporation flux and decrease in separation factor as the

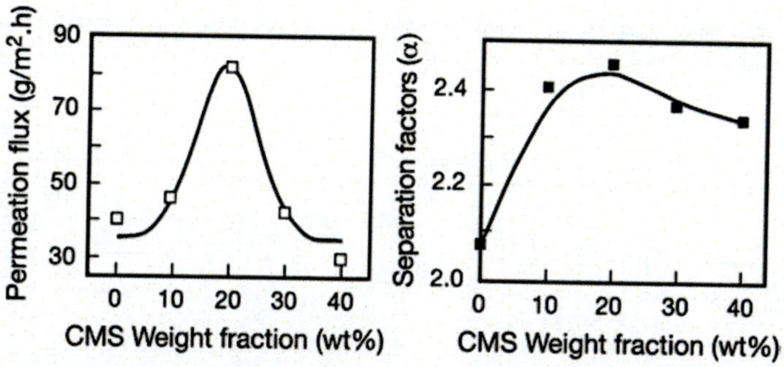

Fig. 9.2 Flux and separation factor versus weight fraction of CMS; concentration of acetic acid in feed, 5.2 mol%, temperature 45°C. (From [8])

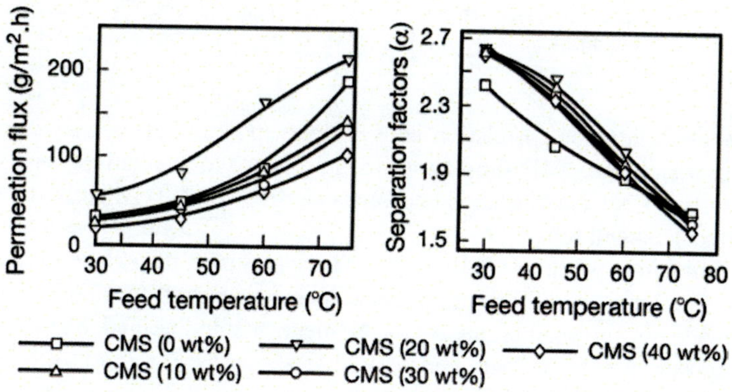

Fig. 9.3 Flux and separation factor versus feed temperature for various CMS loadings; concentration of acetic acid in feed, 5.2 mol%. (From [8])

pervaporation temperature is increased. Figure 9.4 shows that increase in total flux, which is mainly attributed to the flux increase of acetic acid and near constant water flux throughout the concentration range studied. The water flux remains almost constant. As a result the separation factor increases with the feed acetic acid concentration and levels off at high acetic acid concentrations.

Sakata et al. made porous carbon membrane plates (PCMP) to separate benzene-cyclohexane azeotrope by pervaporation [9]. PCMPs were prepared from phenol resin powder (BELLPEARL S-870, Kanebo) by pressurizing the powder in a hydraulic press. The disk was heated at 300°C for 1 h in air stream and then carbonized at 800°C in nitrogen stream (A) or CO_2 stream at 800°C (B) or at 850°C (C). A had only micro-pores, B had much wider pore size distribution than A and C had both micro- and meso-pores. The separation factor (benzene over cyclohexane) of 2.8 was obtained by pervaporation at 60°C. Flux of pure benzene and pure cyclohexane was 0.34 and 0.57 kg/m^2 h, respectively.

Peng et al. prepared novel nanocomposite membranes (PVA-CNT(CS)) by incorporating chitosan (CS)-wrapped MWNT into poly(vinyl alcohol)(PVA) [10]. To

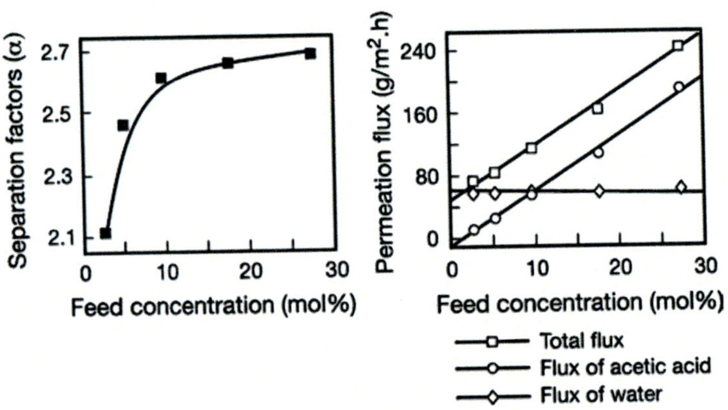

Fig. 9.4 Flux and separation factor versus feed acetic acid concentration; CMS loading 20 wt%, temperature 45°C. (From [8])

further explore intrinsic correlation between pervaporation performance and free volume characteristics, MD simulation was introduced to qualitatively analyze the contribution of CNT incorporation on improving free volume characteristics of the nanocomposite membrane.

MD simulations were carried out on Materials Studio developed by Accelrys Software Inc. Four amorphous cell models including pure PVA membrane, PVA-CNT nanocomposite membrane, PVA-CS blend membrane and PVA-CNT(CS) nanocomposite membrane were constructed as shown in Fig. 9.5 [10]. CNT was used for the simulation. The calculation results concerning the free volume of these four membranes are shown in Fig. 9.6. The authors state that the free volume of PVA-CNT and PVA-CNT(CS) nanocomposite membranes are distinctively larger than corresponding PVA and PVA-CS blend membranes, which indicates that the incorporation of CNT has made the PVA chain packing looser. The fractional free volume of PVA-CTN(CS) was 6.25% which is significantly larger than that of PVA-CNT membrane (2.54%), indicating that chitosan is acting not only as a dispersant but also enhances the fractional free volume. The structure of chitosan wrapping the CNT was also simulated as shown in Fig. 9.7. Chitosan stays along the nanotube axis, enabling the dispersion of CNT in chitosan aqueous solution. The TEM image shown in Fig. 9.8 also confirms the chitosan wrapping on the CTN.

A chitosan solution of 1 wt% was prepared by dissolving chitosan in 2 wt% aqueous acetic acid solution and stirring at 353 K for about 1 h. A predetermined amount of CNT was added to the chitosan solution that was kept under mechanical stirring for about 24 min at 90°C. Then ultrasonic agitation was applied for few minutes. A black suspension of CNTs wrapped by chitosan was obtained. PVA was dissolved in distilled water at 90°C to obtain 10 wt% solution. Glutaraldehyde, the cross-linker, and 0.1 mL HCl (0.5 M) and a predetermined amount of chitosan-wrapped CNT were added to the PVA solution. The mixture was mechanically stirred for about 1 h at 90°C and then ultrasonically agitated for a few minutes. The solution was cast on

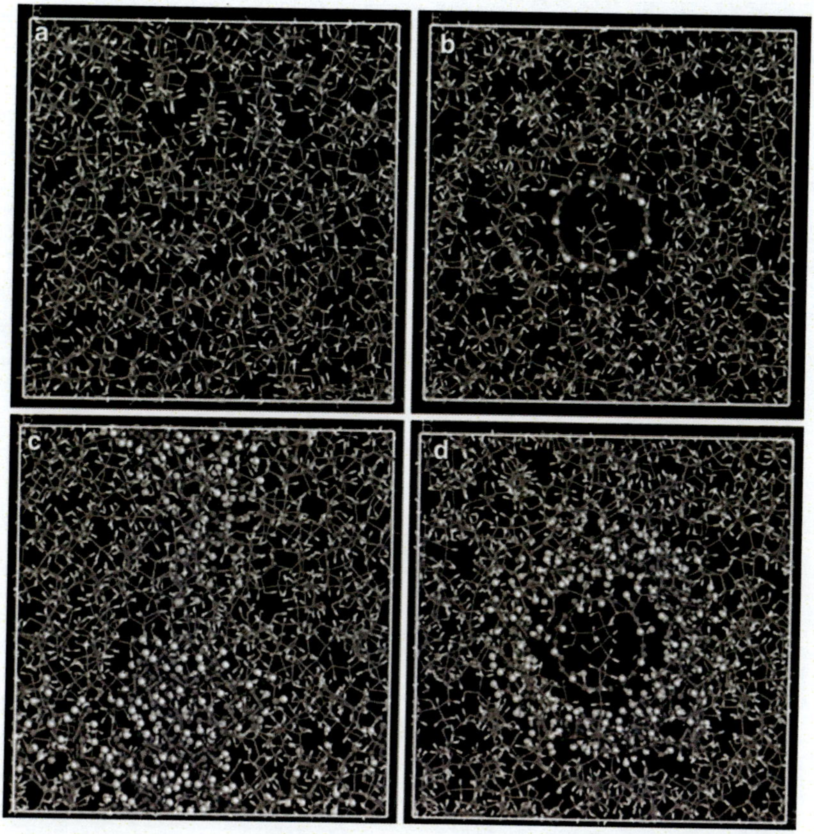

Fig. 9.5 Amorphous cell model of (**a**) pure PVA membrane, (**b**) PVA-CNT nanocomposite membrane, (**c**) PVA-CS blend membrane, (**d**) PVA-CNT(CS) nanocomposite membrane. (From [10])

a glass plate and dried in the air for about 24 h. Membrane thickness obtained was 80 μm. Figure 9.9 displays the cross-sectional SEM pictures of the PVA-CNT(CS) nanocomposite membrane with a CNT content of 2 wt%. The pictures show that the chitosan-wrapped CNTs were well dispersed in the membrane.

The pervaporation results are summarized in Table 9.1.

It is obvious that both flux and separation factor were increased by the incorporation of chitosan wrapped CNT. The best results were obtained when the CNT content was 2.0 wt%. Both flux and permeability decreased at a higher loading CNT than 2.0 wt%. The solubility selectivity for benzene kept increasing by the addition of CNT. This is because of the stronger interaction between CNT and benzene due to the π–π bond than between CNT and cyclohexane. The optimum performance of the membrane at the CNT content of 2.0 wt% is therefore due to the diffusivity change. This is explained by the presence of net-work and aggregate pore. The flux and the separation factor versus feed benzene concentration is plotted in Fig. 9.10.

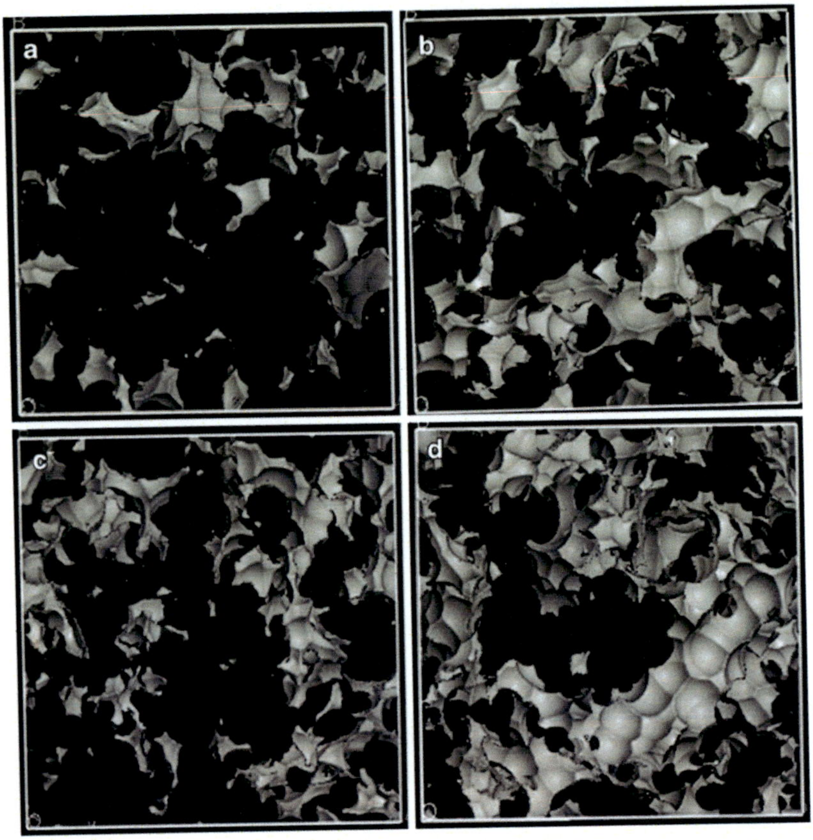

Fig. 9.6 Free volume (shown in *gray color*) simulation of (**a**) pure PVA membrane, (**b**) PVA-CNT nanocomposite membrane, (**c**) PVA-CS blend membrane, (**d**) PVA-CNT(CS) nanocomposite membrane. (From [10])

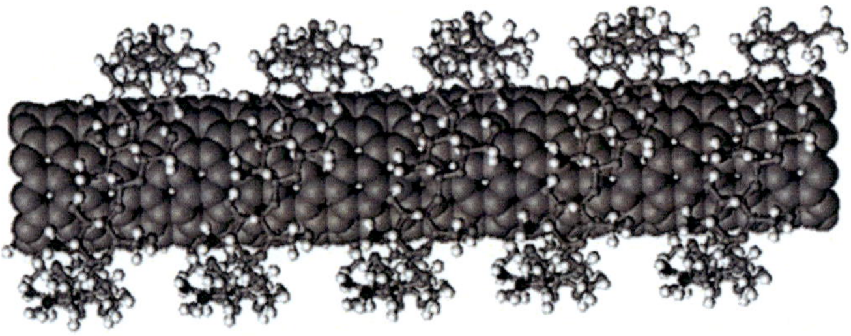

Fig. 9.7 Chitosan-wrapped carbon nanotube. (From [10])

The flux increases while the separation factor decreases rapidly with an increase in benzene concentration. This is the effect of the membrane swelling at the higher benzene concentration. The effect of the incorporation of chitosan-wrapped CNT is presumably two fold: (1) tuning the packing of polymer chains in the vicinity of

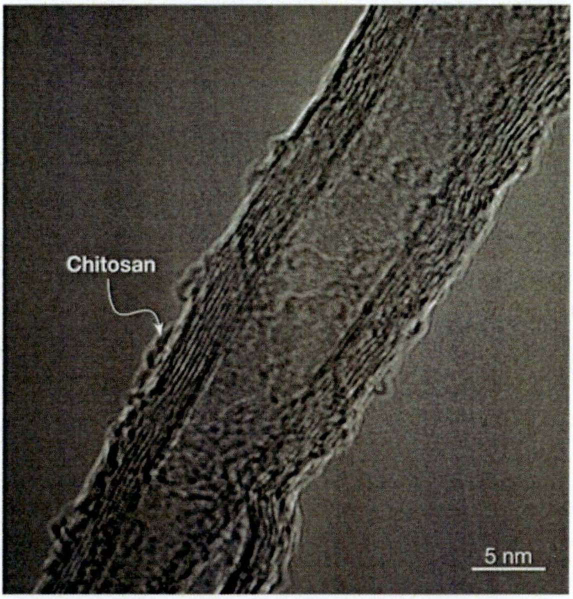

Fig. 9.8 TEM Image of chitosan-wrapped carbon nanotube. (From [10])

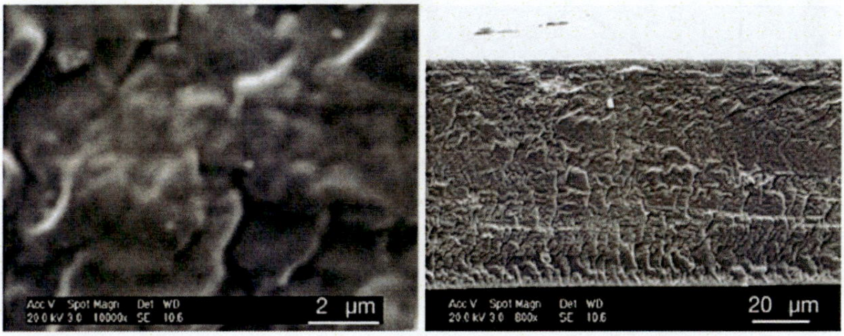

Fig. 9.9 The cross-sectional SEM micrographs of PVA-CNT(CS) nanocomposite membrane. (From [10])

the inserted particles. The fractional free volume is increased and the free volume size is adjusted. (2) providing nanochannels through added CNTs. The authors have suggested that, most likely, the first effect is the main reason for the enhanced flux and selectivity when the CNT content is less than 2.5 wt%. To further confirm this concept, positron annihilation life time spectroscopy (PALS) characterization was performed. Apparent free volumes f_{app} obtained from PALS was for PVA and PVA-CNT(CN) membrane, 0.56 and 0.71, respectively. Hence the increase in free volume by the addition of CN-wrapped CNT was confirmed. The authors distinguish pores present in the membranes as net work and aggregate pore. By coincidence, the pore size in the PVA-CNT(CS) nanocomposite membrane was 0.269 nm, which is between the molecular size of benzene (0.263 nm) and cyclohexane (0.303 nm).

Table 9.1 The pervaporation flux and separation factor of nanocomposite membranes for benzene/cyclohexane (50/50, w/w) mixtures at 523 K

Membrane	Permeation flux (g/m^2 h)	Separation factor	Solubility selectivity	Diffusion selectivity
PVA	20.3	9.6	7.56	1.27
PVA-CNT(CS) (0.5 wt%)[a]	53.0	23.1	12.66	1.82
PVA-CNT(CS) (1.0 wt%)[a]	60.8	30.4	15.00	2.02
PVA-CNT(CS) (1.5 wt%)[a]	67.3	37.6	16.86	2.23
PVA-CNT(CS) (2.0 wt%)[a]	65.9	53.4	18.56	2.88
PVA-CNT(CS) (2.5 wt%)[a]	58.9	46.5	21.27	2.18

[a] CNT content

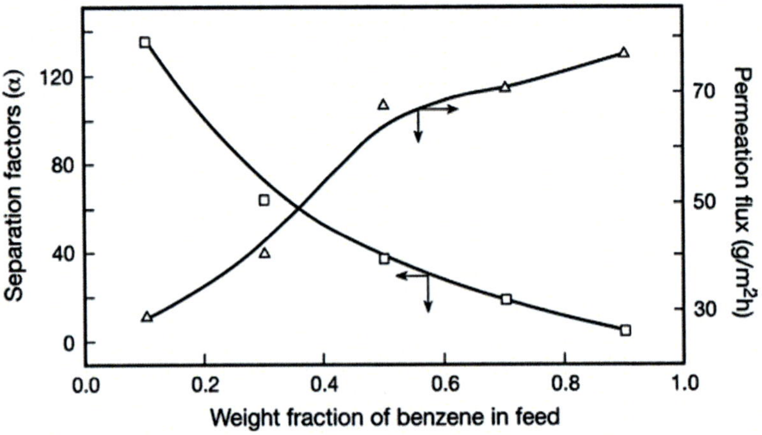

Fig. 9.10 The effect of feed benzene concentration on pervaporation performance for the separation of benzene/cyclohexane at 323 K by PVA-CNT(CS) nanocomposite membrane (CNT content 2 wt%). (From [10])

9.4 Fuel Cell Application

Polymer electrolyte membrane fuel cells (PEMFCs) receive much attention these days for their potential to produce clean energy for automotive, stationary, and potable applications. Nafion based membranes are widely used as the polymer electrolyte membrane (PEM) in fuel cells that operate in the temperature range 60–80°C. Above this temperature range the proton conductivity decreases significantly. As well, although the Nafion membranes show good proton conductivities of 0.1–0.01 S cm^{-1} in a humid environment, they may still have limitations: (1) dependence of conductivity on water, (2) high methanol permeability, especially for the direct methanol fuel cell (DMFC) applications, (3) tendency to be disintegrated in the presence of hydroxyl radicals, an intermediate in the cathode reaction, and (4) moderate mechanical and chemical stability.

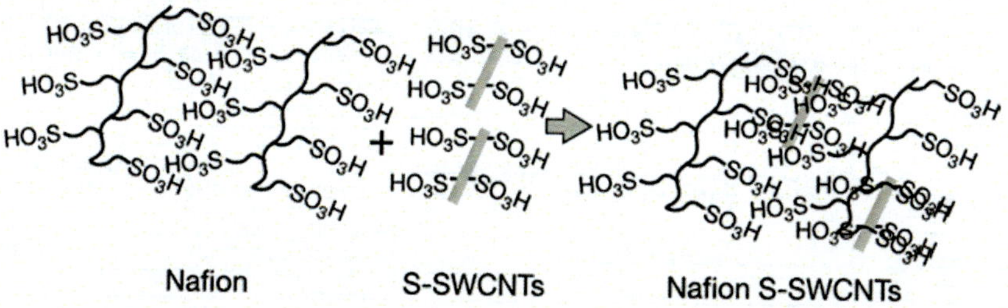

Fig. 9.11 Preparation of S-SWCNT/Nafion composite membrane by mixing solutions of each component. (From [11])

An attempt was made by Kannan et al. to increase the sulfonic acid content of Nafion membranes by incorporating sulfonic acid functionalized single walled carbon nanotubes (S-SWCNTs), and to demonstrate the use of this composite membrane as a PEM [11].

By the addition of SWCNT, the formation of improved channel-like network for proton transfer is expected as depicted by Fig. 9.11. Figure 9.12a shows the temperature dependence of the proton conductivity of Nafion/S-SWCNT composite membrane to be compared with that of Nafion 1,135 membrane. The conductivity of the composite membrane is an order of magnitude higher even at temperatures above 100°C. This seems to be due to an increase in sulfonic acid content in the composite membrane. The activation energies were 0.072 and 0.12 eV for the Nafion/S-SWCNT composite membrane and Nafion 1,135 membrane, respectively, indicating that the proton transfer is more facile in the composite membrane. This improved proton conductivity can be explained by the Grotthus-type mechanism, in which the reorganization of the hydrogen bonds plays a key role. Since additional sulfonic acid moieties are anchored on the CNTs' surface, these could provide more chance for the proton hopping, thus accounting for the increase in proton conductivity. In contrast, when nonfunctionalized CNTs are added, no additional sulfonic acid groups are provided and no improvement of the proton conductivity is shown. Figure 9.12b shows the thermogram from DSC experiments for both Nafion 1,135 and the Nafion/S-SWCNT composite membrane. For both membranes the weight loss of about 5% takes place up to 300°C. Nafion membrane decomposed at 600°C while the composite membrane decomposed at 580°C. This is most likely due to the catalytic activity of S-SWCNT that takes part in the decomposition of Nafion backbone. The tensile strength of Nafion 1,135 is 10.3 MPa while that of the composite membrane is 7.6 MPa indicating slight weakening of mechanical strength by the incorporation of S-SWCNTs.

Figure 9.13a shows the typical impedance spectra for the membrane electrode assembly (MEAs) fabricated with both Nafion 1,135 and the Nafion/S-SWCNT composite membrane using a standard formulation of 20% Pt/C along with Nafion binder in the catalyst layer as electrodes. Obviously, the Nafion/S-SWCNT composite membrane shows less electrolyte resistance. The enhanced conductivity of MEAs based on the composite membranes is solely due to the enhanced conductiv-

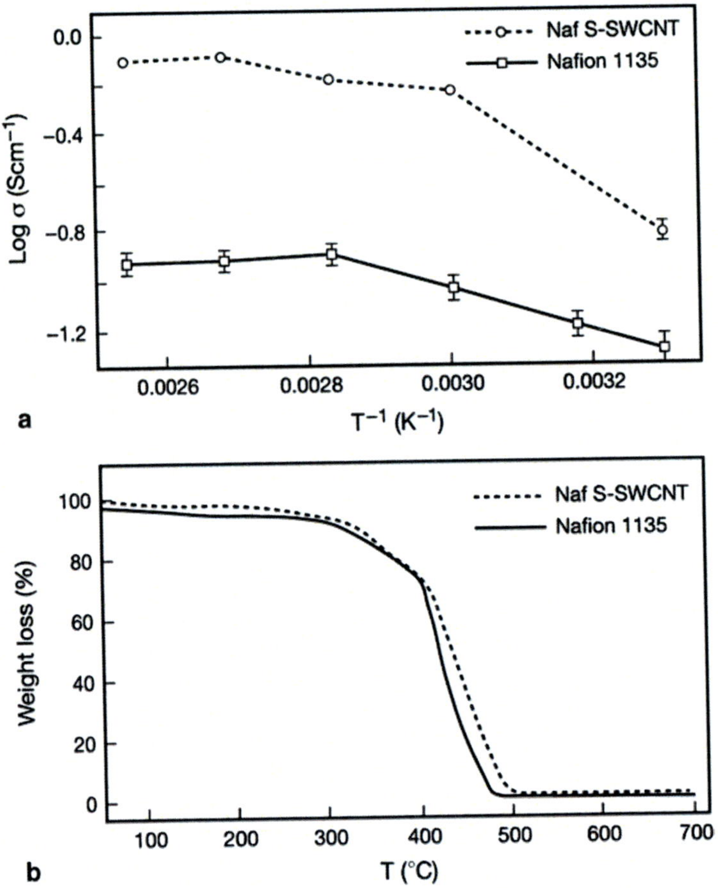

Fig. 9.12 (**a**) Proton conductivity versus temperature for Nafion 1,135 and Nafion/S-SWCNT composite membrane at 100% relative humidity, (**b**) thermogravimetric analysis of Nafion 1,135 and Nafion/S-SWCNT composite membrane in air at the heating rate of 10°C/min. (From [11])

ity of the Nafion/S-SWCNT composite membrane, since other components were maintained identical. The conductivity data were 0.0155 and 0.0101 S cm^{-1} for the composite membrane and Nafion membrane, respectively. Figure 9.13b displays the polarization plot for a MEA fabricated using 20% Pt/C both for cathode and anode in a single cell experiment at 60°C with a humidified H_2/O_2 gas flow rate of 0.4 standard liters per minute. The composite membrane seems significantly better than the Nafion 1,135 membrane. The open-circuit voltage (OCV) obtained from the composite membrane is 0.9 V, slightly lower than that of Nafion 1,135 membrane (0.96 V). However, the activation loss and ohmic loss are much less for the composite membrane. As a result, the Nafion/S-SWCNT composite membrane showed a maximum power density of 260 mW cm^{-2} at 0.42 V, whereas the Nafion 1,135 membrane gives 210 mW cm^{-2} at 0.39 V. Several more studies are required

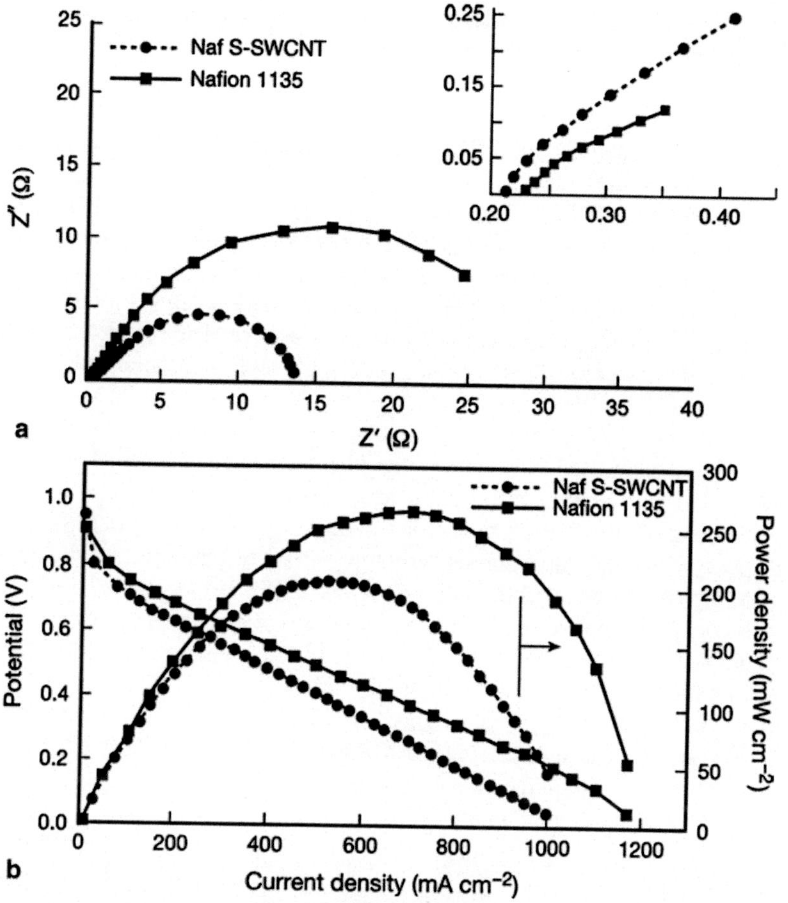

Fig. 9.13 (**a**) Nyquist Impedance (Z) plots for MEAs made with the Nafion 1,135 and the Nafion/S-SWCNT composite membrane. The measurements carried out at room temperature with a flow of humidified H_2 and O_2. Insert: the high-frequency range at higher magnification, (**b**) polarization curve for the Nafion 1,135 and the Nafion/S-SWCNT composite membrane obtained at 60°C with a flow (0.4 slpm) of humidified H_2 and O_2. (From [11])

to exploit these superior features of CNT incorporated composite membranes. They are: Rigorous evaluation of chemical stability and durability of the membranes; Life time studies of MEAs; Studies on the possible corrosion problems of the electrode materials in the presence of higher sulfonic acid content.

The commercialization of proton exchange membrane fuel cells (PEMFCs) is now challenged by the high cost of noble metal catalysts such as Pt. For H_2 fueled PEMFCs, H_2 oxidation is rapid at the Pt anode whereas the oxygen reduction is slow at the Pt cathode. In addition to the attempts to facilitate the electrocatalytic reaction, alloying of Pt as well as replacement of Pt have been pursued. Another attempt is to increase the efficiency of the Pt catalyst and thus to decrease the amount of this expensive metal. Especially, support materials are required to distribute and stabilize the catalyst particles in the catalyst system. They play an important role in the catalytic activities by determining the size, degree of dispersion, degree of alloying of the catalyst, and also the diffusion kinetics of the reactant and product. Synthesis of Pt with uniform size and dispersion in the support nanoparticles is of paramount importance. Hence, the ideal support should have the characteristics such as stability in harsh chemical and electrochemical conditions, high surface area and high electrical conductivity, high porosity to allow efficient diffusion of reactants and products, adequate water-handling capability and low cost. Recently, CNTs have been investigated as a new candidate for the catalyst support for PEMF-Cs and direct methanol fuel cells (DMFCs). It was considered to replace traditional carbon black particles. For example, simply by replacing carbon black particles with disordered MWNTs 20% higher power density could be achieved than Pt/carbon black electrode for hydrogen fuel cell [12]. However, this superior performance is compromised by several difficulties involved in the use of MWNTs such as synthesis, purification, dispersion and surface activation in the practical PEMFC applications. Hollow core/mesoporous shell/(HCMS) carbon (HCMSC) capsules are newly developed nanoporous carbon materials, which have an interconnected bimodal pore structure, composed of hollow macroporous pore and mesoporous shell. Highly developed pores, uniform particle size and narrow pore size distribution make them promising candidate for the efficient catalyst support in low temperature fuel cells. Fang et al. examined the suitability of HCMSC for the Pt catalyst support in PEMFC [13].

Spherical carbon capsules with HCMS structures were synthesized by using the submicrometer sized solid core/mesoporous shell silica spheres as templates. A total of 40 mL of aqueous ammonium (32 wt%) was added to a solution containing 1 L of ethanol and 80 mL of deionized water. After stirring about 15 min 60 mL of tetraethoxysilane (TEOS) (purity 98% from ACROS) was added to the above solution and the reaction mixture was stirred for about 6 h to yield uniform silica spheres. A solution containing 50 mL of TEOS and 20 mL octadecyltrimethoxysilane (C_{18}-TMS) was added to the colloidal solution containing the silica spheres and further reacted for 1 h. The resulting octadecyl group incorporated silica shell/solid core nanocomposite was retrieved by centrifugation, dried at room temperature and further calcinated at 823 K for 6 h under an oxygen atmosphere to produce the final solid core/meso-porous shell (SCMS) silica material. A total of 1.0 g of SCMS silica was added to an aqueous solution containing 0.27 g of $AlCl_3$-$6H_2O$ in 0.3 mL of water, and the resulting slurry was stirred for 30 min. The powder was dried in air at 353 K. Finally Al-impregnated SCMS silica was calcinated at 823 K for 5 h in air to yield SCMS aluminosilicate.

The synthetic route for the preparation of HCMSC capsules is as follows. A total of 0.374 g of phenol was incorporated into the mesopores of 1.0 g of SCMS template by heating at 100°C for 12 h under vacuum. The resulting phenol-incorporated SCMS template was reacted with paraformaldehyde (0.238 g) under vacuum at 130°C for 24 h to yield a phenol resin/SCMS aluminosilicate composite. The composite was heated at 1 K/min to 160°C at which the temperature was held for

5 h in nitrogen stream. The temperature was then increased to 950°C and held for 7 h to carbonize the cross-linked phenol resin inside the mesopores of the SCMS structure. The dissolution of the SCMS template was done by 2.0 N NaOH and washing in EtOH-H_2O (1:1) solution, yielding HCMSC. HCMSC supported Pt (20 wt%) catalysts were synthesized at room temperature through the impregnation method using H_2PtCl_6-$6H_2O$ as a metal precursor and $NaBH_4$ as a reducing agent. The performance of HCMSC supported catalyst was compared with its commercial counterpart E-TEK catalyst in which Pt (20 wt%) is supported on the Vulcan XC-72 (VC) (most widely used commercially available catalyst support). Pt support on VC was made in the laboratory. The latter was used in most cases for comparison with the HCMSC supported catalyst, unless VC supported catalyst is specified as E-TEK.

Appropriate amounts of the catalyst inks, containing various carbon supported dispersed Pt nanoparticles, were painted uniformly on Teflonized carbon paper and dried at 70°C overnight. The MEA was employed to construct a single fuel cell, which was fabricated by hot pressing a pretreated Nafion 112 sandwiched by the anode and cathode. The performance test of the fuel cells was conducted under constant current and constant voltage using a unit cell of 6.25 cm^2 (cross-sectional catalyst area). Hydrogen and oxygen were supplied to the anode and cathode at the flow rates of 200 and 1,000 mL/min, respectively.

Figure 9.14 shows the SEM (a) and TEM (b) images of the HCMSCs. Figure 9.14a shows that the HCMSCs are uniform and spherical (diameter ca. 360 nm). The TEM image in Fig. 9.14b shows the core size and the thickness of 280 and 40 nm, respectively. High resolution TEM image showed mesopores of ca. 3 nm in the shell. Nitrogen adsorption experiment also revealed that the average pore size is 3 nm. The surface area is 917 m^2/g and the total pore volume is 1.1 cm^3/g, among which the mesopore volume is 0.8 cm^3/g.

Table 9.2 shows the pore characteristics of HCMSC compared with commercial VC. It can be noticed that HCMSC has much higher pore volume as well as surface area.

Most of Pt particles were dispersed homogeneously as small, spherical and uniform dark spots on the HCMSC surface. On the other hand, Pt nanoparticles supported by VC shows more aggregation and less uniform dispersion (Fig. 9.15c).

Figure 9.16 shows the XRD patterns for the HCMSC and the commercial carbon black VC supported Pt (20 wt%) catalyst. The Pt size is 2.3 nm on the HCMSC as compared with 2.9 nm on the in-house VC supported Pt (laboratory prepared) and close to 2.2 nm of E-TEK Pt (20 wt%) VT. Thus HCMSC ensures the smaller particle size and more uniform pore size distribution.

Figure 9.17 displays the constant-current cell polarization and power density plots for PEMFCs with various carbon-supported cathode catalysts, Fig. 9.17a, b, c for O_2-fed at 60°C, O_2 fed at 80°C and air-fed at 60°C, respectively. In all cases, the HCMSC-supported Pt cathode catalyst shows lower ion voltage loss than the VC supported Pt catalysts. The maximum power density for O_2-fed at 60°C is 444 mW/cm^2, corresponding to the enhancement of 83% compared with Pt/VC and of 59% compared with Pt/VC (E-TEC). Similar power density enhancement by the

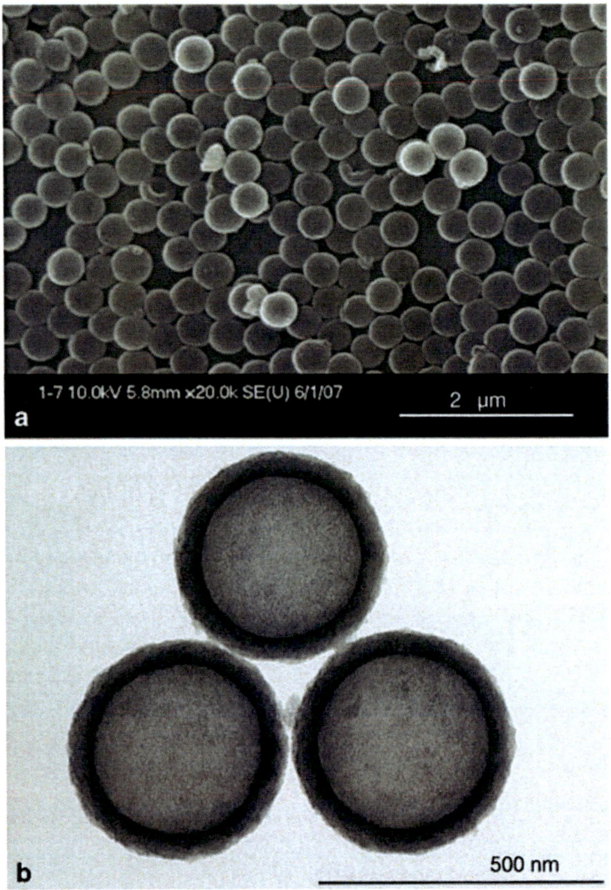

Fig. 9.14 Representative (**a**) SEM and (**b**) TEM images for the HCMSC capsules. (From [13])

Table 9.2 Structural parameters for the HCMSC and commercial VC

Sample	BET area (m^2/g)	V_{total} (cc/g)	V_{meso} (cc/g)	V_{micro} (cc/g)	Pore size (nm)
HCMSC	917	1.10	0.80	0.30	3.0
VC	230	0.31	0.22	0.09	–

HCMSC-supported Pt cathode catalyst was observed for other conditions. As well, Fig. 9.18 shows the highest initial and final current of the HCMSC-supported Pt cathode catalyst among all the carbon-supported cathode catalysts. The enhancement in the electrocatalytic activity is attributed to the unique structural properties of the HCMSC. The electrochemical active layer was also calculated to be 106, 43 and 55 m^2/g for the HCMSC-supported Pt cathode catalyst, for the Pt/VC cathode catalyst and the Pt/VC (E-TEC) cathode catalyst, respectively.

Liao et al. presented a novel one-step preparation of functionalized MWCNTs by free radical modification. MA-POA, i.e. maleic acid (MA) grafted on molecular weight 400 and 2,000 poly(oxialkylene)amines (POA400 and POA2000) was

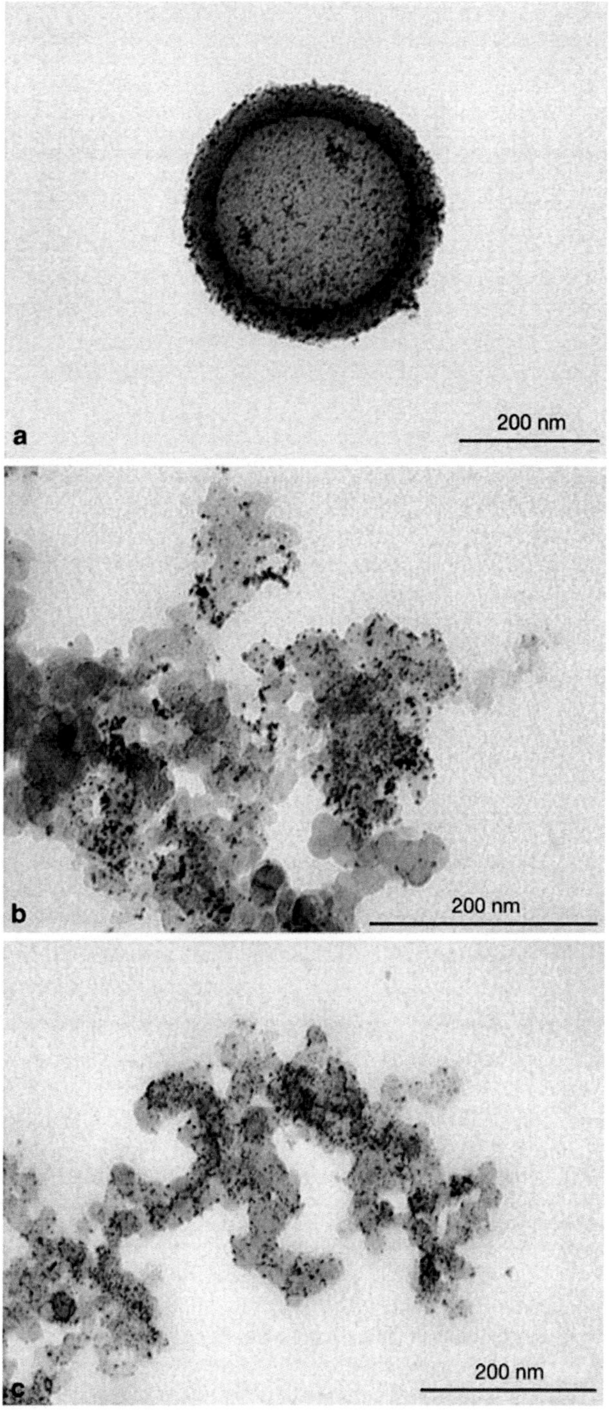

Fig. 9.15 TEM images for (**a**) Pt (20 wt%) supported on HCMSC (**b**) Pt (20 wt%) supported on commercial carbon black VC (in-house pepared) (**c**) Pt (20 wt%) supported on VC (E-TEC). (From [13])

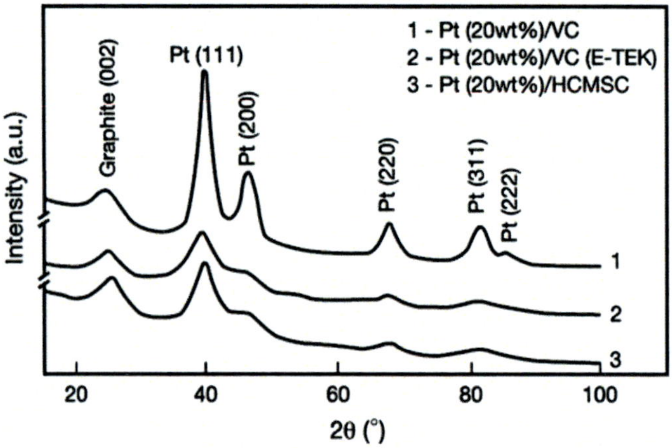

Fig. 9.16 Typical XRD patterns for the HCMSC-supported Pt cathode catalyst, for the Pt/VC cathode catalyst (in-house prepared) and the Pt/VC (E-TEC) cathode catalyst. (From [13])

attached onto the MWCNTs, forming MWCNTs/MA-POA400 and MWCNTs/MA-POA2000. Furthermore, this study also investigated mechanical, electrical and single fuel cell properties of functionalized MWCNT nanocomposite bipolar plates for use in polymer electrolyte membrane fuel cells (PEMFCs).

PEMFCs are promising power sources for stationary, mobile, and potable applications such as cellular phones, PDAs, laptop computers, military back power packs, etc. The main components of PEMFCs are the electrolyte membrane, catalyst and bipolar plates. The bipolar plates electrically connect successive cells in a fuel cell stack, and also provide a gas flow field. This has to fulfill the following basic requirements; i.e., good electrical conductivity, (ii) good thermal conductivity to achieve stack cooling, (iii) good mechanical stability at low thickness, (iv) low weight especially for transport applications and (v) low cost. Although graphite bipolar phases have been used for many years, graphite suffers from being brittle, expensive, bulky and difficult to machine. Hence bipolar plates have to be made several millimeters thick, causing heavy and voluminous fuel cell stack. In place of graphite bipolar plates, graphite based polymer composite bipolar plates are made from a combination of graphite, carbon powder filler or various fibers and a polymer resin with conventional polymer processing methods such as compression or injection molding. They offer low cost, higher flexibility, and greater ease of manufacturing than graphite plates. However, the composite bipolar plates posed an electrical conductivity problem. A high loading of carbon or graphite fillers, typically about 60% has to be incorporated to meet the requirement, which reduces the mechanical strength and ductility of the polymer composite. Nevertheless, there are many reports that polymer composites in which CNTs are incorporated possess excellent electrical conductivity and mechanical properties. Often, however, those composite materials suffer from the poor compatibility between the CNTs and polymeric material. In this work, functionalized MWCNTs attached to linear MA-POA chains have been prepared by free radical modification for the fabrication of nano-

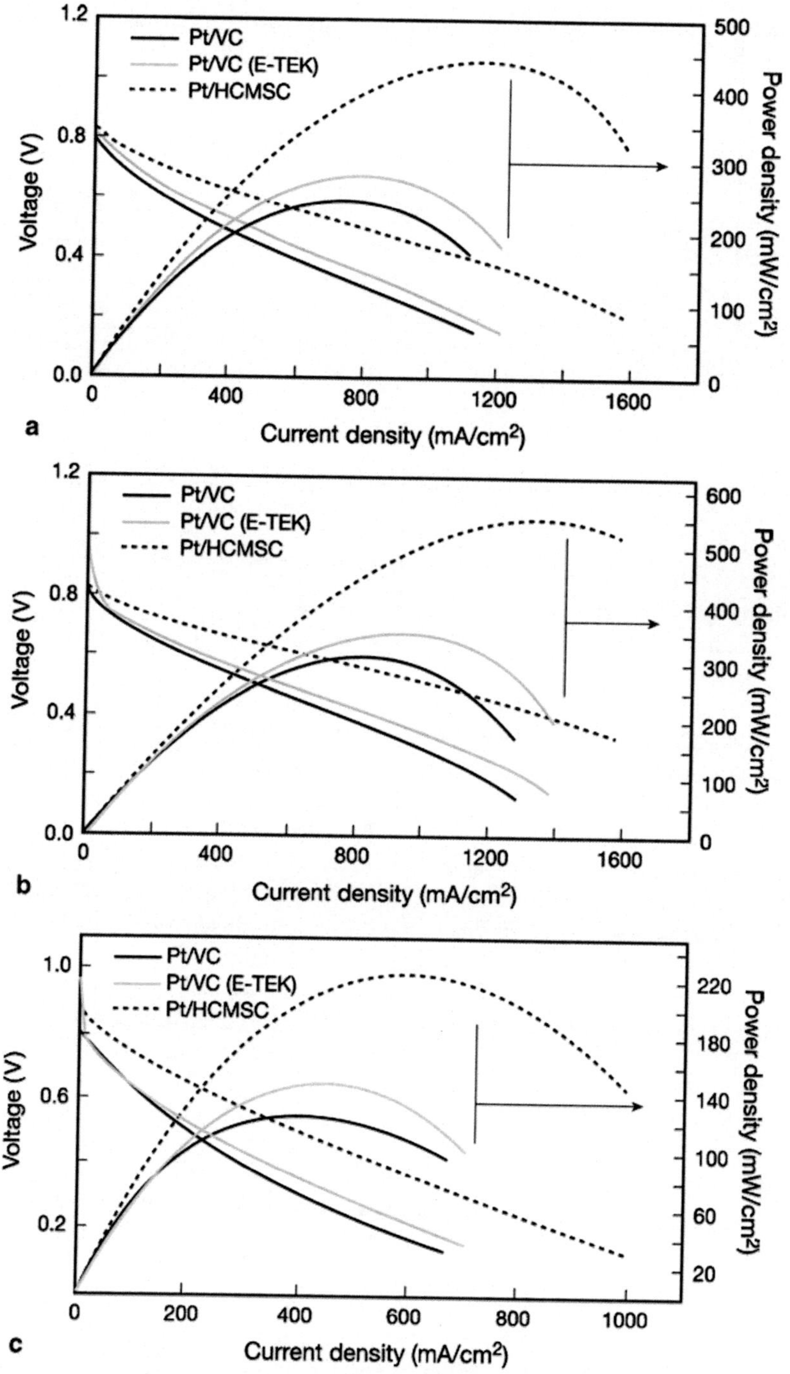

Fig. 9.17 Cell polarization and power density plots for the H_2 fueled PEMFCs using various carbon supported (Pt 20 wt%) cathode catalysts (**a**) O_2 fed at 60°C (**b**) O_2 fed at 80°C and (**c**) air fed at 80°C. (From [13])

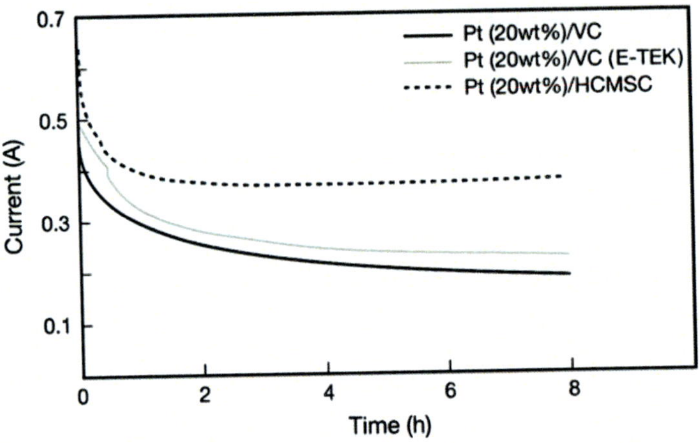

Fig. 9.18 Representative chronoamperograms obtained at 0.75 V and 60°C with O_2-fed cathode mode for the H_2-fueled PEMFCs using various carbon-supported (Pt 20 wt%) cathode catalysts. (From [13])

composite bipolar plates by the bulk molding compound (BMC) process. Surface modification by the free radical is supposed to minimize the damage on the CNTs during the CNT treatment process.

After the preparation of POA bearing one MA, MWCNTs/MA-POA400 and MWCNTs/MA-POA2000 were made as follows: First 200 mg MWCNTs was suspended in 100 mL THF by mixing for 1 h followed by sonication for 15 min. Subsequently, 2.67 mmol of either MA-POA400 or MA-POA2000 dissolved in 10 mL THF was slowly added into the MWCNT suspension and the mixture was further shear mixed at 80°C by refluxing for 8 h. Free radical reaction was initiated by adding benzoyl peroxide (BPO), which was followed by filtration, washing and drying.

To prepare the MWCNTs/Phenolic-novolac (an epoxy-based vinyl ester (VE) nano composites) 6.75 g of VE and 0.162 g of initiator (tert-butyl perbenzoate, TBPB) were dissolved in 2.25 g of styrene monomer and 0.1 g of pristine MWCNTs or the corresponding amount of MWCNTs/MA-POA400 or MWCNTs/MA-POA2000 was added and the mixture was subjected to sonication for 10 min. The prepared MWCNT/VE mixture (pristine MWCNTs, MWCNTs/MA-POA400 or MWCNTs/MA-POA2000) was slowly poured onto an aluminum plate and then was cured by slowly increasing the temperature from 80 to 120°C. Finally, MWCNT/VE nanocomposites were completely cured under vacuum at 140°C for 24 h.

The bulk molding compound (BMC) was prepared by mixing the MWCNT/VE mixtures, low profile agent (a series of polystyrene resins dissolved in styrene monomer), styrene monomer, thickening agent (MgO), release agent (zinc stearate) and graphite in a kneader for 30 min. The BMC formulation is summarized in Table 9.3.

The BMC was thickened for 36 h prior to the hot-pressing process. The processing temperature was 140°C and the processing time was 5 min. The dimensions of composite bipolar plate are 30 mm × 30 mm and 3 mm thick. The dimensions of the channel depth and width are 1 and 1 mm.

Table 9.3 The BMC formulation

Components	Composition	
	Resin composition	MBC composition
Vinyl ester (wt%)	75	30 (resin)
Low profile agent (wt%)	8	
Styrene monomer (wt%)	17	
TBPB (phr)[a]	1.8	
Zinc stearate (phr)	3.5	
Magnesium oxide (phr)	1.8	
MWCNTs (phr)	0–2	
Graphite powder (wt%)		70
Total		100

[a] Parts per hundred parts of resin based on the amount of vinyl ester, low profile agent and styrene monomer

When the MWCNT concentration was kept at 1 wt%, the in-plane electrical conductivity for MWCNT, MWCNTs/MA-POA400 and MWCNTs/MA-POA2000 were 513, 956 and 1,340 S cm^{-1}, respectively. For the pristine MWCNT the formation of local aggregates tend to increase the number of filler to filler hops to traverse a given distance, which causes the decrease in in-plane electrical conductivity when the MWCNT content is further increased. The higher conductivity of MWCNTs/MA-POA400 and MWCNTs/MA-POA2000 than the pristine MWCNT is probably due to the better dispersion of the MWCNTs in the VE matrix because of the introduction of long MA-POA chains on the MWCNT surface.

Figure 9.19 displays the through-plane electrical conductivity of bipolar plates corresponding to the 1% MWCNT content. Here again, the conductivity increased in the following order: without MWCNT (19.1 S cm^{-1})<with pristine MWCNT

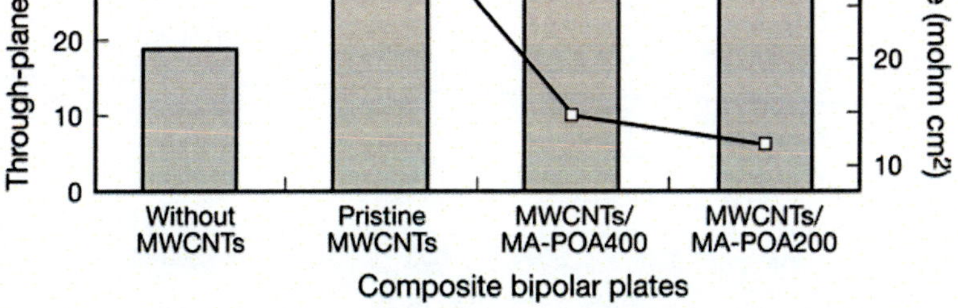

Fig. 9.19 The through-plane conductivity and half-cell resistance of nanocomposite bipolar plates with 1 wt% MWCNTs. (From [14])

(27.9 S cm^{-1})<MWCNTs/MA-POA400 (42.1 S cm^{-1})<MWCNTs/MA-POA2000 (51.2 S cm^{-1}). The through-plane electrical conductivity is significantly lower than the in-plane conductivity. This might be attributed to the orientation of the graphite particles in a plane perpendicular to the direction of the compression force. Thus only a few conducting paths are formed, leading to low through-plane conductivity.

9.5 Water Treatment

Polyvinylidene chloride (PVDC) and polyvinyl alcohol (PVA) microspheres were carbonized on ceramic membrane to fabricate activated carbon membrane for coke furnace wastewater treatment [15]. A ceramic tube was dipped into a polymer latex containing 70 wt% PVDC and PVA microspheres of 0.10–0.15 μm to form aggregates of polymeric microspheres on and within (within pores of) the ceramic pipe. The precursor was heated at 300°C and further to 750°C for carbonization. Major decomposition of the polymeric precursor seems to occur at 300°C. By nitrogen adsorption applying Horvath and Kowazoe method the membrane was found to have micropores of 0.7–0.8 nm in diameter and meso-pores of 2–20 nm. Hence, the membrane has bimodal pore size distribution. The molecular weight cutoff of the membrane was ca. 10,000 Dalton.

Coke furnace wastewater containing dissolved organic carbon (DOC) of 69 ppm and two peaks at the molecular weight of 400 and 10,000 Da was subjected to the filtration by the activated carbon membrane. Figure 9.20 displays (permeate concentration in chemical oxygen demand (COD)/feed concentration in COD = *C/Co*) versus effluent volume. Each curve indicates the typical breakthrough curve but interestingly the curve never exceeds (*C/Co*) ratio of 0.68, which means 32% of the COD is removed by the sieve mechanism. Hence, the COD removal by the

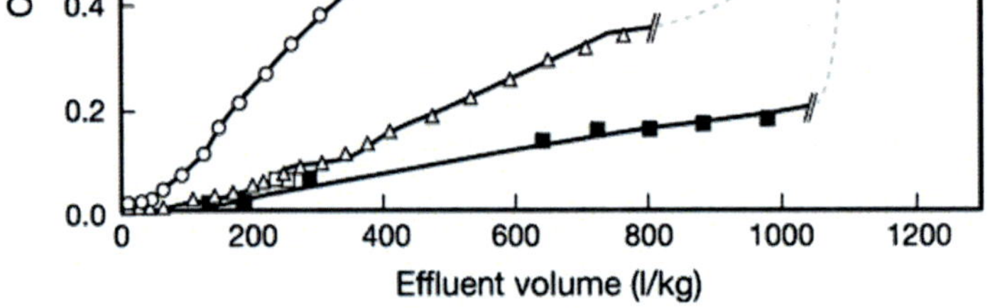

Fig. 9.20 DOC changes in permeant (normalized) versus time. (From [15])

activated carbon membrane exhibits the combination of adsorption and filtration phenomena. Most likely, the adsorption occurs in the micropores in the carbon particles while the filtration takes place by the pores formed as the interstitial spaces between the carbon particles.

The immobilization of biofilms on permeable membranes, for the biodegradation of pollutants has drawn increasing interest for applications, where conventional treatment technologies are difficult to apply. As one of such examples, Hu et al. demonstrated the usefulness of membrane aerated biofilm reactor (MABR) for the wastewater treatment [16]; a carbon-membrane aerated biofilm reactor (CMABR) was constructed to remove organics and nitrogen containing compounds simultaneously in one reactor.

As shown in Fig. 9.21 the bacterial adherence to the membrane surface results in the formation of biofilm. Oxygen pumped into the lumen (left side) of the membrane passes through the membrane wall and is utilized by the bacteria within the biofilm to oxidize the pollutants as it penetrates into the biofilm. The oxygen and substrate concentration profiles within the biofilm may be quite different from the conventional biofilm, allowing simultaneous nitrification and heterotropic oxidation. The region near the membrane shell (right side of membrane before entering into the biofilm) would be in a favorable condition for nitrification because of sufficient oxygen and organic carbon depleted condition. However, in the region near the bulk liquid (right side of the biofilm) is in a favorable condition for heterotrophic denitrification, because organic carbon, nitrite and nitrate sufficient and oxygen depleted condition is created.

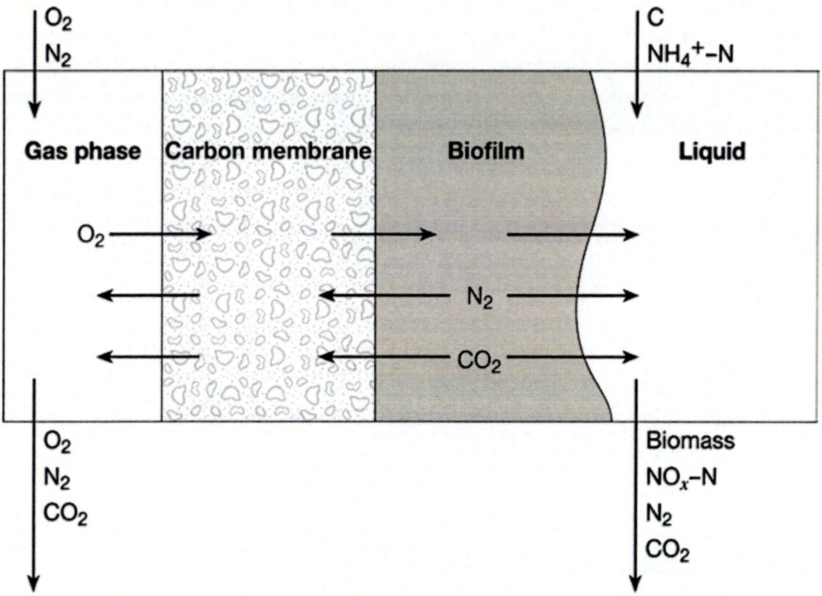

Fig. 9.21 Schematic diagram of MABR. (From [16])

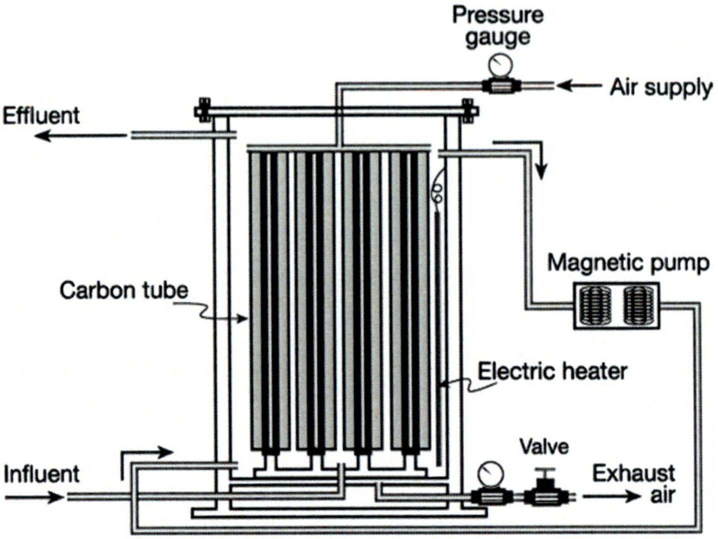

Fig. 9.22 Schemaic diagram of CMABR. (From [16])

The reactor feed contained glucose and ammonium chloride as the sources of organic carbon and nitrogeneous component. In addition, the feed contained a small amount of minerals as element nutrition. The experimental membrane bioreactor contained 16 carbon tubes of length 20 cm with inner and outer diameter of 4.7 and 8.9 mm, respectively. The pore size was 2 μm. To enhance the bacterial attachment the shell side surface of the carbon membrane was covered with nonwoven material,

The CMABR (2.4 L) is schematically shown in Fig. 9.22. Complete mixing of the feed is ensured by the recirculation of the feed mixture by a magnetic pump, while the influent was supplied from the reactor bottom by a peristaltic pump. The effluent is discharged from the top of the reactor. Air was supplied from the top of the reactor into the lumen side of the carbon tubes. A part of the air permeated through the carbon membrane tube to the shell side and reacted with the biofilm. The rest was emitted from the bottom of the reactor. The reactor was maintained at 32°C by heating and also kept dark. The reactor was operated according to the operational scheme given in Table 9.4.

To start the reactor, activated sludge from the Municipal Wastewater Treatment Plant (Dalian, China) was inoculated into the bioreactor. Sludge suspension of 2.4 L with mixed liquor suspended solids (MLSS) concentration of 4,500 mg/L was seeded. The organic loading was gradually increased when a steady state, in terms of COD removal, was reached after a certain period of operation.

Table 9.4 Operational conditions

Run	Period (day)	HRT (h)	Influent COD (mg/L)	Influent ammonium (mg NH_4^+-N/L)
1	0–68	20	253	49
2	69–105	12	348	77
3	106–125	8	410	102

The oxygen supply rate is the most important parameter in the operation of CMABR. If the excess oxygen is provided, heterotrophic bacteria will grow aerobically and the organic carbon for heterotrophic denifitrication will be in short supply. On the other hand, if the oxygen supply rate is low, the heterotrophic bacteria will compete oxygen with nitrifying microorganisms. Therefore, the supply of the oxygen should be controlled for the optimum operation of CMABR. The oxygen transfer was controlled by changing the oxygen pressure. Under the aeration conditions, assumption was made that all the carbon and nitrogen in the feed water are oxidized into CO_2 and NO_3^-. With this assumption the maximum oxygen supply rate (OSR) at each run (1–3 in Table 9.4) was evaluated to be 1.38, 3.43 and 6.17 g/day, respectively. The corresponding air pressure was 7.5, 18.8 and 30 kPa, respectively. In fact in run 3, the pressure is supposed to be more than 30 kPa, but because of the operational restriction, the pressure was limited to 30 kPa.

The performance of CMABR is shown in Fig. 9.23a and b. The figure shows that COD and NH_4^+-N were removed effectively over a period of 121 days. A stable

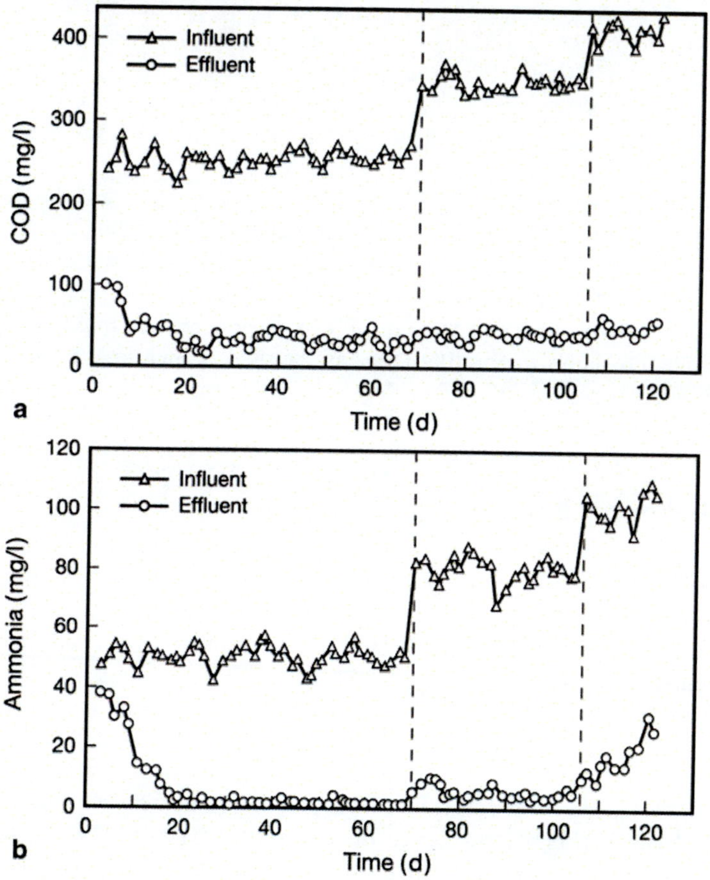

Fig. 9.23 Removal of (**a**) COD and (**b**) NH_4^+-N during the operation of CMABR. (From [16])

Table 9.5 Total nitrogen (TN) removal by CMABR

Run	Period (day)	Influent TN[a] (mg/L)	Effluent TN[b] (mg/L)	TN removal (%)
1	0–18	49.0	29.1	40.6
	19–68	49.0	10.1	77.3
2	69–105	77.0	12.4	83.9
3	106–125	102.0	38.7	62.1

[a] Influent TN contains only NH_4^+-N
[b] Effluent TN contains NH_4^+-N and NO_3^--N

Table 9.6 Biofilm thickness

Run number	Biofilm thickness (μm)
1 (day 15)	221
2 (day 82)	1,465
3 (day 125)	3,307

operation was achieved for COD removal within 8 days. Even when the feed COD was increased twice, on 69th and 106th day, the effluent COD was stable. The COD removal was 86.0, 87.5 and 87.1% during the run 1, 2 and 3, respectively. The NH_4^+-N removal did not reach the steady state so quickly. Its removal was as high as 99 and 94% during the run 1 and 2, when the steady state was reached.

As for the removal of the total nitrogen (TN) the data are summarized in Table 9.5.

The biofilm formation was not quite sufficient during the run 1 and the NO_3^--N concentration was quite high. There was a significant growth of biofilm from 68th day and the NO_3^--N concentration as well as TN concentration decreased considerably. The TN removal increased up to 83.9%. During the run 3 the removal of both NH_4^+-N and NO_3^--N went down and the TN removal decreased to 62.1%. As the data in Table 9.6 shows the biofilm thickness became as high as 3.3 mm, which started to hinder the supply of oxygen through the biofilm.

9.6 Membrane Reactor

The combination of reaction and separation at high temperature in a membrane reactor offers interesting new possibilities. In a membrane reactor, the separate product and feed compartments allow more ways to optimize both selectivity and conversion [17].

Zhang et al. constructed a carbon membrane reactor (CMR) for methanol steam reforming and found CMR gave a higher methanol conversion than the conventional fixed bed reactor (FBR) and produced a CO free hydrogen stream [18]. Carbon membrane was prepared by spraying a DMF solution containing phenol formaldehyde novolac resin (PFNR) and polyethylene glycol (PEG) on a PFNR based green membrane support, the details of which are given elsewhere. The sprayed layer was then carbonized under an inert gas stream to 1,073 K at a heating rate of 0.5 K/min. As the catalyst for the steam reforming reaction, laboratory made Cu/ZnO/Al_2O_3 composite catalyst was used. FBR consists of a stainless steel tube with a diameter of 6 mm. CMR consists of a tubular carbon membrane with the same inner diameter. The permeation mechanism through the membrane is by the sieve mechanism since the gas permeance decreases with an increase in the kinetic diameter of the gas molecule.

The steam reforming consists of the following chemical reactions:

$$CH_3OH + H_2O = CO_2 + 3H_2 \quad \text{endothermic reaction}$$

$$CO + H_2O = CO_2 + H_2 \quad \text{exothermic reaction}$$

$$CH_3OH = CO + 2H_2 \quad \text{endothermic reaction}$$

The first and the third reactions are highly endothermic and the second reaction is highly exothermic.

Table 9.7 shows the experimental membrane reactor data.

From Table 9.7, CH_3OH conversion increases with an increase in temperature. This is because both CH_3OH conversion reactions (first and third reaction) are endothermic. CH_3OH conversion is higher for CMR than for FBR. This is because the equilibrium shifted to the right side in the first and the last reaction when H_2 is preferentially removed from the reactor through the membrane wall. There is practically no change in H_2 yield.

Table 9.8 shows that the permeate is mostly H_2 gas. There is a small amount of CO_2. There is practically no CO and methanol found in the permeate, because they do not permeate through the carbon membrane.

Table 9.7 Conversion and yields of reaction in the CMR and FBR at different temperatures

	473 K		498 K		523 K	
	FBR	CMR	FBR	CMR	FBR	CMR
CH_3OH conversion (%)	57.57	64.35	89.21	93.27	96.20	99.87
H_2 yield (%)	75.31	75.75	75.42	73.97	75.49	75.77
CO_2 yield (%)	24.55	24.16	24.10	25.62	23.33	23.16
CO yield (%)	0.14	0.09	0.48	0.41	1.18	1.07

Data for the condition, WHSV=1.0/h, H_2O/CH_3OH molar ratio=1.5 and pressure 0.2 MPa
CH_3OH conversion=% of methanol converted, H_2 yield=% of H_2 in the total reaction product (sum of H_2, CO and CO_2)

Table 9.8 Mole percent of gas in the permeate (shell side) stream of CMR at different temperatures

	473 K	498 K	523 K
H_2	97.6	97.4	96.9
CO_2	2.4	2.6	3.1

9.7 Miscellaneous Applications

Biomimetic membranes with embedded nanopores can be ideal platforms for bulk chemical separations, drug delivery and sensor applications, provided that molecular transport through these membranes is precisely regulated with selective and reversible chemical interactions. Nednoor et al. developed a novel synthetic nanopore membrane which mimics protein channels regulated by phosphorylation/dephosphorylation by using aligned array of CNTs impregnated in a polystyrene (PSt) matrix [19]. The opened tips (7 nm inner core diameter) of the CNTs in this membrane were plasma-treated to have carboxyl functionalities. The latter can be easily derivatized through carbodiimide chemistry with a molecule that binds to a receptor, which can thus regulate the flow through the pore entrance. A nine residue synthetic peptide containing a serine residue [G-R-T-G-R-N-S-I-NH_2], a specific substrate of protein kinase A, was attached at the tip of CNTs to obain a biomimetic system where phosphorylation regulates ligand gated protein channels. The basal CNTs impregnated PSt membrane was fabricated using a conventional method, and was functionalized with the peptide by attaching amine terminus of the [G-R-T-G-R-N-S-I-NH_2] peptide to the carboxylate-functionalized CNTs. An amount of 20 mg of CNT membrane was added to 4 mL of 2-(N-morpholino)ethanesulfonic acid-NaOH buffer (MES/NaOH), pH 6.5, containing 2 mg of peptide, and 10 mg of 1-ethyl-3-(3-dimethylaminopropyl) carbodiimide (EDC). CNT membrane was incubated in this reaction mixture at 4°C on a shaker overnight. Afterwards, it was washed three times with MES buffer to remove any unreacted peptide. The membrane was subjected to FT-IR to confirm that the peptide was bound to the CNT.

Protein bound to the CNT was further phosphorylated using protein kinase A. The protein kinase mixture was prepared by adding 22 μg of ATP and 30 μL protein kinase A (1 mg/mL) to 4 mL of 20 mM Tris–HCl, 100 mM NaCl, 12 mM $MgCl_2$, pH 7.5. The peptide-functionalized CNT membrane was incubated in the reaction mixture at 4°C on a shaker overnight. Phosphorylation of the peptide was confirmed by using anti-phosphoserine (mouse IgG_1 isotype) monoclonal antibody. This antibody reacts with phospholinated serine, but not with non-phospholinated serine, phospholinated tyropsine and threonine or ATP. After phosphorylation of the serine residue, peptide functionalized CNT membranes were incubated in a monoclonal anti-phosphoserine solution prepared by diluting 200 μL of monoclonal antiphosphoserine solution (purchased as a 1 mg/mL solution, Sigma P4757) in 3.8 mL of 100 mM phosphate buffer, pH 7.4. The reaction was carried out at 4°C on a shaker overnight. The CNT membranes were washed repeatedly with phosphate buffer to remove the unreacted antibody.

A scheme depicting the binding of the monoclonal anti-phosphoserine antibody to the phosphorylated peptide functionalized CNT membrane is shown in Fig. 9.24. In the presence of alkaline phosphatise, the peptide is dephosphorylated and antibody fails to bind.

Transport studies were conducted to test the hypothesis that enzyme-catalyzed conversion can modulate ionic diffusional flux through a CNT synthetic membrane.

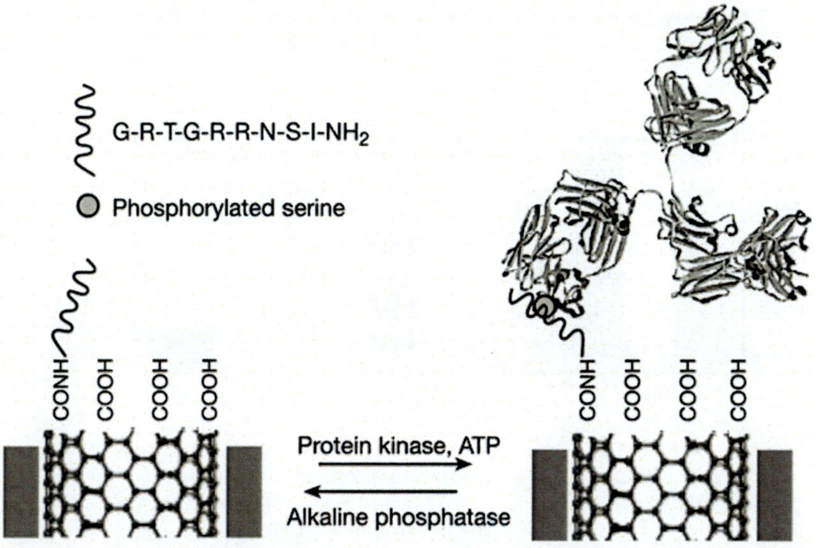

Fig. 9.24 Schematic of phosphorylation based biomimetic system. Also shown is the synthetic peptide with the phosphorylated serine. The phosphorylated peptide tethered to the CNT core entrance binds the antibody to modulate ionic flux through the CNT cores across the membrane structure. *Grey* region represents the polystyrene membrane that forces ionic flow through the CNT cores. (From [19])

The ionic diffusional flux through a CNT synthetic membrane was carried out using a U-tube cell described earlier. The exposed membrane area separating the two chambers of the U-tube cell was 0.3 cm^2. Permeation of both negatively charged (8-anilinonaphthalane-1-sulfonic acid (ANS)) and positively charged marker molecules (methyl viologen (MV^{2+}), tri(2,2′-bipyridyl)ruthenium ($Ru(bpy)_3^{2+}$)) through the membrane was analyzed to evaluate the effects of peptide functionalization and antibody binding. Transport rate relative to that of the unmodified carbon nanotube is shown in Fig. 9.25 for ANS, (MV^{2+}), and ($Ru(bpy)_3^{2+}$).

The trend in the normalized rate of two positively charged marker molecules is similar; i.e., the rate decreases after peptide funtionalization but increases after the antibody binding. On the contrary, the trend is the opposite for the negatively charged marker molecule (ANS). The rate increases remarkably after peptide functionalization but decreases after antibody binding. The effect of the peptide functionalization (rate decrease for the positively charged molecular marker and rate increase for the negatively charged molecular marker) indicates the net positive charge of the peptide. Looking closely into the positively charged molecular marker, the decrease in normalized rate is more for $Ru(bpy)_3^{2+}$ than for MV^{2+}. Noting that $Ru(bpy)_3^{2+}$ is much bulkier than MV^{2+}, this indicates that the rate is controlled not only by the electrostatic charge effect but also by the steric effect. Thus the attachment of peptide seems to have dual effect on the transport rate. As a result, the rate is controlled both by the charge and the size of the marker molecules.

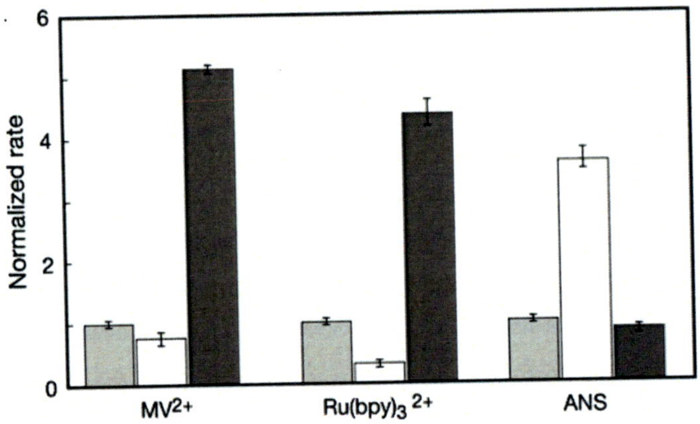

Fig. 9.25 Relative flux data of marker molecules at each stage of functionalization; (*gray bars*) unmodified membrane, (*white bars*) after peptide functionalization, (*hatched bars*) after antibody binding to the peptide functionalized CNT membrane; normalized rate means (mass transfer rate/mass transfer rate for the unmodified CNT membrane). (From [19])

Indeed, the effect of antibody binding indicates that the region of the antibody close to the CNT pores is negatively charged. Thus the rate is enhanced for the positively charged markers by antibody binding while it is reduced for the negatively charged marker molecule.

The charge effect can be minimized by increasing the ionic strength of the solution, since high ionic strength screens the charge at the pore entrance and thus minimizes the electrostatic effect. Figure 9.26 displays the results of experiments where the ionic strength was increased. The flux through the peptide functional-

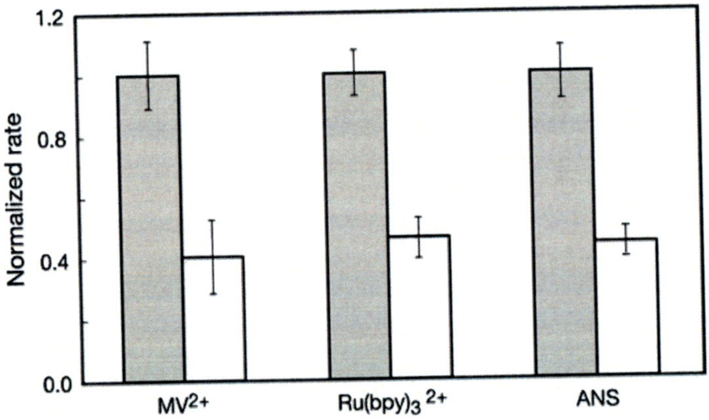

Fig. 9.26 Relative flux data of marker molecules in the presence of 10 mM KCl; (*gray bars*) after the peptide functionalizaton, (*white bars*) after antibody binding to the peptide functionalized CNT membrane. (From [19])

ized membrane in the presence and absence of antibody was investigated by using solutions that contained 10 mM of KCl. The transport rates of all marker molecules were reduced by 54–59% regardless of the charge and size by antibody binding. The fact that the reduction is independent of size of the marker molecule indicates that the antibody binding is not inducing significant steric hindrance through much smaller channels near the pore entrance.

Phospholization of peptide may also affect the flux due to its additional negative charge. However, the transport rates of ANS were 1.6, 1.4 and 1.5 mmol/cm^2 per h for the peptide, phosphorylated peptide and dephosphorylated peptide, respectively, indicating that phosphorylation and dephosphorylation has practically no effect on the transport rate.

In conclusion, synthetic nanopore membranes were used to mimic biological ion channels regulated by phosphorylation and dephosphorylation of proteins. Phosphorylation and dephosphorylation of synthetic peptides at the entrance of CNT cores are shown to modulate ionic flux through the membrane by antibody coordination to the phosphrylated state.

There is a growing interest in preparing new nano-materials composed of CNTs and organic binders for electrochemical and materials science applications. The coupling of polymers with CNTs and biorecognition elements forming a composite is of increasing importance due to the simplicity of construction and its ability to incorporate conducting materials into porous polymers in order to form electrochemical biosensors. Sánchez et al. described CNT/polysulfone (PS) composite thick-film screen printed electrodes for amperometric sensing and biosensing [20]. The combination of MWCNT and PS results in a novel composite material consisting of an interconnected CNT-polymer network, and processing mechanical flexibility, high toughness, and high porosity, while retaining the attractive electrochemical behavior of CNT electrodes and biocompatibility of PS.

In the next section, it is described that such CNT/PS functional modules have favorable electrochemical properties which lead to sensitive detection system based on thick-film screen-printed electrode.

The amperometric sensors used in this investigation consisted of a single working screen-printed electrode deposited onto polycarbonate (PC) substrate. Silver ink acting as a conductive layer was printed and cured in a furnace at 60°C overnight. Carbon paste ink was printed and cured at the same temperature overnight. A non-conductive isolating ink was applied and cured at 60°C overnight. The reaction area of the working electrode was 20 mm^2. Figure 9.27 displays the schematic of the structure of the electrode.

The carbon/PS composites were fabricated as follows. MWCNTs or graphite suspension was mixed with the 7.5 wt% PS solution in DMF for 10 min under continuous stirring. Mixing ratios (carbon/PS-DMF suspension) were 6.5, 9.6, 12.5, 15.0, and 17.6 wt%. Serigraphy is applied to print the composite onto the reaction region of 12 working electrodes. The electrodes were then immersed into bidistilled water for the phase inversion for 5 min followed by rinsing for 1 min and then the electrodes were dried at room temperature.

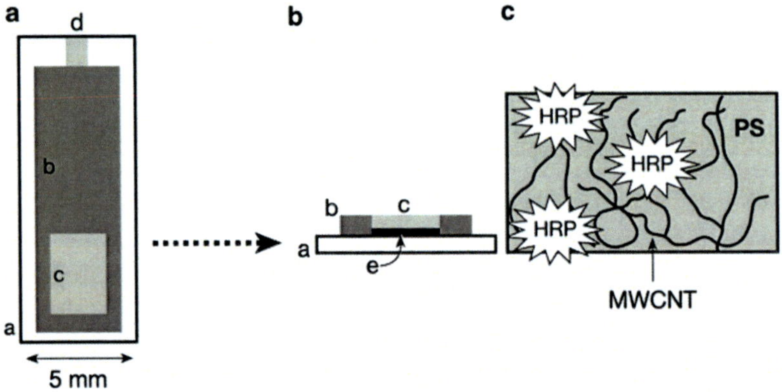

Fig. 9.27 (**a**) The enzyme/MWCNT/PS screen-printed thick film electrochemical detector, *top view*, (**b**) cross-section of the detection area of enzyme/MWCNT/PS screen-printed detector, (**c**) schematic drawing showing structure of HRP/MWCNT/PS composite; (*a*) polycarbonate (PC) substrate, (*b*) insulator layer, (*c*) HRP/MWCNT/PS conducting composite, (*d*) silver contact for the working electrode, (*e*) carbon ink contact layer. (From [20])

For the preparation of horseradish peroxidase enzyme (HRP)/MWCNT/PS, the same method as described above was used. But this time, the MWCNT/PS electrodes prepared above were dipped into HRP of 5.34 Units/mL for 5 min and rinsed for 1 min with bidistilled water and dried at room temperature.

Cyclic voltametric experiments were performed to study the conducting composite at a scan rate of 50 mV/s, adding 0.1 mM ferricyanide to the PBS buffer solution. Hydrodynamic voltammograms were carried out for 4.6 mM hydrogen peroxide in PBS buffer solution using the graphite/PS, MWCNT50/PS and MWCNT200/PS screen printed electrodes. The carbon/PS composition ratio was 6.5 wt% and 17.6 wt% for CNT and graphite, respectively.

SEM images of MWCNT50/PS (b), MWCNT200/PS (c) are compared in Fig. 9.28 with that of graphite/PS (a). Figure 9.28a indicates a very rigid structure where graphite granules are compactly distributed yielding dense surface. On the

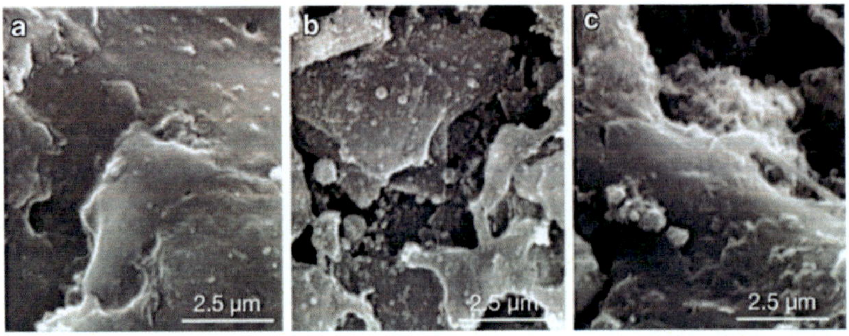

Fig. 9.28 SEM Images of the surfaces of (**a**) graphite/PS, (**b**) MWCNT50/PS, (**c**) MWCNT200/PS modified screen-printed thick film electrodes. (From [20])

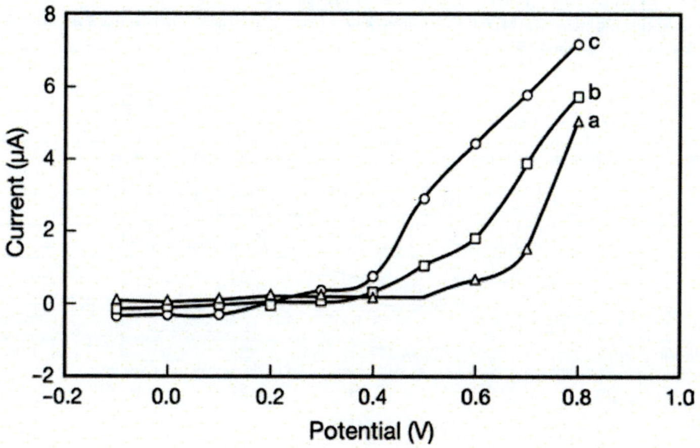

Fig. 9.29 Hydrodynamic voltammograms for 4.6 mM hydrogen peroxide using the (*a*) graphite/PS, (*b*) MWCNT50/PS, (*c*) MWCNT200/PS screen printed electrode. (From [20])

other hand, (b) and (c) show more porous and spongy structures. Hence, MWCNT incorporated membranes show a larger surface area and their interior are more easily accessible, resulting in higher electrolyte mass transfer.

From the current peak heights obtained from the cyclic voltammograms for ferricyanide (recorded at 50 mV/s) at various loadings of MWCNT50/PS, MWCNT200/PS and graphite/PS, it was concluded that 6.5 wt% of MWCNT50/PS and MWCNT200/PS were enough while for graphite/PS, loading of 17.6 wt% was necessary. Figure 9.29 compares the hydrodynamic voltammograms (HDV) for 4.6 mM hydrogen peroxide for graphite/PS (a), MWCNT50/PS (b) and MWCNT200/PS (c) composite electrode. No redox activity was observed for graphite/PS electrode below +0.5 V. MWCNT200/PS and MWCNT50/PS showed amperometric response at +0.3 and +0.4 V, respectively. A rapid increase in current was observed for the latter two electrodes when the potential was higher than the above values. This indicates that hydrogen peroxide can be detected at lower potentials when MWCNT/PS electrodes are used than graphite/PS electrode.

Thus, different carbon materials used for the construction of carbon/PS electrodes has a profound effect on their electrochemical behavior. Figure 9.30 compares current-ferricyanide concentration plots obtained from the electrodes with different carbon materials. All plots exhibit linear relationships. The sensitivity was the lowest for graphite/PS electrode while it was the highest for MWCNT200/PS with MWCNT50/PS that follows. These results indicate that MWCNTs maintain their conducting properties even when they are immersed in a relatively hydrophobic PS matrix. Moreover, the experimental data scattered least for the MWCNT200/PS electrode, indicating that the latter electrode is the best choice among the studied electrodes.

Long term stability of the electrodes was tested for a period of 2 months and the results are shown in Fig. 9.31. The MWCNT200/PS showed stable response for the

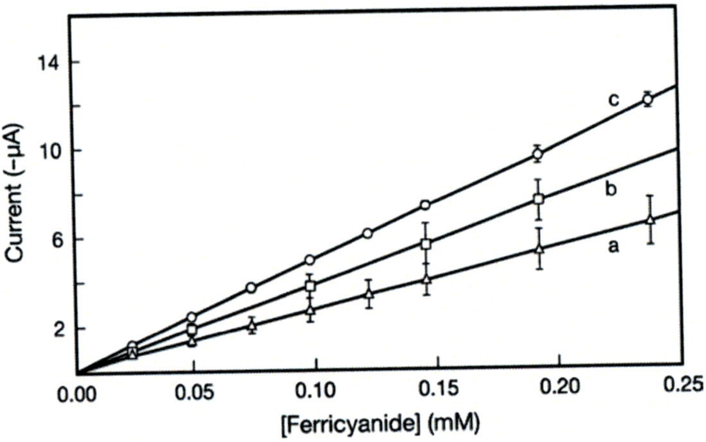

Fig. 9.30 Calibration plots for potassium ferricyanide using PS-carbon Screen-printed electrodes of different compositions; (*a*) graphite/PS, (*b*) MWCNT50/PS, (*c*) MWCNT200/PS composite; operating potential –0.2 V. Error bars represent standard devation for measurements on different electrodes ($n=6$). (From [20])

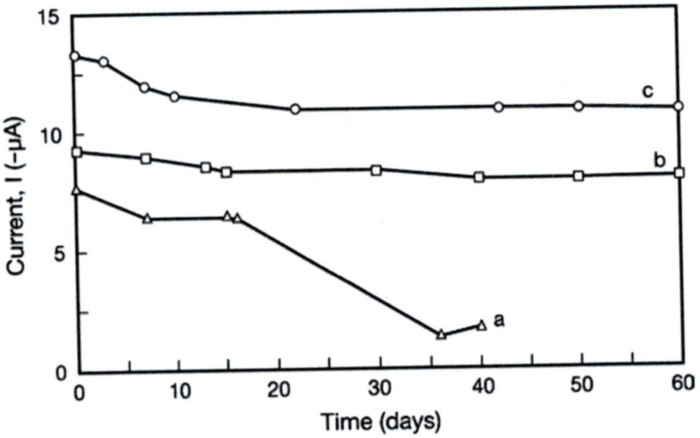

Fig. 9.31 Long-term stability of the response to 0.1 mM potassium ferricyanide using (*a*) graphite/PS, (*b*) MWCNT50/PS, (*c*) MWCNT200/PS screen printed electrode; operating potential –0.2 V. (From [20])

2 months period. MWCNT50/PS was also stable even though the response was a little bit lower. On the other hand, the response of the graphite/PS composite deteriorated after 40 days. This can be attributed to the enhanced mechanical properties of MWCNT/PSs in the polymer matrix as compared to graphite/PS. The electrode was also used for the preparation of enzyme biosensors by the incorporation of HRP into the composite matrix by the phase inversion method. It is known that HRP catalyzes the reduction of hydrogen peroxide to water.

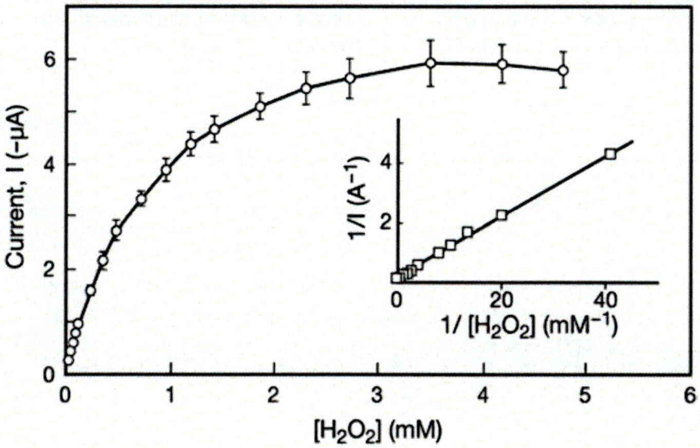

Fig. 9.32 Current dependence upon hydrogen peroxide concentration for HRP/MWCNT200/PS biocomposite screen-printed electrode (6.5 wt% loading); conditions; 5.3 U/mL HRP in the biocomposite, 1.8 mM hydroquinone used as mediator, buffer PBS pH 7.0. (From [20])

Figure 9.32 displays the amperometric response at −200 mV for HRP/MW-CNT200/PS biocomposite sensor to H_2O_2. A linear response is found in the H_2O_2 concentration range of 0.02–0.5 mM and the detection limit is 25 μM. The apparent Michaelis-Menton constant $K_M{}^{app}$ is calculated from the inserted linear plot to be 0.71 mM, which is lower than 0.46 mM for the sol–gel based biosensor, indicating a much lower diffusion barrier of this composite biosensor.

References

1. Strano MS, Zydney AL, Barth H, Wooler G, Agarwal H, Foley HC (2002) Ultrafiltration membrane synthesis by nanoscale templating of porous carbon. J Membr Sci 198 (2): 173-186
2. Strano MS, Agarwal H, Pedrick J, Redman D, Foley HC (2003) Templated pyrolytic carbon: The effect of poly(ethylene glycol) molecular weight on the pore size distribution of poly(fufuryl alcohol)-derived carbon. Carbon 41 (13): 2501-2508
3. Fuertes AB, Centeno TA (1999) Preparation of supported carbon molecular sieve membranes. Carbon 37 (4): 679-706
4. Rao MB, Sircar S (1993) Nanoporous carbon membranes for separation of gas mixtures by selective surface flow. J Membr Sci 85 (3): 253-264
5. Okamoto K, Kawamura S, Yoshino M, Kita H, Hirayama Y, Tanihara N, Kusuki Y (1999) Olefin/paraffin separation through carbonized membranes derived from an asymmetric polyimide hollow fiber membrane. Ind Eng Chem Res 38 (11): 4424-4432
6. Suda H, Haraya K (1997) Alkene/alkane permselectivities of a carbon molecular sieve membrane. J Chem Soc Chem Commun: 93-94
7. Dagan G, Agam G, Krakov V, Kaplan L (2000) Carbon membrane separator for elimination of SF_6 emissions from gas-insulated electrical utilities. In Proc. of the EPA Conference on SF_6 and Environment Emission Reduction Strategies, San Diego, CA, USA (www.epa.gov/highgwp1/sf6/agenda.html).

8. Li L, Xiao Z, Zhang Z, Tan S (2004) Pervaporation of acetic acid/water mixtures through carbon molecular sieve-filled PDMS membranes. Chem Eng J 97 (1): 83-86
9. Sakata Y, Muto A, Uddin MA, Suga H (1999) Preparation of porous carbon membrane plates for pervaporation separation applications. Sep Purif Technol 17 (2): 97-100
10. Peng F, Pan F, Sun H, Lu L, Jiang Z (2007) Novel nanocomposite pervaporation membranes composed of poly(vinyl alcohol) and chitosan-wrapped carbon nanotube. J Membr Sci 300 (1-2): 13-19
11. Kannan R, Kakade BA, Pillai VK (2008) Polymer electrolyte fuel cells using Nafion-based composite membranes with functionalized carbon nanotubes. Angew Chem Int Ed 47 (14): 2653-2656
12. Girishkumar G, Retter M, Underhile R, Binz D, Vinodgopal K, McGinn P, Kamat P (2005) Single-wall carbon nanotube-based proton exchange membrane assembly for hydrogen fuel cells. Langmuir 21 (18): 8487-8494
13. Fang B, Kim JH, Lee C, Yu J-S (2008) Hollow macroporous core/mesoporous shell carbon with a tailored structure as a cathode electrocatalyst support for proton exchange membrane fuel cells. J Phys Chem C 112 (2): 639-645
14. Lio S-H, Yen C-Y, Hung C-H, Weng C-C, Tsai M-C, Lin Y-F, Ma C-CM, Pan C, A Su (2008) One-step operation of carbon nanotubes by free radical modification for the preparation of nanocomposite bipolar plates in polymer electrolyte membrane fuel cells. J Mater Chem 18 (33): 3993-4002
15. Sakoda A, Nomura T, Suzuki M (1996) Activated carbon membrane for water treatments: Application to decolorization of coke furnace wastewater. Adsorption 3 (1): 93-98
16. Hu S, Yang F, Sun C, Zhang J, Wang T (2008) Simultaneous removal of COD and nitrogen using a novel carbon-membrane aerated biofilm reactor. J Environ Sci 20 (2): 142-148
17. Keizer K, Verweij H (1996) Progress in inorganic membranes. Chemtech 26 (1): 37-41
18. Zhang X, Hu H, Zhu Y, Zhu S (2006) Methanol steam reforming to hydrogen in a carbon membrane reactor system. Ind Eng Chem Res 45 (24): 7997-8001
19. Nednoor P, Gavalas VG, Chopra N, Hinds BJ, Bachas LG (2007) Carbon nanotube based biomimetic membranes: Mimicking protein channels regulated by phosphorylation. J Mater Chem 17 (18): 1755-1757
20. Sánchez S, Pumera M, Cabruja E, Fàbregas E (2007) Carbon nano-tube/polysulfone composite screen-printed electrochemical enzyme biosensor. Analyst 132 (2): 142-147

Chapter 10
Economic Evaluation

10.1 Recovery of Hydrogen from the Natural Gas Network

The establishment of hydrogen distribution system would be lengthy and costly, hence a transitional approach was proposed in an EU 6FWP (Integrated Project) named NaturalHy to use existing natural gas (NG) networks to transport mixtures of hydrogen and NG [1].The fraction of hydrogen in the pipeline ranges 5–30 vol.%. It is hence necessary to separate hydrogen from the pipeline gas for the end use components requiring relatively pure hydrogen, such as fuel cells. One of such technologies is membrane gas separation technology with carbon molecular sieve membranes (CMSMs)—this is currently being tested within this project. Carbon molecular sieves have been shown to achieve excellent performance, with respect to hydrogen permeability and selectivity, in the separation of hydrogen from light hydrocarbons including methane. CMSMs derived from the carbonization of cellulosic films have been shown by Grainger and Hägg to perform better than polymeric membranes for hydrogen/methane separation in terms of the Robeson type plot [2, 3].

10.1.1 Introduction

10.1.1.1 Local Permeance and the Temperature Effect on the Permeability

The permeation rate of gas species i through a segment of the membrane is calculated by

$$dF_i = dA\,\frac{P_i}{l}\,(p_{i,f} - p_{i,p}) \tag{10.1}$$

where P is the permeability, A is the membrane area, l is the effective membrane thickness, p is the partial pressure and subscript f and p indicate the feed and permeate. Permeance, P_i/l, is the permeability normalized by membrane thickness.

A. F. Ismail et al., *Carbon-based Membranes for Separation Processes*,
DOI 10.1007/978-0-387-78991-0_10, © Springer Science+Business Media, LLC 2011

If Henry's law holds for the sorption of gases into the carbon pore network, the permeability of gases in the carbon molecular sieve membranes can be written as

$$P_i = D_{0i} S_{0i} \exp\left(\frac{-E_{s,i}}{RT}\right) \tag{10.2}$$

where D_0 and S_0 are the diffusivity and solubility corresponding to $T=\infty$, T is absolute temperature, R is the universal constant and $E_s = E_d + H_s$, where E_d is the activation energy of diffusion and H_s is the heat of sorption. The experimental data were obtained in a range of 25–90°C to calculate the activation energy of permeability, E_s.

10.1.1.2 NaturalHy Design Basis

The NaturalHy concept involves the injection of hydrogen from a large source to a network, to be conveyed to the end users. In this project two end-use pressure levels are considered; one is industrial segment, 40 bar, and the other the residential segment, 8 bar. The typical natural gas composition used in the experiments as well as simulation is given in Table 10.1. In the simulation the feed gas temperature was assumed to be 25°C, which was the lowest temperature used in the experiments. The hydrogen at the end use point was assumed to be consumed in a polymer electrolyte membrane (PEM) fuel cell and the specification for the total impurity was set to 5 mol%. Carbon monoxide and other PEM poisons were assumed to be absent and the delivery pressure of hydrogen was set to 1.5 bar (abs). The proposed hydrogen production rate was in the range from 100 N m^3 h^{-1} (suitable for hospital fuel cell) to 1,000 N m^3 h^{-1} (suitable for industrial use). The membrane unit performance depends on the following variables:

1. The hydrogen mole fraction in the pipe line, $y_{H_2 i\, pipe}$
2. Pipe line gas pressure, p_{feed}
3. The ratio of end-use hydrogen flow rate to the pipeline hydrogen flow rate, $F_{H_2 use}/F_{H_2 pipe}$
4. The price and area of the membrane
5. The module temperature
6. Permeate pressure.

10.1.1.3 Performance Data

It was shown that the hydrogen permeability in the cellulose-derived CMSMs can be reduced by membrane pore blocking via chemical adsorption of oxygen from air.

Table 10.1 Representative natural gas composition

Gas	i-Butane	N_2	CO_2	Propane	Ethane	Methane
Mole %	0.3	1.5	2.0	4.5	10.0	81.7

Table 10.2 Gas transport data (average from H_2-natural gas runs for H_2 and CH_4). [2]

Gas	H_2	i-Butane	N_2	CO_2	Propane	Ethane	Methane
Permeability (25°C) (m^3(STP) m/m^2 bar h)	1.40×10^{-6}	0	4.1×10^{-9}	7.1×10^{-8}	0	0	5.5×10^{-10}
Permeability (90°C) (m^3(STP) m/m^2 bar h)	2.49×10^{-6}	0	4.1×10^{-9}	2.2×10^{-7}	0	0	2.7×10^{-9}
Permeance (25°C) (m^3(STP)/ m^2 bar h)	1.4	0	0.004	0.07	0	0	0.00055
Permeance (90°C) (m^3(STP)/ m^2 bar h)	2.5	0	0.004	0.20	0	0	0.00270

The flux decreases about 50% after 1 year. The strategy to cope with this problem is as follows:

1. Preventative: Blanketing of the module with nitrogen
2. Adaptive: Add new modules, increase pressure, increase operating temperature
3. Conservative: Overdesign the process
4. Regenerative: Periodically apply electric current.

The combination of preventive and conservative was adopted and the data obtained after 7 months of exposure to air was used in the simulation work. The membrane used in the experiments was prepared by doping a cellulose–hemicellulose precursor derived from wood pulp with 4 wt% copper (II) nitrate, and carbonizing it at 650°C to form a flat carbon sheet. In the simulation however hollow fiber configuration was used with a wall thickness and outer layer selective thickness of 16 and 1 μm, respectively. The permeability and permeance data used for the simulation studies are given in Table 10.2 for various gases. These values were obtained either from the experiments with the mixed gas of the composition given in Table 10.1 or from the pure gas, for methane and nitrogen, experiments. Since ethane and propane were not found in the permeate side permeability of zero was assigned for those gases. Experiments showed that the permeance is not affected by the feed pressure very much. Therefore, the permeance was assumed to be constant regardless of the pressure. The parameters associated with temperature effect on the permeability are listed in Table 10.3.

Table 10.3 Activated transport parameters for carbon membrane

Gas	E_s (kJ/mol)	P_0 (m^3(STP) m/m^2 bar h)	Temperature range
H_2	8.0	4.1×10^{-5}	25–90
CH_4	24.3	9.2×10^{-6}	25–90

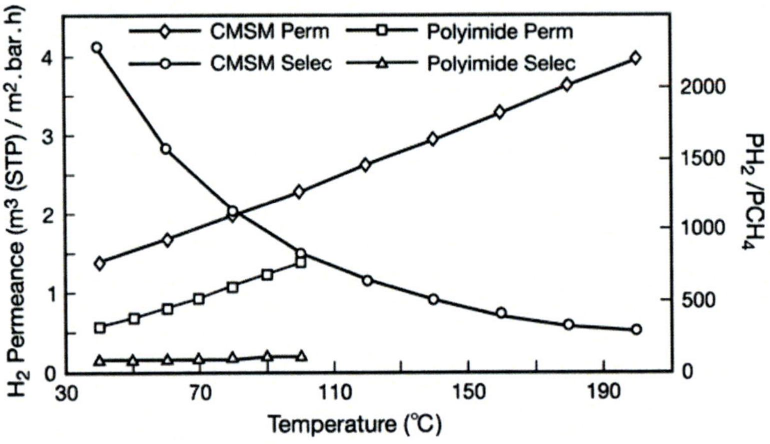

Fig. 10.1 Effect of temperature on membrane performance. (From [1])

10.1.1.4 Comparison of Membrane Performance with Commercial Polyimide Membrane

The performance of CMSM is compared in Fig. 10.1 with polyimide membrane manufactured by Ube Industries, Japan. At 25°C, the hydrogen permeance of Ube membrane is about one half of CMSM and the selectivity is an order of magnitude lower. However, the selectivity of CMSM decreases quickly as the temperature is increased. The temperature limit of Ube membrane was 100°C.

10.1.1.5 Carbon Membrane Module Cost

Polymeric hollow fiber membranes cost approximately \$ 20–50/m^2. It is difficult to evaluate the CMSM module cost because no commercial CMSM module is available. Koros and Mahajan [4] stated that carbon, glass, zeolite and other inorganic membranes could cost from one to three orders of magnitude higher than polymeric membranes. Hence, the cost of the CMSM module was chosen to be \$ 50–500/m^2 for the simulation work.

10.1.2 Methodology

The flow sheet of the process is given in Fig. 10.2. Pretreatment of the natural gas is not included. Hence, the major instalment cost comes from the compressor. If heating of the feed gas is necessary gas-fed heater has also to be installed.

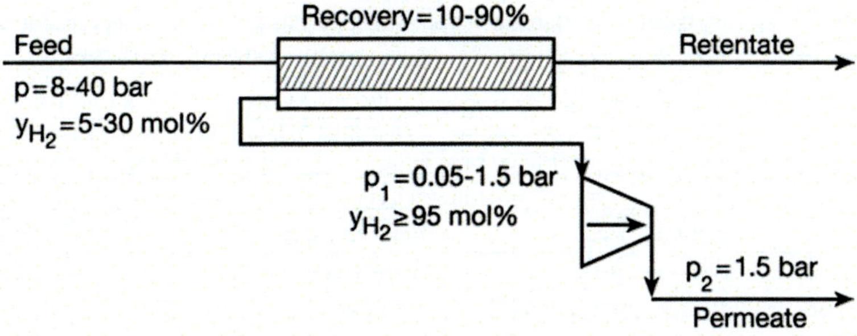

Fig. 10.2 Simple one-stage counter-current membrane process. (From [1])

10.1.2.1 Sensitivity Analysis and Optimization

When all process variables such as feed pressure, permeate pressure, temperature, hydrogen feed fraction and hydrogen recovery, are considered the volume of data becomes too large. Hence, some simplification was made.

1. Feed compression is not made. The membrane module is operated at pipeline or off-gas stream pressure.
2. The feed mixture consists of hydrogen and methane only. This assumption is justified by the high methane concentration (82%) in natural gas. As a result, the hydrogen mole fraction in the permeate is overestimated.

A Microsoft Office Excel spreadsheet can deliver answers based on Mulder's analytical solution for binary gas system under cross-flow condition [5].

Regarding the economic parameters, the membrane cost was evaluated from the cost per unit area of the membrane. The compressor cost was evaluated from adiabatic power of the compressor. Other economic parameters are summarized in Table 10.4.

Table 10.4 Economic factors for excel-based performance analysis

Installation factor for purchased equipment	4
Annual operating time	8,000 h
Fixed cost	
Capital charge	15% of installed capital cost
Module replacement frequency	Every 5 years
Maintenance	2% of installed capital cost excluding module replacement
Insurance	1% of installed capital cost
Variable cost	
Electric power	60.035 kWh^{-1}
Natural gas price	$ 7/MMBTU
$ to € exchange rate	1.3

Excel was then used to generate the membrane area, mass balance, energy consumption and operating cost simultaneously over the range of recovery (10–90%), temperature (25–90°C) and permeate pressure (up to 1.5 bar). The permeate that produced the lowest specific operating cost (€/kg H_2 separated) and met the purity specification for a certain recovery was selected and the corresponding membrane area and energy consumption recorded. The objective function for the optimization (minimization) was

$$\text{Annual specific separation cost} = \frac{(C_{\text{capital charge}} + C_{\text{module replacement cost per year}} + C_{\text{maintenance}} + C_{\text{insurance}} + C_{\text{electricity consumption}} + C_{\text{natural gas combusted per year}})}{\text{mass } H_2 \text{ recovered per year}}$$

where C is the cost. Capital charge was calculated as 15% of the total installed cost of equipment per year. Module replacement cost was taken as 20% of the module purchased cost per year, based on the replacement frequency.

More rigorous simulation was made by using Aspen Hysys® 2004.2 using the full feed composition.

10.1.3 Recovery from a Mixed Hydrogen-NG Network

Figures 10.3 and 10.4 show examples of the simulation results corresponding to 40 bar feed pressure with 30% feed H_2 mole fraction and 8 bar with 5% feed H_2 mole fraction, respectively. Optimum values are recorded all as the function of H_2 recovery. The end-use is assumed to be a hospital fuel cell requiring 100 N m^3 H_2 h^{-1} (ca 4.5 kmol h^{-1}). For the first case (Fig. 10.3) H_2 feed partial pressure is high enough to maintain the permeate pressure at 1.5 bar (150 kPa) in most cases. The hydrogen purity is high but decreases slightly over the studied recovery range. Both area and cost follow upward trend. When the feed partial pressure is low (Fig. 10.4) the low driving force causes membrane area to significantly influence separation cost. Overall, the following conclusions are obtained.

1. CMSMs meet the PEM fuel cell requirement of H_2 purity more than 95%, even when the feed hydrogen concentration is as low as 5 mol%.
2. The lowest cost is obtained for the low H_2 recovery from a feed of 40 bar and 30 mol% hydrogen.
3. Product H_2 purity decreased as the H_2 recovery increased. When the module is operated a 25°C, however, purity requirement (H_2>95 mol%) was met up to the recovery of 90%. This was not the case for the operating temperature of 90°C.
4. The optimum permeate pressure decreased as H_2 recovery increased and as feed pressure or feed hydrogen concentration decreased.

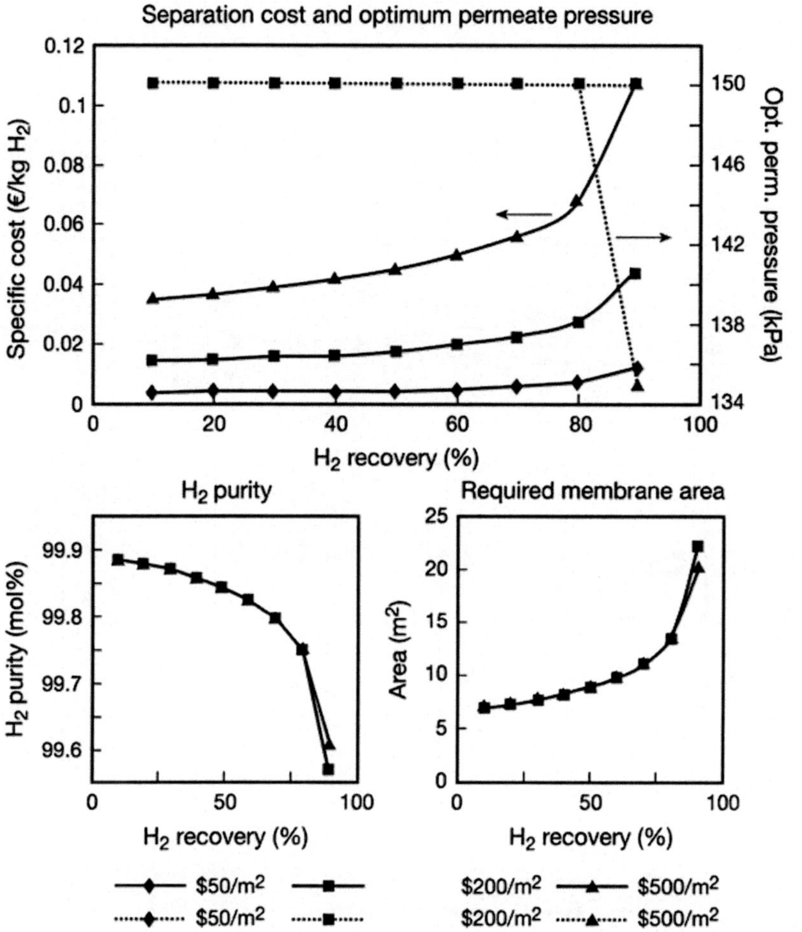

Fig. 10.3 Specific cost, optimum permeate pressure, H_2 purity in the permeate and membrane area at the optimum condition as the function of H_2 recovery; feed pressure, 40 bar; feed hydrogen mole fraction 30%, temperature 25°C. (From [1])

5. The cost of membrane influences the optimum permeate pressure. If the membrane costs more, the optimum permeate pressure becomes less. This is to increase the driving force for the gas permeation and reduce the membrane area requirement.

The effect of the heating of the feed gas was also examined. An extra heating cost is added to the total cost, but, on the other hand the membrane cost is reduced due to the enhanced permeation rate and resulting membrane area reduction. Thus, the cost benefit of feed gas heating is obtained only when the membrane cost is very high. For example, for the 40 bar and 30% hydrogen case, there is no benefit of heating the feed gas to 90°C until the membrane cost becomes as high as \$ 4,000/m^2, while

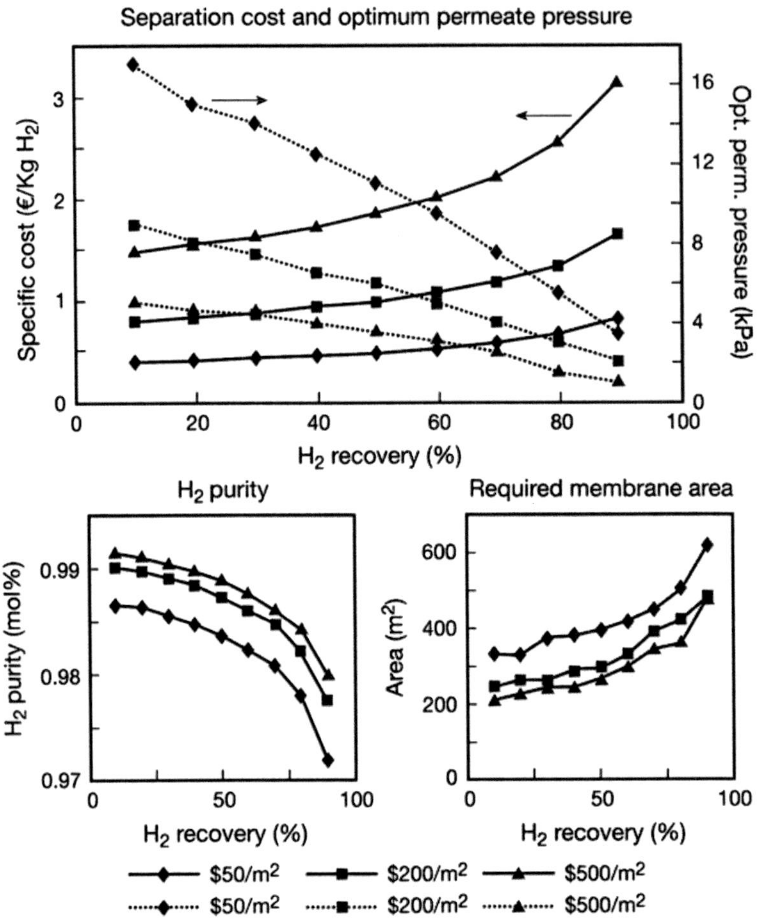

Fig. 10.4 Specific cost, optimum permeate pressure, H_2 purity in the permeate and membrane area at the optimum condition as the function of H_2 recovery; feed pressure, 8 bar; feed hydrogen mole fraction 5%, temperature 25°C. (From [1])

in the case of 8 bar and 5% hydrogen, the heating becomes advantageous when the membrane cost is \$ 500/m^2 or more.

There is also practically no scale merit in the product capacity range of 100–1,000 Nm^3 h^{-1}.

The equipment cost is often scaled by the following equation,

$$C_1 = C_2\left(\frac{S_1}{S_2}\right)^n \tag{10.3}$$

where C is the purchase cost and S in the capacity. Cost is proportional to membrane area and also nearly proportional to power required for the centrifugal compressors (n is ca 0.93). Only the heater shows the economy of scale with n=0.8.

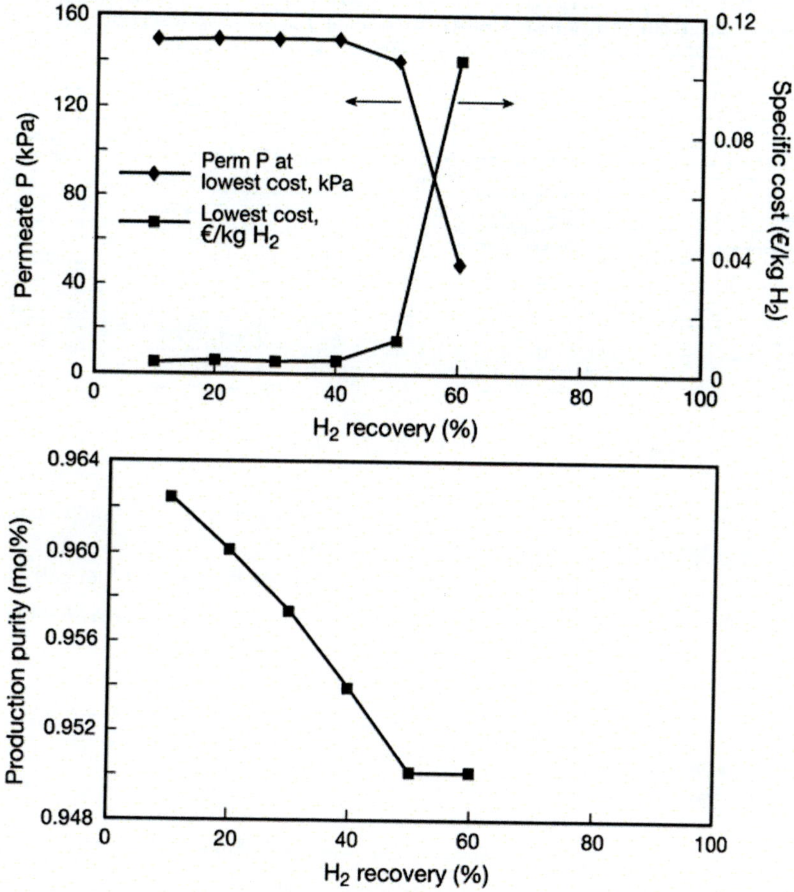

Fig. 10.5 Performance of polyimide membrane at 40 bar, 30% H_2 feed (T=25°C). (From [1])

10.1.3.1 Comparison with a Polyimide (PI) Membrane

The performance of the polyimide membrane is shown in Fig. 10.5. Despite the low polyimide module cost, CMSM is more economical at hydrogen recovery higher than 40%. The reason is that the permeate pressure should be low for the PI membrane to maintain the high hydrogen purity of 95 mol%, whereas the permeate pressure from CMSM is high enough so that recompression is not necessary for the full range of recovery (Fig. 10.3). However, if the hydrogen recovery is less than 30%, PI membrane module is cheaper. The hydrogen purity requirement of>95 mol% can not be met by the one stage PI membrane process when the hydrogen recovery is higher than 60% and the product must be further purified by attaching the second stage (Fig. 10.6).

Regarding the two stage system, the permeate pressures of both stages were set equal to simplify the calculation. On the other hand, the feed pressure for the second

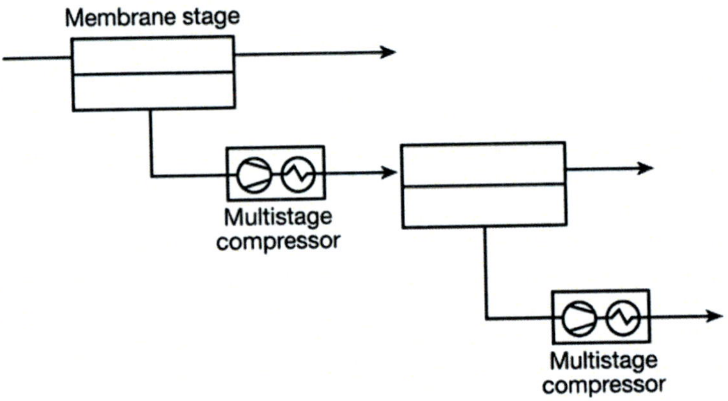

Fig. 10.6 Two-stage polyimide membrane process. (From [1])

stage was varied to minimize the lowest specific cost and it was found that 400 kPa was the optimum feed pressure for the second stage.

Since the product rate is fixed at ca 4.5 kmol/h for the hospital fuel cell, the feed flow rate will vary according to the over all recovery, $R = F_{H_2product}/F_{H_2feed}$, since

$$F_{feed} = F_{product}\, y_{H_2product}/(R\, y_{H_2feed}) \tag{10.4}$$

In Eq. (10.4) all quantities on the right side are fixed other than R. Hence, F_{feed} increases as R decreases.

Similar to the above equation, applying to the first stage module

$$F_{feed} = F_{feedstage2}\, y_{H_2feedstage2}/(R_1\, y_{H_2feed}) \tag{10.5}$$

(overall $R = R_1 R_2$).

Keeping all other quantities constant, $F_{feedstage2}$ will increase as R_1 increases. R_1 has a large impact on the cost since increase in $F_{feedstage2}$ results in heavier compression duty. It is therefore better to minimize R_1.

I was found that setting $R_1 = 1.05R$ satisfies this requirement.

Heating the feed stream is not beneficial since the membrane is cheap and saving in the membrane area does not have economic impact. For a feed of 8 bar and 5 mol% hydrogen the specific cost of two stage PI process is similar to the CMSM as shown in Fig. 10.7 where the comparison of two stage PI and CMSM are made. This is because CMSM membrane area is large and the membrane is expensive while the PI process requires extra compression for the second stage.

Figure 10.8 compares CMSM and PI membranes in terms of the specific energy required to produce one mole of hydrogen. These are thermal equivalent values. The consumption of electrical power by the compressor is converted to the thermal consumption. Heating duty is also added if any. The energy consumption of the CMSM process is low for both 30 and 5 mol% hydrogen feed cases, because less permeate recompression is required. In particular, in the 30 mol% case the permeate in CMSM process is above 1.5 bar and does not require recompression at all.

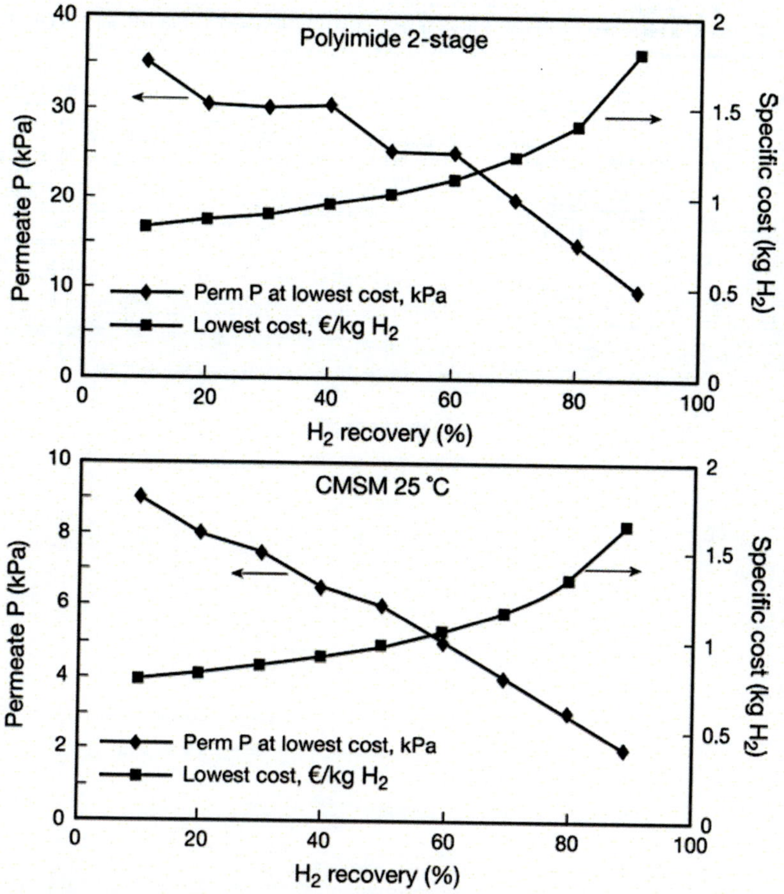

Fig. 10.7 Comparison of membranes for the feed mixture of 8 bar and 5% H_2; CMSM module cost is set equal to \$ 200/m². (From [1])

10.2 Applications in Landfill Gas Energy Recovery

10.2.1 Introduction

Prosser et al. discussed on the performance and economics of carbon molecular sieve membranes in landfill gas energy recovery applications [6].

CMSMs look promising for landfill gas (LFG) processing since they may enhance the BTU values of LFGs. Shallow landfills, landfills located in arid environments, or the landfills sites that extract landfill gas higher than it is generated in order to protect the environment are excellent candidates for this technology. Removal of CO_2 gas from the LFG will restore the methane quality to an acceptable level for fuel. It is estimated that this process could benefit 5–15% of all landfills

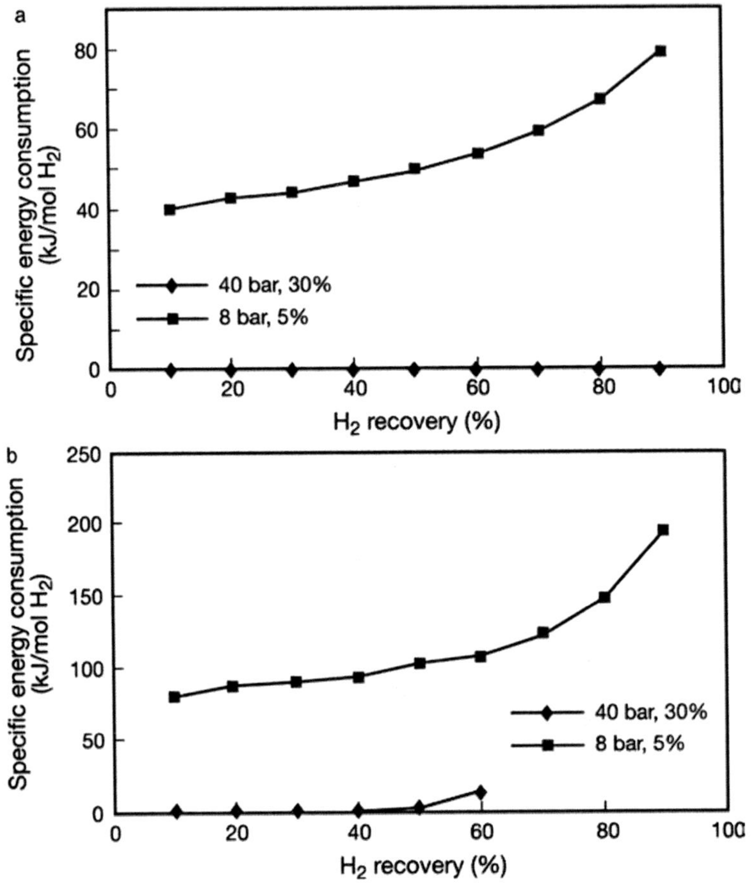

Fig. 10.8 Energy consumed per mole of hydrogen produced by (*top*) CMSM, (*bottom*) polyimide membrane. (From [1])

by enhancing the quality of the collected LFG. The study deals with the use of CMSMs in two applications. In one application spark-ignited engine generators to produce electricity is utilized while in the second CMSMs are used to purify LFG to pipeline quality methane. The focus is on smaller landfills, i.e., 1–5 million tons of municipal solid waste (MSW) in place, since there are more landfills in this size range that have not yet been developed.

10.2.1.1 Membrane Development

US Filter mesoporous tubular substrates were used as the support to study the membrane feasibility in laboratory scale. Polyetherimide (PEI) was used as the polymeric precursor. The support substrate (7 mm ID, 10 mm OD) was dip coated in PEI solution followed by drying in air and carbonization in an argon atmosphere.

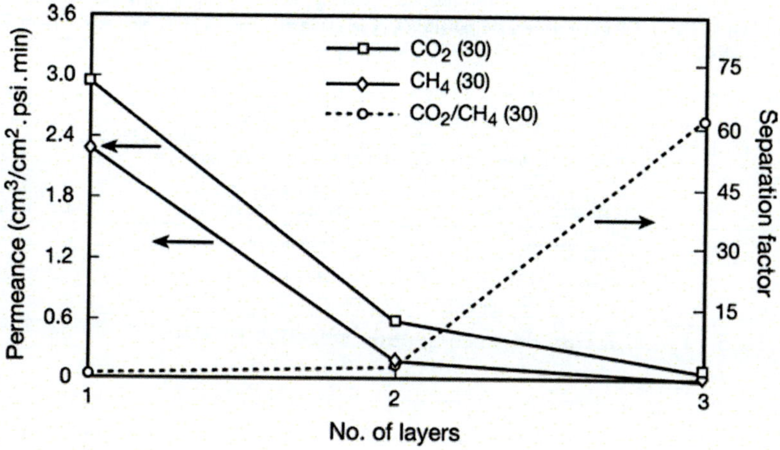

Fig. 10.9 Effect of number of carbon layers on permeance and separation factor for single gases. (From [6])

The coating-carbonization cycle was repeated until desired performance data were obtained. Figure 10.9 illustrates the permeance and ideal separation factor as the function of the number of coating. The actual separation factor for an equimolar binary mixture of CO_2/CH_4 is about twice as high as the ideal separation factor [7].

10.2.2 Application

10.2.2.1 Internal Combustion Engine Power Generation

LFG from the interior of a landfill may not require CO_2 removal. In this case the value added will be to utilize perimeter gas or poor quality interior gas that may otherwise be flared. In arid climates, LFG collected even in the interior part of a landfill may not be suitable for use in engines. The value added in this case is to treat the entire gas stream to a level acceptable to the engines.

10.2.2.2 Upgrade Landfill Gas to Pipeline Quality

Upgrading the LFG to the level suitable for injection to the natural gas pipeline is another application. With the aid of CMSMs this application may be economical even for small scale (1–5 million tons of MSW) landfills. No modification is necessary for the equipment at the end user. The application requires relatively expensive treatment of LFG and also, in order to inject into the natural gas pipeline, there are strict specifications that impose additional quality control and compression requirements.

Table 10.5 Methane generation and collection rates

Waste in place (million t)	1	5
Methane gas per pound refuse per year (cf/lb/year)	0.05	0.05
Annual methane gas generation (mmcf/year)	100	500
Methane gas generation per hour (scfh)	11,400	57,100
LFG flow at 35% methane and 80% collection efficiency (scfh) (internal combustion application)	26,100	130,500
LFG flow at 57% methane and 70% collection efficiency (scfh) (injection into pipeline)	14,000	70,100

10.2.2.3 Landfill Gas Generation and Collection Efficiency

The quantity of methane generated varies from site to site. The quantity of 0.05 cf/lb is used in this calculation. This value is somewhere in the intermediate range of the scattered data and is based on the LFG containing 50% of methane. Dry landfills produce less methane and wet landfills more methane. The methane generation and collection rates are shown in Table 10.5.

To make Table 10.5, it was assumed that the quality of the product being collected for the internal combustion (IC) engine application is 35% CH_4, 35% CO_2 and 30% N_2 and O_2 mixture. The quality of the product collected for the injection into the pipeline is 57% CH_4, 42% CO_2 and 1% N_2 and O_2 mixture. The difference in these gas compositions is due to the difference in the amount of air drawn into LFG during its collection and the collection efficiency.Collection efficiency of 80% was applied for the IC engine application. The product used for the high BTU gas application (pipeline injection) should have little air, therefore a lower collection efficiency of 70% was applied. The hourly product flow was calculated by multiplying the methane generation rate with collection efficiency and dividing by the methane concentration.

10.2.2.4 Carbon Molecular Sieve Membrane Performance

For the IC combustion application, the feed and permeate gas pressures are assumed to be 40 psig and 0 psig, respectively. The feed composition is 35% CH_4, 35% CO_2 and 30% N_2 and O_2. The processed gas quality that is collected on the retentate side should meet the requirement of 50% CH_4, 10% CO_2 and 40% N_2 and O_2. Methane yield of the process is 83.9%. The CO_2/CH_4 selectivity of the membrane is set equal to 10. The same selectivity is assumed for CO_2/N_2 and CO_2/O_2. The permeance of CO_2 is set equal to 1.47 m^3/m^2 h bar. Using these data the permeate composition can be calculated to be 15% CH_4, 72% CO_2 and 13% N_2 and O_2. Mass balance tells us the stagecut required to achieve the above retentate and permeate composition. From the feed and retentate flow rate and the CH_4 concentration in the feed and retentate, the CH_4 yield can finally be obtained. The results of the calculation are summarized in Table 10.6.

For upgrading the LFG to pipeline quality, the feed reject pressure is assumed to be 120 psig. The feed LFG composition is 57% CH_4, 42% CO_2 and 1% N_2 and

Table 10.6 Membrane simulation results for internal combustion engine and injection into pipe line applications

	IC engine		Injection into pipe line	
	1 MM t	5 MM t	1 MM t	5 MM t
Feed pressure (psig)	40	40	120	120
Retentate pressure (psig)	40	40	120	120
Permeate pressure (psig)	0	0	0	0
Feed CO_2 mole fraction	0.35	0.35	0.57	0.57
Feed CH_4 mole fraction	0.35	0.35	0.42	0.42
Feed N_2, O_2 mole fraction	0.3	0.3	0.01	0.01
Retentate CO_2 mole fraction	0.1	0.1	0.02	0.02
Retentate CH_4 mole fraction	0.5	0.5	0.96	0.96
Retentate N_2, O_2 mole fraction	0.4	0.4	0.02	0.02
Permeate CO_2 mole fraction	0.72	0.72	0.88	0.88
Permeate CH_4 mole fraction	0.15	0.15	0.117	0.117
Permeate N_2, O_2 mole fraction	0.13	0.13	0.003	0.003
Overall yield (%)	59.9	59.9	36.1	36.1
CH_4 yield (%)	83.9	83.9	82.5	82.5
Area (m^2)	270	1350	280	1420
Selectivity	10	10	30	30
CO_2 permeance (m^3/m^2 h bar)	1.47	1.47	0.588	0.588
BTU out (mmBTU/h)	6.96 LHV	34.8 LHV	6.66 HHV	33.3 HHV
Gross kW output-engine at 12,500 (BTU LHV/kWh)	556	2,782	–	–
Net kW output (kW)	484	2,421	–	–
Net BTUs produced (mmBTU/day)	–	–	160	800
Annual capacity factor	0.9	0.9	0.9	0.9
Net energy sold (kWh/year)	3,816,000	19,085,000	–	–
Net energy sold (mmBTU HHV/year)	–	–	52,500	262,500

O_2 mixture. The required product gas composition is 96% CH_4, 2% CO_2 and 2% N_2 and O_2 mixture. The CH_4/CO_2 selectivity and the CO_2 permeance are assumed to be 30 and 0.59 m^3/m^2 h bar, respectively. A similar calculation as above yields the CH_4 yield of 82.5%.

Table 10.6 summarizes the results of all the above calculations.

10.2.3 *Economic Analysis of Applications*

10.2.3.1 Internal Combustion Engine Power Generation

The figures given in the first column of the table can be calculated as follows. From the production rate of methane

$$\text{The BTU out} = \text{Landfill gas flow (cfh)} \times \text{Energy content of the Landfill gas (BTU/cf)}$$

Table 10.7 Cost of electricity generated by internal combustion engine

	Internal combustion engine	Internal combustion engine
Waste in place (million t)	1	5
Installed capital cost ($/kW)	630	480
Membrane capital cost ($/kW)	242	242
Capital charge rate (1/year)	0.136	0.136
Annual capacity factor (¢/kWh)	90	90
Levelized capital cost (¢/kWh)	1.50	1.25
O&M expenses (¢/kWh)	1.80	1.80
Cost of electricity generated (¢/kWh)	3.30	3.05

The paper does not give the energy content of the landfill gas but the total BTU out is given as 6.96 mm BTU LHV/h. This is divided by electric to heat energy conversion factor of 12,500 kWh/BTU to obtain Gross kW output, 556 kW. When 13% of parasitic loads (including compressors and other auxiliary equipment) are deducted the net kW output becomes 484 kW. The annual capacity factor (the fraction of the year in which the equipment is producing electricity at its rated capacity) is considered to be 0.9. Then, 484 × 24 × 365 × 0.9 = 3,816,000 kWh/year of net energy can be generated to be sold. Similar calculation can be done for the second column for 5 MM t operation. LFG power projects are selling electricity to the utilities for 3–4 ¢/kWh currently. When power is sold directly to the retail customer, higher power rates can be obtained.

The costs to generate the electricity by internal combustion engine are summarized in Table 10.7. The levelized capital costs (¢/kWh) in column 2 can be calculated as follows.

$$\text{Levelized capital cost (LCC)} = \frac{\text{(Installed capital cost + membrane capital cost)} \times \text{capital charge rate}}{\text{(annual capital factor} \times 8{,}760\ \text{h/year}} = 1.50\,¢/\text{kWh}$$

Further, adding operation and maintenance (O&M) expenses (1.80 ¢/kWh), the resulting cost of electricity generated becomes 3.30 ¢/kWh.

10.2.3.2 Injection into Pipeline

Upgrading gas to pipeline quality is substantially more expensive than the other alternative because of the high demand in gas quality. Hence the gas price becomes much larger. Table 10.8 summarizes the annual cost in $ involved in the treatment system. The calculation was done for two capacities of 1 and 5 million ton. For each capacity, the calculation was done for two cases, i.e. one for annual interest of 0.08 and the other 0.05. Annual payment was calculated based on the annual interest rate and number of annual payment. Annual energy content came from Net energy

Table 10.8 Cost of pipeline quality gas generated

Waste in place (million t)	1		5	
Installed capital cost ($)	938,400	938,400	2,050,800	2,050,800
Annual interest rate	0.08	0.05	0.08	0.05
Number of annual payment	10	10	10	10
Annual annuity payment	139,850	121,500	305,600	265,600
Annual O and M expenses ($)	142,000	142,000	314,000	314,000
Cost of pipeline gas generated ($)	281,850	263,500	619,600	579,600
Annual energy content (mmBTU/year)	57,600	57,600	288,000	288,000
Value of pipeline gas generated ($/mmBTU)	4.89	4.58	2.15	2.01

sold in Table 10.6. Instead of 52,500 and 262,500 mmBTU HHV/year, 57,600 and 288,000 mmBTU/year were used, respectively in Table 10.8.

As a conclusion the projects are not feasible unless the price of such gas is greater than 4.89 and 4.58 $/mmBTU at the interest rate of 8 and 5%, respectively, for the 1 million t landfill. Te price should be higher than 2.15 and 2.01 $/mmBTU, respectively, at the interest rate of 8 and 5% for the 5 million t landfill.

References

1. Granger D, Hägg M-B (2008) The recovery by carbon molecular sieve membranes of hydrogen transmitted in natural gas network. Int J Hydrogen Energy 33 (9): 2379-2388
2. Grainger D, Hägg M-B (2007) Evaluation of cellulose-derived carbon molecular sieve membranes for hydrogen separation from light hydrocarbons. J Membr Sci 306 (1-2): 307-317
3. Robeson LM (1991) Correlation of separation factor versus permeability for polymeric membranes. J Membr Sci 62 (2): 165-185
4. Koros WJ, Mahajan R (2000) Pushing the limits on possibilities for large-scale gas permeation: which strategies? J Membr Sci 175 (2): 181-196
5. Mulder M (1997) Basic principles of membrane technology, 2nd ed. Springer, Berlin, p 496
6. Prosser D, Ackerman MJ, Liu PKT, Sedigh MG, Tsotsis TT, Sahimi M (1999) An economic evaluation of carbon molecular sieve membranes in landfill gas applications. GC Environemtal Inc.: 1-9
7. Sedigh MG, Xu L, Tsotsis, TT, Sahami M (1999) Transport and morphological characteristics of polyetherimide-based carbon molecular sieve membranes. Ind Eng Chem Res 38 (9): 3367-3380

Chapter 11
Current Research and Future Direction

11.1 Inorganic Membranes and Carbon Membrane

In the beginning of this century, more than 20 large membrane plants with 100 million standard cubic feet (scf) have been built, as well as many smaller systems [1]. Increasing interest in gas separation by organic membranes had further led to exploitation of inorganic membranes for high temperature or corrosive gas separation applications [2]. Nowadays, inorganic membrane producers are generally in the start-up and technology push stage. Meanwhile, the end-user industry has exhibited a "wait-and-see" attitude when it comes to adopting advanced inorganic membrane applications. Industries currently have a major interest in basic quantitative knowledge of inorganic membranes. They are interested to know the performance of inorganic membranes in separation process and their stability in aggressive environment [3].

In general, inorganic membranes for gas separation can be classified into three categories; i.e., (1) microporous, amorphous membranes, (2) microporous, crystalline membranes and (3) dense, high temperature membranes. Efforts will be made to realize complex, however well-defined porous architectures and all these three types of membranes will be combined to exhibit new separation properties with improved long-term stability in these few years. In the near future a gradual shift will take place from the exploration of new membrane concepts toward better control of membrane preparation and understanding of performance, long-term stability and process integration in the applications [3].

Today, 26 firms are manufacturing inorganic membranes and most of them were introduced to the market over the last 5 or 10 years. The inorganic membranes produced by these firms are summarized in the Table 11.1. The oldest inorganic porous membranes manufacturers, SCT/US Filter and TECH-SEP are now dominating the market for the inorganic membranes. As well, it is porous ceramic membranes which dominate commercialized inorganic membranes. The most prevalent membrane materials are metallic oxides, more specifically alumina and zirconia, but other materials, such as titania, carbon or glass are also available [4].

A. F. Ismail et al., *Carbon-based Membranes for Separation Processes*,
DOI 10.1007/978-0-387-78991-0_11,

Table 11.1 Commercial porous inorganic membranes

Manufacturer	Trade name	Material	Pore size[a]	Geometry
USF/SCT	MEMBRALOX®	ZrO_2/Al_2O_3	20–100 nm	Monolith
		Al_2O_3/Al_2O_3	5–12 nm	
	CERAFLO®	Al_2O_3/Al_2O_3	0.2–1.0	Monolith
TECH-SEP	CARBOSEP®	ZrO_2/C	10–300 kD	Tube
		TiO_2/C	0.14	
	KERASEP®	$TiO_2/Al_2O_3+TiO_2$	0.1–0.45	
		$ZrO_2/Al_2O_3+TiO_2$	15–300 kD	Monolith
Le Carbone Lorraine		C/C	0.1–1.4	Tube
CERASIV		Al_2O_3/Al_2O_3	0.1–1.2	Tube/monolith
		$TiO_2/ZrO_2/Al_2O_3$	5–100 nm	
NGK		$Al_2O_3/Al_2O_3+SiO_2$	0.2–5	Tube/monolith
Whatman	ANOPORE®	Al_2O_3	20–0.2 nm	Disc
Gaston Country	UCARSEP®	ZrO_2/C	4 nm	Tube
Du pont/Carre		$Zr(OH)_4/SS$	0.2–0.5	Tube
TDK	DYNACERAM®	ZrO_2/Al_2O_3	10 nm	Tube
		Al_2O_3/Al_2O_3	50 nm	
ATEC		SiC/SiC	0.05–1	Tube/monolith
		Al_2O_3		
Asahi glass		Glass	0.1–1.4	Tube
Fuji filters		Glass	4–90 nm	Tube
		Glass	0.25–1.2	
Fairey	STRATA-PORE®	Ceramics/ceramics	1–10	Tube/plate
	MICROFIL-TREX®	SS	0.2–1	Tube/plate
Osmonics	HYTREX®	Ag/none	0.2–5	Tube/plate
	CERATREX®	Ceramics/ceramics	0.1	
Ceramen		Ceramics/cordierite	0.05–0.5	Honeycomb
Trideltafiltration		Al_2O_3/Al_2O_3	0.1–7	Tube/monolith
Hoogovens		Al_2O_3/Al_2O_3	0.1–1	Tube
Steenecker		Al_2O_3	0.4	Tube
NOK		Al_2O_3/Al_2O_3	0.2–6	Tube
TOTO		Al_2O_3/Al_2O_3	0.1–0.2	Tube
		ZrO_2/Al_2O_3	5–30 nm	Monolith
Carre		ZrO_2/SS		Tube
Jiangsu Jiuwu Hitech Co., Ltd.	CERAMFIL®	α-Alumina/zirconia	50–500 nm	Tubular
Sterlitech Corporation		$ZrO_2–TiO_2$	1 KD–1.4 μm	Disc
		Glass	0.4 μm	Disc
Pall Corporation	ACCUSEP®	Ceramic		Tube
Kerafol		Ceramic		Capillary/disc
Mempro	nCeramic	Ceramic		

[a] μm unless otherwise stated

Commercialized inorganic membranes exist in three configurations: disks or sheets, tubes and multichannels/honeycombs. Usually, flat disks or sheets are limited to small scale industrial, medical and laboratory applications. They are used almost exclusively in flow-through filtration in contrast to cross-flow filtration in tubes and multichannel monoliths. Meanwhile, tubes and monoliths are used for various industrial applications [2, 4].

As we can see in Table 11.1, there are only a few manufacturers involved in the production of carbon membranes. *It is because higher costs in production and packaging these carbon membranes in modules have prevented their use in large-scale membrane modules* [5]. A large amount of efforts should be made to know the optimum conditions to prepare an excellent carbon membrane with excellent performance. However, the production of carbon membranes involves a very high cost. The cost of carbon membrane per unit of membrane area is reported to be between one and three orders of magnitude higher than the polymeric membrane [6].

Thus, *investigation should be made on material selection of more suitable carbon membrane precursor*. Finding a more economical material than polyimide and other polymeric materials currently used is still necessary in carbon membrane production. This indicates that *the discovery of the method to manufacture carbon membranes of excellent separation properties without losing the processability of polymeric membrane materials would be a major breakthrough* of this field [5].

Other challenges faced by the current membrane technology are: *achieving higher permselectivity with acceptable productivity, maintaining these properties in the presence of complex and aggressive feeds and preventing the need for recompression of the desired product* [6].

In the future, the polymer blend carbonization method will become an important preparation method to *produce a membrane with mixed-matrix materials*, which will probably be able to overcome the challenges mentioned above. Mixed matrix materials comprising molecular sieve entities embedded in a polymer matrix offer the potential to combine the processability of polymers with the superior gas separation properties of rigid molecular sieving materials [6]. By the carbonization of polymer blend ideal pore structure will be formed since the thermally unstable polymer (pyrolyzing polymer) will decompose to leave pores in the carbon matrix formed from the stable polymer (carbonizing polymer). Although the polymer blend has been now widely used as an important industrial raw material or as a highly functional material, this carbonization method is just at the stage of the beginning [7]. A large amount of studies are needed to find out the suitable pyrolyzing polymer and carbonizing polymer for blending.

It is clear from the above discussion, carbon membrane still requires much improvement and a long journey to go through before it will become a dominant commercialized inorganic membrane in this century. However, carbon membrane has a great potential to replace other inorganic membranes in the market because it has a number of unique characteristics and is able to separate gas mixtures, which have similar size of gas molecules, efficiently.

In short, the increasing number of researches and investigations of carbon membrane technology indicates that the carbon membrane will become another alternative membrane to be used for the industrial separation process. They consist of four major configurations; i.e., flat, supported on tube, capillary and hollow fiber. Carbon membranes, especially carbon hollow fiber membranes, have a great potential to be used widely in the gas separation processes. Nowadays, hollow fibers are the most popular membrane geometry due to their high surface area per unit volume in the membrane module [1]. Carbon membranes with their unique characteristics and advantages will make them competitive with other porous inorganic membranes and even with currently dominant polymeric membranes. However, many research and development efforts are needed in order to commercialize the carbon membrane in the international market.

In the following sections, the above general concept is discussed more thoroughly.

11.2 Current Research and Future Direction of Carbon Membrane Development for Gas Separation

There is a growing interest in the development of gas separation membranes based on materials that provide better selectivity, thermal and chemical stability than those which already exist (i.e. polymeric membranes). Carbon membranes offer the best candidates for the development of new membrane technologies, because of their stability and molecular sieving capabilities. The most notable advantages of carbon membranes have been recently reviewed [8] in comparison to those of polymeric membranes, in order to highlight the factors that make carbon membranes very attractive and useful as separation tools. Table 11.2 inserted in the end of the chapter summarizes the performance of carbon membranes for the separation of mixtures of permanent gases as reported in the literature.

11.2.1 Advantages of the Carbon Membranes

Selectivities of the carbon membranes are much greater than those of other inorganic membranes and the polymeric membranes. The selectivities of typical highly selective carbon membranes are 10–20 times larger than Vycor glass and silicone rubber membranes. In addition, the average permeabilities are an order of magnitude higher than that of Vycor glass [9].

Koros and Mahajan summarized three main challenges faced by current membrane technology, which are (1) achieving higher permselectivity with acceptable productivity, (2) maintaining these properties in the presence of complex and aggressive feeds and (3) preventing the need for recompression of the desired product [6].

Carbon membranes have the potential to overcome these three challenges with their numerous advantages compared with polymeric membranes.

Table 11.2 Performance of carbon membranes for the separation of mixtures of permanent gases

Precursor	Pyrolysis	P (O_2)	P (CO_2)	P (H_2)	a (O_2/N_2)	a (CO_2/N_2)	a (CO_2/CH_4)	a (H_2/N_2)	a (H_2/CH_4)	Reference
6-FDA/ BPDA-DAM	550°C, 2 h		25–30 GPU				73–85			[42]
Matrimid 5218	550°C, 2 h		11–13 GPU				69–83			
6FDA/ BPDA-TMPD	500°C, 120 min	20–50 GPU			8.5–11.5	55–56				[16]
	550°C, 120 min	14–40 GPU			11–14		140–190		400–500	
BPDA-ODA/DAT	600°C, 0 min		180 Barrer				60			[44]
BPDA-pp' ODA	600°C		330 Barrer			30	80			[45]
	700°C		160 Barrer			55	60			
	800°C		7.8 Barrer			30	100			
BPDA-pp' ODA	700°C, 0 min				9.7	47				[46]
	CVD				14	73				
BPDA-pp' ODA	700°C, 0 min		90–222 Barrer			14–40				[47]
BPDA-pp' ODA	700°C, 0 min					35				[12]
BPDA-pPDA	550–700°C	1.68–50.1 Barrer	2.52–71.4 Barrer		1.3–14	1.8–21	1.4			[23]
BPDA-pPDA	550°C		18 Barrer		5.5		37.4			[48]
BTDA-TDI/MDI	700°C, 60 min				6	22–29				[49]
Cellulose	800°C, Ar,			300 Barrer					1,200	[50]
	600°C	98 Barrer			12					
	400°C	56 Barrer			11					
Coal tar pitch	900°C							3.33		[17]
COPNA resin	400–1,000°C					35–45				[51]
Kapton	800°C	1.2 Barrer	13 Barrer		4.62	50				[52]
Kapton	950°C, 120 min			0.95 GPU				10		[19]
Kapton	600°C, 120 min	300 GPU	1,200 GPU	1,500 GPU	3	12				[30]
	800°C, 120 min	60 GPU	200 GPU	790 GPU	6	20				
Kapton	600°C, 2 h	383 Barrer	1,820 Barrer	1,600 Barrer	4.7	22.2		19.6		[53]
	800°C, 2 h	34.8 Barrer	128 Barrer	669 Barrer	11.5	42.2		220		
	1,000°C, 2 h	0.96 Barrer	4.15 Barrer	59.4 Barrer	23.4	101		1,450		

Table 11.2 (continued)

Precursor	Pyrolysis	P (O_2)	P (CO_2)	P (H_2)	a (O_2/N_2)	a (CO_2/N_2)	a (CO_2/CH_4)	a (H_2/N_2)	a (H_2/CH_4)	Reference
Matrimid and Kapton	475–650°C	1–7 Barrer	2–12 Barrer		2.9–5.9		5.4–32.9			[54]
PAN	500°C, 10–180 min	6.76–71.66GPU			1.28–3.74					[55]
PDMS	600°C			16.2 Barrer				20		[56]
PEI	600°C						91			[57]
PEI	700°C, 60 min				7.4	15	25			[58]
PFA	600°C	1.2–7.2			2.7–3.7					[59]
PFA	300°C, 120 min	0.064 Barrer		0.828 Barrer	13.5			175		[60]
	450°C, 120 min	0.334 Barrer		3.63 Barrer	30.4			331		
	600°C, 120 min	0.762 Barrer		8.136 Barrer	2.8			29.8		
PFA	150–600°C	0.06–81.6 Barrer		0.83–245.4 Barrer	2.1–30.4			6.3–331		[61]
	450°C, He	1.67 Barrer		10.5 Barrer	9.9			62.4		
	450°C, H_2	55.26 Barrer			4.1					
PFA	450°C, 120 min	1.452–2.742 Barrer		9.3–20.34	2.2–6.4			12.5–37.7		[62]
PFA	600°C	4.65–5.07 Barrer	34.92–16.02 Barrer	153–36.24 Barrer	10.6–12.7		92–82			[63]
Phenol formaldehyde	800–950°C, 120–180 min	2,300 Barrer		5,100 Barrer	10.65			23.61		[64]
Phenol formaldehyde	900°C, 60 min	0.351 GPU	0.597 GPU	4.95 GPU	5.24	8.91	20.58	73.88	170.69	[65]
Phenolic resin	700°C, 1 coat	13.4 Barrer	46.1 Barrer		10.7	36.9	85.3			[66]
	700°C, 3 coat	4.5 Barrer	16.1 Barrer		11	39.5	115.7			
Phenolic resin	700°C		11.7 Barrer		10	45				[67]
Phenolic resin	700°C				11.4	38	120			[10]
Phenolic resin	700°C				13.2	39	101			[68]
Phenolic resin	700°C				1.5–9.2	2.4–39				[69]

Table 11.2 (continued)

Precursor	Pyrolysis	P (O_2)	P (CO_2)	P (H_2)	a (O_2/N_2)	a (CO_2/N_2)	a (CO_2/CH_4)	a (H_2/N_2)	a (H_2/CH_4)	Reference
Phenolic resin	700°C	22.2–295.2 Barrer	127.2–960 Barrer		3.3–9	17–31	13–67			[70]
Polyamic acid	750°C, 3 h						200			[43]
Polyamic acid	750°C, N_2		189.6 Barrer				42			[71]
Polyimide	550°C, vacuum	25–50 GPU		372–473 GPU	7.4–9.0			64–110		[72]
	550°C, Ar	1–284 GPU		451–713 GPU	2.8–6.1			6.8–31.2		
	550°C, He	73–140 GPU		428–676 GPU	4.7–6.1			15.2–5.7		
	550°C, CO_2	75–306 GPU		400–654 GPU	2.6–6.1			10.4–30		
	800°C, vacuum	1.6–4.2 GPU			6.8–10.3					
Polyimide	535°C	952 Barrer			4.6					
	550°C	239 Barrer			9.9					[73]
	800°C	23 Barrer			12.3					
Polyimide	700°C, 3.6 min			1,000 GPU					132	[21]
	850°C, 3.6 min			180 GPU					631	
Polyimide	600°C	280 GPU	590 GPU		3.2	6.8				[28]
	700°C	170 GPU	420 GPU		3.7	10				
Polyimide	700°C, 3.6 min			420 GPU					540	[14]
Polyimide	500–900°C			30–800 GPU					80–1,000	[74]
Polypyrrolone	500°C, 1 h	674 Barrer	2,970 Barrer	6,580 Barrer	9.1	40	120	88.56	270	[75]
	700°C, 1 h	68.1 Barrer	250 Barrer	1,720 Barrer	11	39	180	267.5	1,200	
	800°C, 1 h	0.183 Barrer	0.178 Barrer	25.4 Barrer	7.8	7.6	7.1	1,080.85	1,000	
PVD-CPVC	700°C				14	25	60			[76]
Sulfonated phenolic resin	500°C, 90 min	0.13–240 GPU	0.29–800GPU	10.5–1,950 GPU	1.8–5.2		4–27		157–65	[77]

1. Carbon membranes are effective for separating gas mixtures with similar molecular sizes such as O_2/N_2, CO_2/CH_4 and CO_2/N_2 [10].
2. Carbon membranes have greater mechanical strength and are able to withstand higher pressure differences for a given wall thickness [11].
3. Feed pressure does not greatly affect the permeation properties of carbon membranes [12, 13]. This is because of the structural stability of carbon membranes, which do not suffer from compaction and swelling problems.
4. The permeation properties of carbon membranes is not time dependent [14]. This means that the operating life of carbon membranes is much larger than that of organic membranes [15].
5. Carbon membranes have the advantage of being able to operate in environments that are generally prohibitive to polymeric materials, and they have superior stability in presence of organic vapour or solvent and non-oxidizing acid or base environments [11, 14, 16, 17]. They are ideal for corrosive applications [18] and are not overly affected by aggressive feeds [6].
6. Carbon membranes have higher thermal stability than polymeric membranes. They are appropriate for application in high temperature separation processes, with temperatures in the range of 500–900°C. In contrast, organic polymer membranes can not resist very high temperatures and they begin to decompose or react with certain components [11, 16, 18–20].
7. The activation energies of the diffusion on the asymmetric carbon membranes are smaller than those on the asymmetric polyimide membrane for the gas especially having larger molecular size. Therefore, the selectivity of polyimide membrane decreases remarkably if the measurement temperature increases [21]. On the other hand, selectivity does not decrease remarkably in the carbon membranes.
8. The same starting material can be served to develop membranes of different permeation properties for different gas mixtures [11].
9. The pore dimension and distribution of the carbon membranes can be finely adjusted by simple thermochemical treatment to meet different separation needs and objective optimally [11, 17].
10. Carbon membranes have a superior adsorptivity for some specific gases, which can enhance its gas separation capacity [17].
11. Carbon membranes have the ability to be back flushed, steam sterilized or autoclaved [2].

11.2.2 Disadvantage of Carbon Membranes

Although carbon membranes have a number of advantages, a few disadvantages of carbon membranes have been identified.

A carbon membrane is very brittle and fragile, and therefore it needs to be handled with care, which can be very demanding.

Besides, the production of carbon membranes currently involves a very high cost. The cost of a carbon membrane per unit of membrane area is reported to be between one or three orders of magnitude greater than that of a typical polymeric membrane [6]. Therefore, carbon membranes must achieve a superior performance in order to compensate for their higher cost. Optimization of fabrication parameters during the pyrolysis process is arguably the best way to achieve this goal. Since there are a large number of parameters involved, using computer programming to optimize the pyrolysis process makes the task more accurate and satisfactory.

It is also difficult to fabricate carbon membranes [6]. In addition, carbon membranes require pre-purifiers for the removal of traces of strongly adsorbing vapors, which clog up the pores [11]. Carbon membranes only demonstrated high selectivities for a certain gas pair, which is limited to gases with molecular sizes smaller than 4.0 to 5.0 Å.

11.2.3 Application of Carbon Membranes

11.2.3.1 Nitrogen Production from Air

The most important application for carbon membranes is in the production of low cost and high purity nitrogen from air. Membranes can produce N_2 with purity of up to 99.5%. It is estimated that membranes currently produce 30% of all gaseous N_2 because many industries and commercial applications do not require ultra high purity nitrogen [6, 22]. It is expected that carbon membranes will offer an effective way of producing N_2.

11.2.3.2 Purification of Methane

Other separation processes include purification of methane as well as the recovery of carbon dioxide in oil fields [15, 23]. Besides that they are useful in the removal of acid gases from natural gas because they can operate in severe environments [15].

Currently, the largest industrial gas separation application is the processing of natural gas. The total worldwide consumption of natural gas is ~95 trillion scf/year. the high consumption of natural gas drives a worldwide market for new gas separation equipment about US$ 5 billion a year, which is < 5% of this market is toward the gas separation using membrane [1]. Although polymeric membranes are able to compete successfully with other technologies, such as amine scrubbing, improvement in membrane fabrication is necessary to grab the broad market opportunity for carbon membranes. This is because some glassy polymeric membranes lose their selectivity and productivity in the presence of trace quantities of condensable heavy hydrocarbon.

Furthermore, extremely high partial-pressure of carbon dioxide can cause the plasticization in the skin layer of a membrane [6].

11.2.3.3 Hydrogen Recovery

Membranes compete with cryogenic, catalytic and pressure swing adsorption processes in the hydrogen recovery process [6]. Carbon membranes can be applied to recover a valuable chemical (H_2) from a waste gas, or can be used for the recovery of hydrogen from gasification gas without further compression of the feed gas, while rejecting a substantial portion of the hydrocarbons [24]. Conventional/polymeric membranes incur additional recompression costs because H_2, as a fast gas, exits the unit at the lower pressure side [5, 6].

11.2.3.4 Light Alkenes/Alkanes

Carbon membranes are promising candidates for separating light alkenes/alkanes, especially propene/propane separation, because carbon molecular sieve membranes possess excellent propene/propane permselectivities.

Up until now, the separation of propene/propane gas mixtures has been investigated mainly by using polymeric membranes. However, their drawbacks include a relatively low permselectivity and the lack of thermal and chemical stabilities of the polymeric membranes. These drawbacks have led to further trials to develop thermally and chemically stable non-polymeric membranes that exhibit better separation performance, specifically for propene/propane separation. One of the choices is carbon molecular sieve membranes [25]. Hayashi et al. [12, 26] reported that a special carbonized membrane is capable of recognizing size differences between alkane and alkene molecules. It is also possible to employ carbon membranes to separate isomers of hydrocarbons into normal and branched fractions [27]. It is expected that they will be superior to other methods, such as distillation, adsorption and absorption-based on energy consumption. Separation of light alkenes/alkanes has been recognized as a key technology for use by the petrochemical industry [25].

11.2.3.5 Olefins and Paraffins

Carbon molecules also can be implemented in processes that are used to separate other olefins and paraffins—another important process in the petrochemical industry (especially for 1,3-butadiene/n-butane separation [28]).

A recent study estimated that about 10,000 BTU of energy is used annually for olefin–paraffin distillation. The distillation process is used commercially in this separation process. However, membrane separation with low energy consumption and with relative ease in operation, can be significantly competitive with the distillation process [6]. Therefore, carbon membranes can contribute greatly to the petrochemical industry.

11.2.3.6 Carbon Membrane Reactor

A carbon membrane reactor constitutes one of the most promising applications of carbon membranes. The performance of a carbon membrane for gas separation and for the dehydrogenation of cyclohexane to benzene was examined by Itoh and Haraya [29]. They concluded that the performance of their carbon membrane reactor for dehydrogenation was fairly good compared with that of a normal reactor, i.e. functioning at equilibrium [29]. On the other hand, Lapkin [30] used a macro-porous phenolic resin carbon membrane as a contactor for the hydration of propene in a catalytic reactor. He found that the use of this porous contactor-type reactor for his high-pressure catalytic reaction is practical.

11.2.3.7 Optimization

The combination of reaction and separation at a high temperature in a membrane reactor offers interesting new possibilities. In reactors based on carbon membranes, separate product and feed compartments allow various ways of optimizing both selectivity and conversion [2, 15].

11.2.4 Challenge in Carbon Membrane Development

Carbon membrane technology has been the focus of research and development in gas separation. In an effort to develop carbon membrane, the main problem that must be overcome before it will be possible to apply the technology broadly is the fabrication of these materials in a manner that is both reproducible and scalable for manufacturing.

11.2.5 Few Manufacturers

As can be seen from Table 11.1, there are only a few manufacturers involved in the production of carbon membranes. This is because greater costs are involved in producing and packaging carbon membranes in modules, and this has prevented their use in large-scale membrane modules [5]. The cost of a carbon membrane is reported to be higher than an equivalent polymeric membrane [6]—between one and three orders of magnitude per unit of membrane area. Since carbon membranes involve high production cost, permeability and permselectivity of carbonized membranes require further refinements and improvements before they can be used on a large industrial scale [27]. High permeability membranes require small membrane area that lower the capital cost of the membrane system. In addition, high selectivity membranes promise to improve separation and make it more efficient. In this way, the operating costs are reduced, because a lower driving force (pressure ratio) is required [6].

11.2.6 Improving Performance

Much research should be carried out to optimize the way in which carbon membranes are produced in order to improve their performance and to render them an important separation tool in industry.

Moreover, an investigation into material selection to find more suitable membrane precursors should be continued. Finding a more economical material than polyimide, which is mainly used by many researchers as a carbon membrane precursor, is another necessary task in carbon membrane production. This indicates that discovering ways to provide carbon membranes with excellent separation properties, without losing the economical processibility of polymeric membrane materials, would be a major breakthrough in this field.

It is the inspiration and the incentive which lies behind the development of carbon hollow fiber membranes, particularly designed for gas separation. It is believed that fundamental knowledge will lead eventually to the development of tailor-made carbon membranes.

11.2.7 Future Directions of Research and Development

Although carbon membranes still require improvement before they can become dominant commercialized inorganic membranes, they have great potential as a replacement for other inorganic membranes in the market place.

This is because they have many useful characteristics and are able to efficiently separate gas mixtures that have molecules of similar sizes. A number of recommendations are made for future investigative work.

11.2.7.1 Choice of Precursor and Optimization of Precursor Preparation Process

The most popular precursor currently used for manufacturing carbon membranes is polyimide. Although polyimide offers the best separation potential for carbon membranes, its cost is too high and its commercial availability is sometimes very limited. Therefore, in order to reduce the overall cost and production time during carbon membrane fabrication, an alternative polymer must be chosen. The choice of polyacrylonitrile (PAN) might be the answer because of many advantages offered by the polymer, i.e. high carbon yield, mechanical and chemical stability, ready availability, etc. However, it is still in a progressing stage aimed at improvement in their separation and permeation properties. It is believed that satisfactory results will be obtained via optimization of the pyrolysis process and combination of suitable pre- and post-treatment methods [23].

Besides pyrolysis, the polymer solution preparation, spinning process, solvent exchange process and others are important steps to ensure the production of excellent carbon membranes.

11.2.7.2 Optimization of Pyrolysis Process

There are abundant parameters during the process that determine the properties and performance of a carbon membrane. A wide coverage of study should be conducted on the pyrolysis process condition, including pyrolysis temperature, soak time, purge gas flow rate, heating rate and the type of the purge gas.

All the experimental results that are obtained should be simulated using computer software to obtain an ideal and practical pyrolysis condition, which can be implemented on a pilot scale.

11.2.7.3 Composite Precursor for Carbon Membranes

Combined usage of low-cost and high-cost polymers to produce composite carbon membranes is an alternative way of obtaining carbon membranes of high performance economically.

The membrane consists of a skin layer with high-yield carbon, and substructure with low-yield carbon. High yield carbon leads to the development of a highly selective skin layer, while low-yield carbon can result in an open porous structure with minimum resistance for gas permeation.

11.2.7.4 Polymer Blend Carbonization

The polymer blend carbonization method will become an important tool for producing carbon membranes, with mixed-matrix materials, to overcome the challenges and limitations of membrane technology used in the gas separation industry.

Mixed matrix materials, comprising molecular sieve entities embedded in a polymer matrix, offer the potential to combine the processibility of polymers with the superior gas separation properties of rigid molecular sieving materials [4, 31].

The carbonization of polymer blend will lead to the formation of porous structure. It is because of the thermally unstable polymer (pyrolysing polymer) that decomposes and pores in the carbon matrix that are formed from the stable polymer (carbonizing polymer).

Polymer blends of phenolic resin and poly(vinyl butyral) [32], as well as poly(diphenylene pyromellitimide) and poly(ethylene glycol) have been used by researchers. Although polymer blends have been used widely as important industrial low materials for highly functional materials, the implementation of this carbonization method is currently in its early stage of development [7]. Further study is still needed to identify suitable pyrolizing polymers and carbonizing polymers for blending.

11.2.7.5 Carbon Membrane Module Design

The efficiency of a membrane separation process in a certain application, depends on the way it is installed in the form of a membrane module.

Many membrane modules have been proposed by researchers. The most challenging task that must be considered during the fabrication of such modules is the poor mechanical stability of carbon membranes. This is more crucial in hollow fiber carbon membranes since they are self-supporting. More effort is now needed to develop a suitable module configuration and module design to accommodate carbon fibers.

11.2.7.6 Fiber Stretching During Spinning or Pre-oxidation Process

The act of fiber stretching is essentially an attempt to improve the mechanical properties of the fiber through molecular orientation.

In general, polyacrylonitrile (PAN) fibers are stretched in two steps that improve the orientation of the molecular chains [33]. An ideal stretching process would allow the removal of surface defects, attenuation of fiber diameter, and molecular orientation prior to heat treatment in order to obtain fibers with a good balance of stiffness and strength [34].

Stretching during spinning offers the advantages that the pore systems (which are obtained after the spinning process) are preserved to larger extent in the carbon membrane.

It also ensures a greater dimensional stability [35, 36]. It is generally thought that better molecular orientation in the original PAN fiber produces better mechanical properties in the resultant carbon fiber [37].

During the pre-oxidation step, a fiber must be held under tension, otherwise samples fuse during heating and undergo weight loss [37].

11.2.7.7 Pre- and Post-treatment

In addition, the effect of pre-treatments and post-treatments during the membrane fabrication process should also be examined. As an example, post-oxidation treatments can be very useful to tailor the properties of a molecular sieving carbon membrane in order to make an adsorption-selective carbon membrane, which is suitable for the separation of hydrocarbon mixtures based on adsorption differences. In fact, there are several pre-treatment methods that can be used to control the uniformity of pore formation during pyrolysis. Hopefully, through manipulation of the available pre- and post-treatment methods, a stable carbon membrane exhibiting high separation properties can be produced.

11.2.7.8 Carbon Membrane Aging and its Regeneration

Once a carbon membrane of high performance is produced, the effect of exposure to water vapor must be considered. It has been reported that the selectivity of a typical membrane decreases as the amount of sorbed water increases [38]. Not many studies have taken into account this phenomenon very seriously, although it becomes quite important when a carbon membrane is commercialized. This is primarily due to the fact that water vapor can be found in a large number of process stream [38]. To date, only Jones and Koros [39, 40] have studied this phenomenon in detail, and they proposed the use of composite membranes (coated DuPont's Teflon AF1600 and AF2400) as a solution to this drawback. On the other hand, the humidity found in the ambient atmosphere can also have an adverse effect on carbon membrane performance. Therefore, the study of storage conditions for carbon membranes is also an important consideration for carbon membrane research in the future.

11.2.8 Chemical Vapor Deposition

The selectivity of a carbon membrane may be increased through the introduction of organic species into the pore system of the membrane's structure, and pyrolyzation (that is chemical vapor deposition or (CVD)) [41].

Generally, to fabricate carbon molecular sieves, the inherent pore structure of the carbonaceous precursor is initially fixed into a suitable pore range by controlling the thermal treatment, followed by final tailoring of the pore structures by CVD. This process is expected to produce asymmetric membranes using homogeneous (symmetric) carbon membranes as the starting membrane structure.

This could be another area of interest to membranologists to further explore the possibility of improving membrane selectivity to bring about the commercial feasibility of carbon membranes for gas separation.

11.2.9 Conclusions

The great potential and numerous advantages of carbon membranes will definitely lead to their wide application in the gas separation industry over the coming years.

It is believed that carbon membrane gas separation technologies will break through the limitation and weakness of conventional membranes. Carbon membranes offer promising separation performance and productivity and can be used in severe environments and in high temperature processes.

This means that intensive research and development work, covering carbon membranes, should be carried out in order to produce tailor-made structures and technologies. These technologies will contribute greatly to the gas separation industry world wide.

References

1. Baker RW and Lokhandwala K (2008) Natural gas processing with membranes: An overview. Ind Eng Chem Res 47 (7): 2109-2121
2. Hsieh HP (1988) Inorganic membranes. AIChE Symposium Series 84 (261): 1-18
3. Keizer K, Verweij H (1996) Progres in inorganic membranes. Chemtech, Jan: 37
4. Soria R (1995) Overview on industrial membranes. Catalysis Today 25 (3-4): 285-290
5. Koros JW (1995) Membranes: Learning a lesson from nature. Chem Eng Progr Oct: 68
6. Koros WJ, Mahajan R (2000) Pushing the limits on possibilities for large-scale gas permeation: which strategies? J Membr Sci 175 (2): 181-196
7. Kyotani T (2000) Control of pore structure in carbon. Carbon 38 (2): 269-286
8. Ismail AF, David LIB (2001) A review on the latest development of carbon membranes for gas separation. J Membr Sci 193 (1):1-18
9. Wang SS, Zeng MY, Wang ZZ (1996) Carbon membrane for gas separation. Sep Sci Technol 31 (16): 2299-2306
10. Fuertes AB (2000) Adsorption-selective carbon membrane for gas separation. J Membr Sci 177 (1-2): 9-16
11. Koresh JE, Soffer A (1987) The carbon molecular sieve membranes. General properties and the permeability of CH_4/H_2 mixture. Sep Sci Technol 22 (2-3): 973-982
12. Hayashi J, Mizuta H, Yamamoto M, Kusakabe K, Morooka S (1996) Separation of ethane/ethylene and propane/propylene system with a carbonized BPDA-pp/ODA polyimide. Ind Eng Chem Res 35 (11): 4176-4181
13. Geiszler VC (1997) Polyimide precursors for carbon membranes. PhD Thesis, University of Texas at Austin, TX, USA
14. Tanihara N, Shimazaki H, Hirayama Y, Nakanishi S, Yoshinaga T, Kusuki Y (1999) Gas permeation properties of asymmetric carbon hollow fiber membranes prepared from asymmetric hollow fiber. J Membr Sci 160 (2): 179-186
15. Fain DE (1991) Technical and economic aspects and prospects for gas separation with inorganic membranes. Key Eng Materials 61/62: 327-336
16. Jones CW, Koros WJ (1994) Carbon molecular sieve gas separation membranes I-preparation and characterization based on polyimide precursors. Carbon 32 (8): 1419-1425
17. Liang C, Sha G, Guo S (1999) Carbon membrane for gas separation derived from coal tar pitch. Carbon 37 (8): 1391-1397
18. Bauer JM, Elyassini J, Moncorge G, Nodari T, Totino E (1991) New development and application of carbon membranes. Key Eng Materials 61/62: 207-212
19. Petersen J, Matsuda M, Haraya K (1997) Capillary carbon molecular sieve membranes derived from kapton for high temperature gas separation. J Membr Sci 131 (1-2): 85-94
20. Koresh JE, Soffer A (1986) Mechanism of permeation through molecular sieve carbon membrane. Part 1. The effect of adsorption and the dependence on pressure. J Chem Soc, Farady Trans 1 82: 2057-2063
21. Kusuki Y, Shimazaki H, Tanihara N, Nakanishi S, Yoshinaga T (1997) Gas permeation properties and characterization of asymmetric carbon membranes prepared by pyrolyzing asymmetric polyimide hollow fiber membrane. J Membr Sci 134 (2): 245-253

22. Roman IC, Ubersax RW, Fleming GK (2001) New directions in membranes for gas separation. Chimica Industria Giug 83 (5): 1-3
23. Fuertes AB, Centeno TA (1999) Preparation of supported carbon molecular sieve membrane. Carbon 37 (4): 679-684
24. Rao MB, Sircar S (1993) Nanoporous carbon membranes for separation of gas mixtures by selective surface flow. J Membr Sci 85 (3): 253-264
25. Suda H, Haraya K (1997) Alkene/alkane permselectivites of a carbon molecular sieve membrane. J Chem Soc Chem Commun 1: 93-94
26. Hayashi J, Yamamoto M, Kusakabe K, Morooka S (1997) Effect of oxidation on gas permeation of carbon molecular sieve membrane based on BPDA-*pp*/ODA poyimide. Ind Eng Chem Res 36 (6): 2134-2140
27. Soffer A, Girlon J, Cohen H (1999) Separation of linear from branched hydrocarbons using a carbon membrane.US patent 5914434
28. Okamoto K, Kawahara S, Yoshino M, Kita H, Hirayama Y, Tanihara N (1999) Olefin/paraffin separation through carbonized membranes derived from an asymmetric polyimide hollow fiber membrane. Ind Eng Chem Res 38 (11): 4424-4432
29. Itoh N, Haraya K (2000) A carbon membrane reactor. Catal Today 56 (1-3): 103-111
30. Lapkin A (1999) Hydration of propene using a porous carbon membrane contactor. Membr Technol 116: 5-9
31. Mahajan R, Koros WJ, Thundyil M (1999) Mixed matrix membranes: Important and challenging. Membr Technol 105: 6-8
32. Ozaki J, Endo N, Ohizumi W, Igarashi K, Nakahara M, Ohya A, Yoshida S, Iizuka T (1997) Novel preparation method for the production of mesoporous carbon fibre from a polymer blend. Carbon 35 (7): 1031-1033
33. Mittal J, Mathur RB, Bahl OP (1997) Post spinning modification of PAN fibers-A review. Carbon 35 (12): 1713-1722
34. Chen JC, Harrison IR (2002) Modification of polyacrylonitrile (PAN) carbon fiber precursor via post-spinning plasticization and stretching in dimethylformamide (DMF). Carbon 40 (1): 25-45
35. Soffer A, Koresh J, Sagee S (1987) Separation device. US patent 4919860
36. Ismail AF, Saufi SM (2003) Development and characterization of PAN-based carbon hollow fiber membrane for oxygen/nitrogen separation, "Regional Symposium and Workshop Membrane Science and Technology", 13-17 January 2003, Songkhla, Thailand
37. Dalton S, Heatley F, Budd PM (1999) Thermal stabilization of polyacrylonitrile fibers. Polymer 40 (20): 5531-5543
38. Jones CW, Koros WJ (1995) Characterization of ultramicroporous carbon membranes with humidified feeds. Ind Eng Chem Res 34 (1): 158-163
39. Koros WJ, Jones CW (1994) Composite carbon fluid membranes. US patent 5288304
40. Jones CW, Koros WJ (1995) Carbon composite membranes: a solution to adverse humidity effects. Ind Eng Chem Res 34 (1): 164-167
41. Soffer A, Azariah A, Amar A, Cohen H, Golub D, Saguee S (1997) Method of improving the selectivity of carbon membranes by chemical vapor deposition. US patent 5695618
42. Vu DQ, Koros WJ, Miller SJ (2002) High pressure CO_2/CH_4 separation using carbon molecular sieve hollow fiber membranes. Ind Eng Chem Res 41 (3): 367-380
43. Ogawa M, Nakano Y (1999) Gas permeation through carbonized hollow fiber membranes prepared by gel modification of polyamic acid. J Membr Sci 162 (1-2): 189-198
44. Yamamoto M, Kusakabe K, Hayashi J, Morooka S (1997) Carbon molecular sieve membrane formed by oxidative carbonization of a copolyimide film coated on a porous support tube. J Membr Sci 133 (2): 195-205
45. Hayashi J, Yamamoto M, Kusakabe K, Morooka S (1995) Simultaneous improvement of permeance and permselectivity of 3,3',4,4'-biphenyltetracarboxylic dianhydride-4,4'-oxydianiline polyimide membrane by carbonization. Ind Eng Chem Res 34 (12): 4364-4370
46. Hayashi J, Mizuta H, Yamamoto M, Kusakabe K, Morooka S (1997) Pore size control of carbonized BPTA-pp'ODA polyimide membrane by chemical vapor deposition of carbon. J Membr Sci 124 (2): 243-251

47. Kusakabe K, Yamamoto M, Morooka S (1998) Gas permeation and micropore structure of carbon molecular sieving membranes modified by oxidation. J Membr Sci 149 (1): 59-67
48. Fuertes AB, Centeno TA (1998) Preparation of supported asymmetric carbon molecular sieve membranes. J Membr Sci 144 (1-2): 105-111
49. Barsema JN, van der Vegt NFA, Koops GH, Wessling M (2002) Carbon molecular sieve membranes prepared from porous fiber precursor. J Membr Sci 205 (1-2): 239-246.
50. Soffer A, Rosen D, Saguee S, Koresh J (1989) Carbon membranes. GB patent 2207666
51. Kusakabe K, Gohgi S, Morooka S (1998) Carbon molecular sieving membranes derived from condensed polynuclear aromatic (COP-NA) resins for gas separation. Ind Eng Chem Res 37 (11): 4262-4266
52. Hatori H, Shiraish M, Nakata H, Yoshitomi S (1992) Carbon molecular sieve films from polyimide. Carbon 30 (4): 719-720
53. Suda H, Haraya K (1997) Gas permeation through micropores of carbon molecular sieve membranes derived from kapton polyimide. J Phys Chem B 101 (20): 3988-3994
54. Fuertes AB, Nevskaia DM, Centeno TA (1999) Carbon composite membranes from matrimid and kapton polyimide. Microporous Mesoporous Mater 33 (1-3): 115-125
55. David LIB, Ismail AF (2003) Influence of the thermostabilization process and soak time during pyrolysis process on polyacrylonitrile carbon membranes for O_2/N_2 separation. J Membr Sci 213 (1-3): 285-291
56. Lee LL, Tsai DS (2001) Synthesis and permeation properties of silicon-carbon-based inorganic membrane for gas separation. Ind Eng Chem Res 40 (2): 612-616
57. Sedigh MG, Xu L, Tsotsis TT, Sahimi M (1999) Transport and morphological characteristics of polyetherimide-based carbon molecular sieve membranes. Ind Eng Chem Res 38 (9): 3367-3380
58. Fuertes AB, Centeno TA (1998) Carbon molecular sieve membranes from polyetherimide. Microporous Mesoporous Mater 26 (1-3): 23-26
59. Acharya M, Foley HC. (1999) Spray-coating of nanoporous carbon membranes for air separation J Membr Sci 161 (1-2): 1-5
60. Shiflett MB, Foley HC (1999) Ultrasonic deposition of high-selectivity nanoporous carbon membranes. Science 285 (5435):1902-1905
61. Shiflett MB, Foley HC (2000) On the preparation of supported nanoporous carbon membranes. J Membr Sci 179 (1-2): 275-282
62. Shiflett MB, Foley HC (2001) Reproducible production of nanoporous carbon membranes. Carbon 39 (9): 1421-1446
63. Wang H, Zhang L, Gavalas GR (2000) Preparation of supported carbon membranes from furfuryl alcohol by vapor deposition polymerization. J Membr Sci 177 (1-2): 25-31
64. Wang S, Zeng M, Wang Z (1996) Asymmetric molecular sieve carbon membranes. J Membr Sci 109 (2): 267-270
65. Wei W, Hu H, You L, Chen G (2002) Preparation of carbon molecular sieve membrane from phenol-formaldehyde novolac resin. Carbon 40 (3): 465-467
66. Centeno TA, Fuertes AB (2001) Carbon molecular sieve membranes derived from a phenolic resin supported on porous ceramic tubes. Sep Purif Tech 25 (1-3): 379-384
67. Centeno T, Fuertes AB (1999) Supportd carbon molecular sieve membranes based on phenolic resin. J Membr Sci 160 (2): 201-211
68. Fuertes AB (2001) Effect of air oxidation on gas separation properties of adsorption-selective carbon membranes. Carbon 39 (5): 697-706
69. Fuertes AB, Menendez I (2002) Separation of hydrocarbon gas mixtures using phenolic resin-based carbon membranes. Sep Purif Tech 28 (1): 29-41
70. Menendez I, Fuertes AB (2001) Aging of carbon membranes under different environments. Carbon 39 (9): 733-740
71. Ogawa M, Nakano Y (2000) Separation of CO_2/CH_4 mixture through carbonized membrane prepared by gel modification. J Membr Sci 173 (2): 123-132
72. Geiszler VC, Koros WJ (1996) Effect of polyimide pyrolysis conditions on carbon molecular sieve membrane properties. Ind Eng Chem Res 35 (9): 2999-3003

73. Ghosal AS, Koros WJ (2000) Air separation properties of flat sheet homogeneous pyrolytic carbon membranes. J Membr Sci 174 (2): 177-188
74. Yoshinaga T, Shimazaki H, Kusuki Y, Sumiyama Y (1991) Asymmetric hollow filamentary carbon membrane and process for producing the same. EU patent 0459623
75. Kita H, Yoshino M, Tanaka K, Okamoto K (1997) Gas permselectivity of carbonized polypyrrolone membrane. J Chem Sco Chem Commun 1051-1052
76. Centeno TA, Fuertes AB (2000) Carbon molecular sieve gas separation membranes based on poly(vinylidene chloride-co vinyl chloride). Carbon 38 (7): 1067-1073
77. Zhou W. Yoshino M, Kita H, Okamoto K (2001) Carbon molecular sieve membranes derived from phenolic resin with a pendant sulfonic acid group. Ind Eng Chem Res 40 (22): 4801-4807

Index

A. F. Ismail et al., *Carbon-based Membranes for Separation Processes*,
DOI 10.1007/978-0-387-78991-0, © Springer Science+Business Media, LLC 2011

D

E

F

R

S